TEST 2-13

LINEAR ALGEBRA

NORMAN J. BLOCH
JOHN G. MICHAELS

Professors of Mathematics
State University of New York College at Brockport

McGraw-Hill Book Company

New York   St. Louis   San Francisco   Auckland   Bogota   Düsseldorf
Johannesburg   London   Madrid   Mexico   Montreal   New Delhi
Panama   Paris   São Paulo   Singapore   Sydney   Tokyo   Toronto

LINEAR
ALGEBRA

1234567890 KPKP 7832109876

This book was set in Times New Roman. The editors were A. Anthony Arthur and Douglas J. Marshall; the designer was Joseph Gillians; the production supervisor was Dennis J. Conroy. The drawings were done by Oxford Illustrators Limited.
Kingsport Press, Inc., was printer and binder.

Library of Congress Cataloging in Publication Data

Bloch, Norman J
    Linear algebra.

    Includes index.
    1. Algebras, Linear. I. Michaels, John G., joint author. II. Title.
QA184.B58   512'.5   76-12464
ISBN 0-07-005906-3

To our wives
Judy and Lois,
and our children
Steven, Susan,
and Margaret

CONTENTS

PREFACE

This text is intended for a one-semester introductory linear algebra course. One of the primary purposes of a linear algebra course at this level is to serve as a transition from computational mathematics to more theoretical mathematics. We have, therefore, tried to strike a reasonable balance between theory and technique.

It is our experience that for many students a course in linear algebra is the first substantial exposure to mathematical structure. Because of this, we have consistently presented the theory as a natural outgrowth of familiar geometric ideas and as a problem-solving tool. As each new concept is presented, it is first related to ideas already familiar to the student. For example, vector spaces are motivated through lines and planes, while the theory of transformations is closely tied to plane and space geometry via projections, reflections, and rotations. We have also presented numerous applications showing the power of linear algebra in dealing with problems arising in the natural and social sciences.

A course in linear algebra is often the first course in which the student is required to carry out theoretical arguments in any detail. In the spirit of developing the student's mathematical maturity, we have taken care to explain to the student how to set up and carry out proofs. For example, the first step in many proofs is the selection of arbitrary elements from the set being considered, a difficult first step for many students; we show in detail how this is accomplished.

Exercises are also a vital part of the learning process. We have included over 500 exercises, some computational, others requiring that the student complete details of proofs and further develop the theory. Where appropriate, hints are provided. Answers to approximately one-half of the problems are given in the back of the book.

Each chapter ends with a review section. The first part of the review section tests the student's recollection of the main ideas of the chapter; following this is a set of review exercises. We recommend that the student first reread the chapter quickly in search of its main points and then do the review.

Exercises marked with an asterisk (*) ask the student to write computer programs. These problems are optional; however, a student with a background in computer science is encouraged to try them.

The text begins with systems of linear equations. In our experience, this topic is the most natural starting point for a linear algebra course, since it ties in nicely with lines and planes, is computational in nature, and yet introduces many of the ideas to be used in later chapters.

Chapters 2 and 3 discuss vector spaces and dimension. The emphasis is on vector spaces of n-tuples of real numbers and vector spaces of matrices.

Chapters 4 and 5 are devoted to linear transformations, their matrices, and the simplification of matrices. The unifying theme in these chapters is the use of transformations to change problems to a form that makes them more readily solvable.

In Chapter 6 many of the geometric ideas that have been touched upon in the previous chapters are brought together and studied systematically using the dot product.

The material in these six chapters provides a reasonable base for a one-semester course. Each chapter concludes with a section on applications; these sections can be dealt with as time and the instructor's taste dictate. If a somewhat briefer course is desired, certain sections can be omitted without loss of continuity: Section 9, which contains some special examples of vector spaces; Section 18, which deals with the concept of isomorphism; and Section 36, which covers inner product spaces in general.

Finally, we are grateful to the many students and colleagues who have helped shape this text. In particular, we would like to thank Frederick Gass of Miami University, Larry Johnson of Fort Lewis College, James Murtha of Marietta College, and George Springer of Indiana University for their many constructive suggestions.

Norman J. Bloch

John G. Michaels

LINEAR ALGEBRA

CHAPTER 1

SYSTEMS OF LINEAR EQUATIONS

Systems of one or more equations arise frequently in mathematics. For example, an equation such as $3x + 2y = 5$ is the equation of a straight line in the xy plane. More generally, an equation $ax + by = c$, where $a, b,$ and c are given constants, represents a straight line. A system of two such equations, such as

$$2x - 3y = 2$$
$$4x + y = 0$$

represents two lines in the plane. To solve a system of equations such as this, we must find particular values for the unknowns x and y that simultaneously satisfy both equations in the system. Geometrically, this would yield all the points (x, y) that lie on both lines, that is, the points where the lines intersect.

Similarly, a system consisting of a single equation such as $3x + 4y - 2z = 5$, or more generally, $ax + by + cz = d$, represents a plane in xyz space. Further details on this will be given in Sec. 4. A system of two such equations, for example

$$x - 3y + 5z = 2$$
$$4x - z = 3 \tag{1}$$

represents two planes. Solving this system entails finding all points (x, y, z) that satisfy both equations simultaneously, that is, all points in the intersection of the two planes. In like fashion, the system

$$
\begin{aligned}
4x - 2y \quad\ \ &= 1 \\
x + 3y - 4z &= 0 \\
5x -\ \ y -\ \ z &= 4
\end{aligned}
\qquad (2)
$$

represents three planes in xyz space, with the solutions being the points (x, y, z) that lie in the intersection of all three planes. Systems of equations, such as (1) and (2) above, can have a unique solution (the planes meet in one point), infinitely many solutions (the planes intersect in a line), or no solution at all (the planes have no common intersection).

The equations in the above systems are called *linear* since the unknowns appear only to the first power; hence we refer to these systems as systems of linear equations. Because of its importance in so many areas of mathematics, we will examine an algorithm for solving systems of linear equations.

1.   GAUSSIAN ELIMINATION

This section will describe a systematic method for solution of systems of linear equations. The method is called *elimination* or *gaussian elimination* since it entails successively eliminating unknowns from all but one of the equations. Eliminating an unknown is done chiefly by adding appropriate multiples of one equation to the other equations. The following two examples illustrate this method.

Example 1-1

We wish to solve the system of linear equations

$$
\begin{aligned}
x +\ \ y + 2z &=\ \ \ 0 \\
3x +\ \ y + 2z &=\ \ \ 2 \\
2x - 2y +\ \ z &= -6
\end{aligned}
$$

We begin by eliminating x from the second equation. To do this we subtract 3 times the first equation from the second equation. When we do this, the second equation becomes $-2y - 4z = 2$; the first and third equations remain unchanged. The system of equations has become

$$
\begin{aligned}
x +\ \ y + 2z &=\ \ \ 0 \\
-2y - 4z &=\ \ \ 2 \\
2x - 2y +\ \ z &= -6
\end{aligned}
$$

In similar fashion we eliminate x from the third equation by subtracting 2 times the first equation from the third, yielding the new system

$$
\begin{aligned}
x +\ \ y + 2z &=\ \ \ 0 \\
-2y - 4z &=\ \ \ 2 \\
-4y - 3z &= -6
\end{aligned}
$$

Thus x has been eliminated from all but one equation.

Next we eliminate y from the first and third equations. This is made easier if we first multiply the second equation by $\frac{1}{2}$:

$$
\begin{aligned}
x + \quad y + 2z &= \quad 0 \\
- \quad y - 2z &= \quad 1 \\
- 4y - 3z &= -6
\end{aligned}
$$

To eliminate y from the first and third equations, we add the second equation to the first, and subtract 4 times the second equation from the third:

$$
\begin{aligned}
x \qquad\qquad &= \quad 1 \\
- y - 2z &= \quad 1 \\
5z &= -10
\end{aligned}
$$

To eliminate z from the second equation, we first multiply the third equation by $\frac{1}{5}$:

$$
\begin{aligned}
x \qquad\qquad &= \quad 1 \\
- y - 2z &= \quad 1 \\
z &= -2
\end{aligned}
$$

and then add 2 times the third equation to the second:

$$
\begin{aligned}
x \qquad\qquad &= \quad 1 \\
- y \qquad &= -3 \\
z &= -2
\end{aligned}
$$

Finally we multiply the second equation by -1:

$$
\begin{aligned}
x \qquad\qquad &= \quad 1 \\
y \qquad &= \quad 3 \\
z &= -2
\end{aligned}
$$

We can now read off the solution to the system of equations: $x = 1$, $y = 3$, $z = -2$; that is, $(1, 3, -2)$. Geometrically, the three equations in the original system represent three planes that meet in the single point $(1, 3, -2)$. ●

Example 1-2

We wish to solve the system

$$
\begin{aligned}
2x + y - \quad z &= \quad 4 \\
x \qquad + 2z &= \quad 9 \\
-3x - y + 2z &= -7
\end{aligned}
$$

To solve this system we again successively eliminate unknowns.

We first interchange the first and second equations (this puts the simplest x term first):

$$\begin{array}{rrcr} x & + 2z = & & 9 \\ 2x + y & - z = & & 4 \\ -3x - y & + 2z = & & -7 \end{array}$$

We now eliminate x from the second equation by subtracting 2 times the first equation from the second, and eliminate x from the third equation by adding 3 times the first equation to the third (the first equation remains the same):

$$\begin{array}{rcr} x \quad + 2z = & & 9 \\ y - 5z = & & -14 \\ -y + 8z = & & 20 \end{array}$$

We eliminate y from the third equation by adding the second equation to the third:

$$\begin{array}{rcr} x \quad + 2z = & & 9 \\ y - 5z = & & -14 \\ 3z = & & 6 \end{array}$$

We multiply the third equation by $\frac{1}{3}$ (this will simplify the next step):

$$\begin{array}{rcr} x \quad + 2z = & & 9 \\ y - 5z = & & -14 \\ z = & & 2 \end{array}$$

Finally we eliminate z from the first and second equations by subtracting 2 times the third equation from the first and by adding 5 times the third equation to the second. We can now read off the solution directly:

$$\begin{array}{rcr} x \quad & = & 5 \\ y \quad & = & -4 \\ z = & & 2 \end{array}$$

Interpreting these three equations as representing three planes, we find that they intersect in the point $(5, -4, 2)$. ●

It is appropriate at this point to discuss some terminology and notation concerning solutions. A symbol such as $(3, 4)$ is called an *ordered pair* (*pair*, for short) of numbers. The word *ordered* is used since the order in which the entries are written is important. For instance, the pairs $(3, 4)$

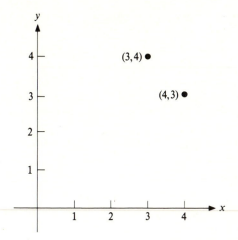

Fig. 1-1

and (4, 3) are different; they certainly correspond to different points in the xy plane (Fig. 1-1). Similarly, $(3, 2, -7)$ is called an (ordered) *triple* of numbers, $(2, 0, 4, 5)$ is an (ordered) *4-tuple*, etc.

The above systems all have two or three unknowns, and the solutions were pairs (x, y) or triples (x, y, z). In many applications, systems of equations with more than three unknowns can arise. If there were four unknowns, say x_1, x_2, x_3, x_4, a solution would be a 4-tuple (r, s, t, u) of numbers that when substituted for x_1, x_2, x_3, x_4 would satisfy all the equations simultaneously. Gaussian elimination can be used no matter how many unknowns there are.

The method of gaussian elimination, used in the above examples, entails only three permissible operations on the equations of a system:

(a)   Interchanging two equations

(b)   Multiplying both sides of an equation by a nonzero number

(c)   Adding a multiple of one equation to another†

The method works since each of these operations leaves the solutions of the system unchanged.

Interchanging two equations certainly does not change the solutions.

Multiplying an equation by a nonzero number can do no harm either.

---

† Note that subtracting a multiple of one equation from another, as is done several times in the examples, is covered by operation (c). Subtracting k times an equation is the same as adding $-k$ times that equation.

For example, $x + 2y = 5$ has exactly the same solutions as $3x + 6y = 15$. (A more general proof is left as Exercise 1-13.)

Finally, adding a multiple of one equation to another leaves the solutions unchanged. We again leave the details to the reader (Exercise 1-14).

Exercises

1-1 Which of $(4, 10)$, $(3, 7)$, or $(9, 1)$ is a solution to the system of equations

$$3x - y = 2$$
$$x + y = 10$$

1-2 Which of $(0, -2, 2)$, $(4, -1, -1)$, $(2, -1, 0)$, or $(2, 0, -1)$ is a solution to the system

$$x + 2y + 2z = 0$$
$$3y + 2z = -2$$
$$4x + y - 5z = 13$$

Solve each of the systems of equations in Exercises 1-3 to 1-8 by gaussian elimination.

1-3 $4x + 6y = 11$
$x - 8y = -2$

1-4 $x + 3y = 0$
$2x + y = 0$

1-5 $x + y = 5$
$-x + 2z = 0$
$2x + y - z = 7$

1-6 $x + 2y + 3z = 0$
$-x + 3y + 2z = 0$
$2x - y - z = 3$

1-7 $x_1 + x_2 - 2x_3 - 4x_4 = -\frac{19}{2}$
$ x_2 + x_3 + x_4 = 2$
$x_1 - 2x_3 - 3x_4 = -\frac{13}{2}$
$ 2x_2 + x_3 + x_4 = 2$

1-8 $x_1 + 2x_2 - x_3 + x_4 = 7$
$ x_2 - x_3 - x_4 = -4$
$x_1 + 2x_2 + 4x_3 - x_4 = -8$
$2x_1 - 2x_3 + 2x_4 = 12$

1-9 Find the point of intersection of the two lines $x - 3y = 1$ and $2x - 5y = 4$.

1-10 Find the intersection of the three planes $-2x - 3y + z = 0$, $-4x + 2y - 4z = -4$, and $x - 5z = 1$.

1-11 The sum of two numbers is 40. The difference between the larger and the smaller is 12. Find the numbers. (Hint: Let $x =$ the larger number and $y =$ the smaller number. Then set up two equations involving x and y and solve.)

1-12 A man can row 2 mi downstream in 15 min and 2 mi upstream in 30 min. How fast is the current in the stream, and how fast would the man be rowing in still water? (Hint: Let x equal the rate of the current and let y equal his rowing rate in still water. Use the relationship "distance = rate × time" to set up two equations.)

1-13 Show that $ax + by + cz = d$ has exactly the same solutions as $kax + kby + kcz = kd$ (where k is a nonzero number). This is the key to showing that operation (b) does not change solutions. [Hint: You must show that a solution (r, s, t) of the first equation is a solution of the second, and vice versa.]

1-14 Show that the systems

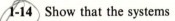

$$ax + by = c$$
$$dx + ey = f$$

and

$$ax + by = c$$
$$(ka + d)x + (kb + e)y = kc + f$$

have exactly the same solutions. This illustrates that operation (c) does not change solutions.

2. A MATRIX METHOD FOR SOLVING SYSTEMS OF EQUATIONS

When solving a system of equations, for example the system in Example 1-1,

$$x + y + 2z = 0$$
$$3x + y + 2z = 2$$
$$2x - 2y + z = -6$$

it is more efficient to write only the *coefficients* of the equations:

$$\begin{bmatrix} 1 & 1 & 2 & \vdots & 0 \\ 3 & 1 & 2 & \vdots & 2 \\ 2 & -2 & 1 & \vdots & -6 \end{bmatrix}$$

We use a dotted line to separate the two sides of the equations. Such an array of numbers is called a *matrix*. Since this one arose from a system of equations, it is called the *matrix of coefficients* of the system. Since it includes the entries on the right sides of the equations, it is referred to as the *augmented* matrix of coefficients. (The *unaugmented* matrix of this system would be

$$\begin{bmatrix} 1 & 1 & 2 \\ 3 & 1 & 2 \\ 2 & -2 & 1 \end{bmatrix}$$

Unaugmented matrices will be dealt with later.)

We now solve the above system, working with its matrix of coefficients. We first eliminate x from the second and third equations by subtracting 3 times the first equation from the second and by subtracting 2 times the first equation from the third. In terms of the matrix this amounts to subtracting 3 times the first row from the second and 2 times the first row from the third. This yields the matrix

$$\begin{bmatrix} 1 & 1 & 2 & \vdots & 0 \\ 0 & -2 & -4 & \vdots & 2 \\ 0 & -4 & -3 & \vdots & -6 \end{bmatrix}$$

whose corresponding system of equations is

$$\begin{aligned} x + y + 2z &= 0 \\ -2y - 4z &= 2 \\ -4y - 3z &= -6 \end{aligned}$$

Continuing, we multiply the second equation by $-\frac{1}{2}$; in terms of the matrix this amounts to multiplying the second row by $-\frac{1}{2}$. Doing this, we obtain

$$\begin{bmatrix} 1 & 1 & 2 & \vdots & 0 \\ 0 & 1 & 2 & \vdots & -1 \\ 0 & -4 & -3 & \vdots & -6 \end{bmatrix}$$

From now on we will consider only the matrix, rather than the equivalent system of equations, bearing in mind that each operation on the matrix is equivalent to a corresponding manipulation on the equations. We eliminate y from the first and third equations by subtracting the second row from the first row and by adding 4 times the second row to the third row, obtaining

$$\begin{bmatrix} 1 & 0 & 0 & \vdots & 1 \\ 0 & 1 & 2 & \vdots & -1 \\ 0 & 0 & 5 & \vdots & -10 \end{bmatrix}$$

We then multiply the third row by $\frac{1}{5}$,

$$\begin{bmatrix} 1 & 0 & 0 & \vdots & 1 \\ 0 & 1 & 2 & \vdots & -1 \\ 0 & 0 & 1 & \vdots & -2 \end{bmatrix}$$

and eliminate z from the second equation by adding -2 times the third row to the second:

$$\begin{bmatrix} 1 & 0 & 0 & \vdots & 1 \\ 0 & 1 & 0 & \vdots & 3 \\ 0 & 0 & 1 & \vdots & -2 \end{bmatrix}$$

From this last matrix we can read off $x = 1$, $y = 3$, $z = -2$; that is, the solution is the triple $(1, 3, -2)$.

We now consider another example, this time working entirely with the matrix of the system

$$\begin{array}{rcl} 2x - 4y & = & 4 \\ 5x - 2y + 10z & = & 0 \\ 3x - y - 5z & = & 1 \end{array}$$

Its augmented matrix is

$$\begin{bmatrix} 2 & -4 & 0 & \vdots & 4 \\ 5 & -2 & 10 & \vdots & 0 \\ 3 & -1 & -5 & \vdots & 1 \end{bmatrix}$$

We want to eliminate x from the last two equations (often referred to as "clearing out" the first column, since it renders all but one entry in that column equal to zero). The following suggestion simplifies the clearing-out process:

A row used to clear out a column should be multiplied so that the row begins with a 1.

We do this to make the numerical calculations needed to clear the column easier. We will use the first row to clear the first column; to make it begin with a 1 we multiply the first row by $\frac{1}{2}$:

$$\begin{bmatrix} 1 & -2 & 0 & \vdots & 2 \\ 5 & -2 & 10 & \vdots & 0 \\ 3 & -1 & -5 & \vdots & 1 \end{bmatrix}$$

Now we can easily see that we must subtract 5 times the first row from the second row and subtract 3 times the first row from the third:

$$\begin{bmatrix} 1 & -2 & 0 & \vdots & 2 \\ 0 & 8 & 10 & \vdots & -10 \\ 0 & 5 & -5 & \vdots & -5 \end{bmatrix}$$

This clears the first column.

We now want to clear out the second column. The next suggestion prevents damage to the column we have just cleared:

Each time a column is to be cleared out, use a different row from any used before.

Thus we cannot use the first row again, but we may use either the second or third row to clear the second column. For convenience, we choose the third row. Again, it helps to make this row begin with a 1. To achieve this, we multiply the third row by $\frac{1}{5}$:

$$\begin{bmatrix} 1 & -2 & 0 & \vdots & 2 \\ 0 & 8 & 10 & \vdots & -10 \\ 0 & 1 & -1 & \vdots & -1 \end{bmatrix}$$

Now we clear out the second column, subtracting 8 times the third row from row 2 and adding 2 times the third row to row 1:

$$\begin{bmatrix} 1 & 0 & -2 & \vdots & 0 \\ 0 & 0 & 18 & \vdots & -2 \\ 0 & 1 & -1 & \vdots & -1 \end{bmatrix}$$

The second column is now cleared.

Lastly, we use a new row—a row not used before—to clear out the third column. We obviously must use the second row. Again, multiplying the second row by $\frac{1}{18}$ helps (by making sure the first nonzero entry in the row is a 1):

$$\begin{bmatrix} 1 & 0 & -2 & \vdots & 0 \\ 0 & 0 & 1 & \vdots & -\frac{1}{9} \\ 0 & 1 & -1 & \vdots & -1 \end{bmatrix}$$

Now we clear the third column and obtain

$$\begin{bmatrix} 1 & 0 & 0 & \vdots & -\frac{2}{9} \\ 0 & 0 & 1 & \vdots & -\frac{1}{9} \\ 0 & 1 & 0 & \vdots & -\frac{10}{9} \end{bmatrix}$$

To avoid any error in reading off the solution, we interchange the last two rows:

$$\begin{bmatrix} 1 & 0 & 0 & \vdots & -\frac{2}{9} \\ 0 & 1 & 0 & \vdots & -\frac{10}{9} \\ 0 & 0 & 1 & \vdots & -\frac{1}{9} \end{bmatrix}$$

We read off $x = -\frac{2}{9}$, $y = -\frac{10}{9}$, $z = -\frac{1}{9}$; thus, the solution is the triple $(-\frac{2}{9}, -\frac{10}{9}, -\frac{1}{9})$.

In general, in simplifying the matrix for a system of equations the two suggestions in this last example facilitate the process:

(a)   A row used to clear out a column should be multiplied so that the first nonzero entry in that row is a 1.

(b)   Each time a column is to be cleared out, use a different row from any used before.

When we simplify a matrix, we are aiming for a matrix that makes the solutions as easy as possible to read off. We now describe precisely what this form should be.

Definition

A matrix is in **reduced echelon form (reduced form,** for short) if:

(a)   The first nonzero entry in any row is a 1 (this "row-leading" 1 may occur after some zeros).

(b)   Every column in which there is a row-leading 1 is cleared out, that is, that column has nothing but zeros apart from the 1.

(c)   The row-leading 1s descend to the right.

Example 2-1

The matrices

$$\begin{bmatrix} \underline{1} & 0 & 5 & 0 \\ 0 & \underline{1} & 2 & 0 \\ 0 & 0 & 0 & \underline{1} \end{bmatrix} \qquad \begin{bmatrix} \underline{1} & 0 & 0 & 0 & 0 \\ 0 & \underline{1} & 0 & 2 & 0 \\ 0 & 0 & \underline{1} & 2 & 0 \\ 0 & 0 & 0 & 0 & \underline{1} \end{bmatrix} \qquad \begin{bmatrix} \underline{1} & 0 & 0 \\ 0 & \underline{1} & 0 \\ 0 & 0 & \underline{1} \end{bmatrix}$$

are in reduced form. The row-leading 1s are underlined in this text to emphasize their importance.

The matrix

$$\begin{bmatrix} \underline{1} & 0 & 0 & 4 & 5 \\ 0 & \underline{1} & 0 & 0 & 2 \\ 0 & 0 & 3 & 2 & 0 \\ 0 & 0 & 0 & 0 & 0 \end{bmatrix}$$

is not in reduced form since the first nonzero entry in the third row is 3, not 1.

The matrix

$$\begin{bmatrix} \underline{1} & 0 & 5 \\ 0 & \underline{1} & 0 \\ 0 & 0 & \underline{1} \\ 0 & 0 & 0 \end{bmatrix}$$

is not in reduced form since the 5 has not been cleared out of the third column.

The matrix

$$\begin{bmatrix} \underline{1} & 0 & 0 & 5 \\ 0 & \underline{1} & 0 & 2 \\ 0 & 0 & 4 & 3 \\ 0 & 0 & 0 & \underline{1} \end{bmatrix}$$

violates conditions (a) and (b) of the definition.   ●

In the examples at the beginning of this section we solved systems of equations by simplifying their matrices, bringing them to reduced form. In fact, any matrix can be put into reduced form by using an appropriate combination of the three operations

(a)   Interchanging two rows

(b)   Multiplying a row by a nonzero number

(c)   Adding a multiple of one row to another

These are the analogs of the corresponding operations on equations listed at the end of Sec. 1. The following two examples illustrate this process of reducing a matrix (*reducing* means bringing to reduced echelon form by means of our three operations).

Example 2-2

Let

$$A = \begin{bmatrix} 4 & 1 & 2 & 1 \\ 0 & 1 & 3 & 0 \\ 1 & 0 & -2 & 1 \\ 2 & 1 & 4 & 0 \end{bmatrix}$$

Here is one way of reducing A (the letters on the arrows refer to the explanation below):

$$\begin{bmatrix} 4 & 1 & 2 & 1 \\ 0 & 1 & 3 & 0 \\ 1 & 0 & -2 & 1 \\ 2 & 1 & 4 & 0 \end{bmatrix} \xrightarrow{(a)} \begin{bmatrix} 1 & 0 & -2 & 1 \\ 0 & 1 & 3 & 0 \\ 4 & 1 & 2 & 1 \\ 2 & 1 & 4 & 0 \end{bmatrix}$$

$$\xrightarrow{(b)} \begin{bmatrix} 1 & 0 & -2 & 1 \\ 0 & 1 & 3 & 0 \\ 0 & 1 & 10 & -3 \\ 0 & 1 & 8 & -2 \end{bmatrix} \xrightarrow{(c)} \begin{bmatrix} 1 & 0 & -2 & 1 \\ 0 & 1 & 3 & 0 \\ 0 & 0 & 7 & -3 \\ 0 & 0 & 5 & -2 \end{bmatrix}$$

$$\xrightarrow{(d)} \begin{bmatrix} 1 & 0 & -2 & 1 \\ 0 & 1 & 3 & 0 \\ 0 & 0 & 1 & -\frac{3}{7} \\ 0 & 0 & 5 & -2 \end{bmatrix} \xrightarrow{(e)} \begin{bmatrix} 1 & 0 & 0 & \frac{1}{7} \\ 0 & 1 & 0 & \frac{9}{7} \\ 0 & 0 & 1 & -\frac{3}{7} \\ 0 & 0 & 0 & \frac{1}{7} \end{bmatrix}$$

$$\xrightarrow{(f)} \begin{bmatrix} 1 & 0 & 0 & \frac{1}{7} \\ 0 & 1 & 0 & \frac{9}{7} \\ 0 & 0 & 1 & -\frac{3}{7} \\ 0 & 0 & 0 & 1 \end{bmatrix} \xrightarrow{(g)} \begin{bmatrix} 1 & 0 & 0 & 0 \\ 0 & 1 & 0 & 0 \\ 0 & 0 & 1 & 0 \\ 0 & 0 & 0 & 1 \end{bmatrix}$$

Explanation:

(a) The first and third rows are interchanged.

(b) The first column is cleared (by subtracting 4 times the first row from the third row and 2 times the first row from the fourth row).

(c) The second column is cleared using a *new* row (by subtracting the second row from both the third and fourth rows).

(d) A row-leading 1 is introduced into the third row—a row not used before—by multiplying the third row by $\frac{1}{7}$.

(e) The third column is cleared.

(f) A row-leading 1 is introduced into the fourth row.

(g) The fourth column is cleared.

Note that each time we cleared a column we chose a row not used before, and a row-leading 1 was introduced at once. The row-leading 1 was then used to clear out its own column.   ●

Example 2-3

Let

$$A = \begin{bmatrix} 3 & 3 & -2 \\ 2 & 1 & 4 \end{bmatrix}$$

The following sequence of operations puts A into reduced form:

$$\begin{bmatrix} 3 & 3 & -2 \\ 2 & 1 & 4 \end{bmatrix} \longrightarrow \begin{bmatrix} 1 & 1 & -\frac{2}{3} \\ 2 & 1 & 4 \end{bmatrix} \longrightarrow \begin{bmatrix} 1 & 1 & -\frac{2}{3} \\ 0 & -1 & \frac{16}{3} \end{bmatrix}$$

$$\longrightarrow \begin{bmatrix} 1 & 1 & -\frac{2}{3} \\ 0 & 1 & -\frac{16}{3} \end{bmatrix} \longrightarrow \begin{bmatrix} 1 & 0 & \frac{14}{3} \\ 0 & 1 & -\frac{16}{3} \end{bmatrix}$$

The reader should check which operation was used at each step in the reduction process.   ●

Using the idea of reducing a matrix, we can summarize our method for solving a system of equations as follows:

(a)   Write its augmented matrix of coefficients.

(b)   Put this matrix into reduced form (using a combination of the three operations).

(c)   Read off the solution to the original system from the reduced matrix.

Example 2-4

To solve

$$\begin{aligned} x_1 + 2x_2 + 4x_3 &= 3 \\ x_1 \qquad\quad + 2x_3 &= 0 \\ 2x_1 + 4x_2 + \quad x_3 &= 3 \end{aligned}$$

for the unknowns x_1, x_2, x_3, we first write its augmented matrix:

$$\begin{bmatrix} 1 & 2 & 4 & \vdots & 3 \\ 1 & 0 & 2 & \vdots & 0 \\ 2 & 4 & 1 & \vdots & 3 \end{bmatrix}$$

We then put the matrix into reduced form:

$$\begin{bmatrix} 1 & 2 & 4 & \vdots & 3 \\ 1 & 0 & 2 & \vdots & 0 \\ 2 & 4 & 1 & \vdots & 3 \end{bmatrix} \longrightarrow \begin{bmatrix} 1 & 2 & 4 & \vdots & 3 \\ 0 & -2 & -2 & \vdots & -3 \\ 2 & 4 & 1 & \vdots & 3 \end{bmatrix}$$

$$\longrightarrow \begin{bmatrix} 1 & 2 & 4 & \vdots & 3 \\ 0 & -2 & -2 & \vdots & -3 \\ 0 & 0 & -7 & \vdots & -3 \end{bmatrix} \longrightarrow \begin{bmatrix} 1 & 2 & 4 & \vdots & 3 \\ 0 & 1 & 1 & \vdots & \frac{3}{2} \\ 0 & 0 & -7 & \vdots & -3 \end{bmatrix}$$

$$\longrightarrow \begin{bmatrix} 1 & 0 & 2 & \vdots & 0 \\ 0 & 1 & 1 & \vdots & \frac{3}{2} \\ 0 & 0 & -7 & \vdots & -3 \end{bmatrix} \longrightarrow \begin{bmatrix} 1 & 0 & 2 & \vdots & 0 \\ 0 & 1 & 1 & \vdots & \frac{3}{2} \\ 0 & 0 & 1 & \vdots & \frac{3}{7} \end{bmatrix}$$

$$\longrightarrow \begin{bmatrix} 1 & 0 & 0 & \vdots & -\frac{6}{7} \\ 0 & 1 & 1 & \vdots & \frac{3}{2} \\ 0 & 0 & 1 & \vdots & \frac{3}{7} \end{bmatrix} \longrightarrow \begin{bmatrix} 1 & 0 & 0 & \vdots & -\frac{6}{7} \\ 0 & 1 & 0 & \vdots & \frac{15}{14} \\ 0 & 0 & 1 & \vdots & \frac{3}{7} \end{bmatrix}$$

From the reduced matrix we read off $x_1 = -\frac{6}{7}$, $x_2 = \frac{15}{14}$, $x_3 = \frac{3}{7}$, which we can rewrite as $(-\frac{6}{7}, \frac{15}{14}, \frac{3}{7})$. ●

Example 2-5

To solve

$$x - 3y = 0$$
$$-2x + 5y = 0$$

we write its augmented matrix

$$\begin{bmatrix} 1 & -3 & \vdots & 0 \\ -2 & 5 & \vdots & 0 \end{bmatrix}$$

We reduce the matrix

$$\begin{bmatrix} 1 & -3 & \vdots & 0 \\ -2 & 5 & \vdots & 0 \end{bmatrix} \longrightarrow \begin{bmatrix} 1 & -3 & \vdots & 0 \\ 0 & -1 & \vdots & 0 \end{bmatrix}$$

$$\longrightarrow \begin{bmatrix} 1 & -3 & \vdots & 0 \\ 0 & 1 & \vdots & 0 \end{bmatrix} \longrightarrow \begin{bmatrix} 1 & 0 & \vdots & 0 \\ 0 & 1 & \vdots & 0 \end{bmatrix}$$

From the reduced matrix we read off $x = 0$, $y = 0$, which we can rewrite as $(0, 0)$. ●

A system of equations, such as the one in Example 2-5, in which all the constant terms are equal to zero, is called a *homogeneous* system. Otherwise the system is called *nonhomogeneous*. The system of equations in Example 2-4 is nonhomogeneous.

The augmented matrix for a homogeneous system will have a column of zeros on the right, and each entry in the right-hand column will remain zero throughout the reduction process (see Example 2-5). Thus, when reducing the matrix for a homogeneous system of equations, we can leave out this right-hand column of zeros, and when we read off the solution, restore the zeros.

For example, the augmented matrix for the homogeneous system

$$x + 2y + 2z = 0$$
$$x + 3y + z = 0$$
$$2x + 4y + 5z = 0$$

is

$$\begin{bmatrix} 1 & 2 & 2 & \vdots & 0 \\ 1 & 3 & 1 & \vdots & 0 \\ 2 & 4 & 5 & \vdots & 0 \end{bmatrix}$$

Leaving out the column of zeros, we have the corresponding unaugmented matrix

$$\begin{bmatrix} 1 & 2 & 2 \\ 1 & 3 & 1 \\ 2 & 4 & 5 \end{bmatrix}$$

This unaugmented matrix reduces to

$$\begin{bmatrix} 1 & 0 & 0 \\ 0 & 1 & 0 \\ 0 & 0 & 1 \end{bmatrix}$$

To read off the solution, mentally restore the column of zeros on the right to get the augmented matrix

$$\begin{bmatrix} 1 & 0 & 0 & \vdots & 0 \\ 0 & 1 & 0 & \vdots & 0 \\ 0 & 0 & 1 & \vdots & 0 \end{bmatrix}$$

and then read off $x = 0$, $y = 0$, $z = 0$; that is, $(0, 0, 0)$.

Exercises†

Find the reduced form for each of the matrices in Exercises 2-1 to 2-4. Remember the two suggestions made in this section:

(a) Introduce row-leading 1s and use them to clear columns.
(b) To clear a column use a row that was not used before.

2-1
$$\begin{bmatrix} 1 & 1 & 7 & 0 \\ 0 & -2 & 9 & 1 \\ 1 & 4 & 5 & 1 \\ 3 & 5 & 17 & 2 \end{bmatrix}$$

2-2
$$\begin{bmatrix} 1 & 5 & -1 & 0 \\ 2 & 10 & -1 & 0 \\ 1 & 5 & 3 & 0 \\ 0 & 0 & 1 & 0 \end{bmatrix}$$

2-3
$$\begin{bmatrix} 2 & 4 & 5 \\ 1 & 0 & 2 \\ 5 & -8 & 4 \\ 1 & 0 & 4 \end{bmatrix}$$

2-4
$$\begin{bmatrix} 2 & 0 & 6 & 8 & 0 \\ 2 & 1 & 8 & 9 & 1 \\ 1 & 0 & 3 & 4 & -1 \end{bmatrix}$$

2-5 Solve the systems in Exercises 1-3 to 1-8 (of Sec. 1) by reducing their augmented matrices.

Solve each of the systems in Exercises 2-6 to 2-11 by reducing its matrix.

2-6
$$\begin{aligned} x + 2y + 3z &= 2 \\ 3x + 7y + 9z &= 0 \\ x + 4y + z &= 2 \end{aligned}$$

2-7
$$\begin{aligned} x \quad\quad - 2z &= 0 \\ y + 3z &= 0 \\ 4x + y + 2z &= 0 \end{aligned}$$

2-8
$$\begin{aligned} 2x_1 + 4x_2 \quad\quad &= 2 \\ x_1 + 3x_2 + 4x_3 &= 3 \\ - 2x_2 + 4x_3 &= 2 \\ 3x_1 + 7x_2 + 4x_3 &= 5 \end{aligned}$$

2-9
$$\begin{aligned} x + y \quad\quad &= 0 \\ 2x + 3y + z &= 0 \\ 3x + 2y + z &= 0 \end{aligned}$$

2-10
$$\begin{aligned} x + 2y + z &= 3 \\ 3x + 8y + 7z &= 1 \\ x + 2y + 5z &= -1 \\ 2x + 6y + 2z &= 2 \end{aligned}$$

2-11
$$\begin{aligned} 2x_1 + x_2 - x_3 + 6x_4 &= 1 \\ x_1 + 2x_2 + x_3 - 2x_4 &= 7 \\ 3x_2 - x_3 + 2x_4 &= -5 \\ x_1 - x_2 \quad\quad - 4x_4 &= 0 \end{aligned}$$

*2-12 Write a computer program for reducing matrices to reduced echelon form.

3. SETS OF SOLUTIONS

Every system of equations solved so far in this chapter has had a unique solution. This is by no means always the case. A system of equations

† The asterisks indicate problems involving computer programs.

might have infinitely many solutions, or it might have no solution at all. Figure 3-1 gives a geometric interpretation of this phenomenon.

Figure 3-1*a* shows three planes that intersect in a single point *P*; Fig. 3-1*b* illustrates three planes that intersect in an infinite set of points, the line ℓ; and Fig. 3-1*c* shows three planes that have no point in common.

In any case the same matrix method can be used to solve the system.

Example 3-1

Consider the system

$$
\begin{aligned}
x + 2y \quad\;\; &= \quad 0 \\
y - z &= \quad 2 \\
x + \; y + z &= -2
\end{aligned}
$$

As we will see, this system has an infinite number of solutions. Its matrix

$$
\begin{bmatrix}
1 & 2 & 0 & \vdots & 0 \\
0 & 1 & -1 & \vdots & 2 \\
1 & 1 & 1 & \vdots & -2
\end{bmatrix}
$$

reduces to

$$
\begin{bmatrix}
1 & 0 & 2 & \vdots & -4 \\
0 & 1 & -1 & \vdots & 2 \\
0 & 0 & 0 & \vdots & 0
\end{bmatrix}
$$

Note that we are unable to clear the third column without doing damage to one of the first two columns. Attempting to read off the solution, we obtain

$$
\begin{aligned}
x + \quad\;\; 2z &= -4 \\
y - \; z &= \quad 2
\end{aligned}
$$

which does *not* yield unique values for *x*, *y*, and *z*. To obtain the possible values for *x*, *y*, and *z* we rewrite this as

$$
\begin{aligned}
x &= -4 - 2z \\
y &= \quad 2 + \; z
\end{aligned}
$$

In this form, when we substitute a number for *z*, we can read out corresponding numbers for *x* and *y*. For example, if we let $z = 1$, then $x = -6$

Fig. 3-1 Three planes can intersect in (*a*) a single point, (*b*) infinitely many points, (*c*) no points at all.

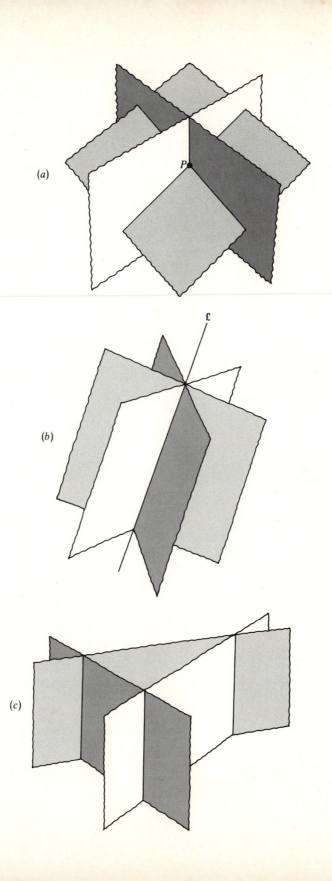

(a)

(b)

£

(c)

and $y = 3$; or if we let $z = 0$, we get $x = -4$ and $y = 2$. More generally, if we let $z = a$, then $x = -4 - 2a$ and $y = 2 + a$. In other words, for any number a we substitute for z, $(-4 - 2a, \ 2 + a, \ a)$ is a solution. The original system of equations has an *infinite* number of solutions, all triples of the form $(-4 - 2a, \ 2 + a, \ a)$.

Geometrically, the three equations we began with represent three planes, and the infinite number of solutions we ended with gives a line. We will return to this idea in Sec. 4. The three planes have this line as their common intersection, as in Fig. 3-1*b*. ●

Example 3-2

Consider the system of equations

$$\begin{aligned} x + 2y + \ z &= 0 \\ y + 2z &= 0 \\ x + \ y - \ z &= 0 \end{aligned}$$

Its unaugmented matrix reduces to

$$\begin{bmatrix} 1 & 0 & -3 \\ 0 & 1 & 2 \\ 0 & 0 & 0 \end{bmatrix}$$

This gives

$$\begin{aligned} x \quad\ - 3z &= 0 \\ y + 2z &= 0 \end{aligned}$$

Observe that the row-leading 1s in the reduced matrix correspond to the unknowns x and y. It is these unknowns that we keep on the left side, shifting all other terms to the right. We thus obtain

$$\begin{aligned} x &= \ 3z \\ y &= -2z \end{aligned}$$

Next we let $z = a$ and write the solutions as

$$\begin{aligned} x &= \ 3a \\ y &= -2a \\ z &= \quad a \end{aligned}$$

In other words, the solutions are all triples of the form $(3a, \ -2a, \ a)$, where a can be any number. ●

Example 3-3

Suppose a certain system of equations with unknowns x_1, x_2, x_3, x_4 has as its reduced matrix

$$\begin{bmatrix} 1 & 0 & 4 & -1 & \vdots & 7 \\ 0 & 1 & -2 & -3 & \vdots & 8 \\ 0 & 0 & 0 & 0 & \vdots & 0 \end{bmatrix}$$

The row-leading 1s correspond to x_1 and x_2. Writing these on the left and shifting all other terms to the right, we obtain

$$x_1 = 7 - 4x_3 + x_4$$
$$x_2 = 8 + 2x_3 + 3x_4$$

We can choose any numbers we wish for x_3 and x_4, say $x_3 = a$ and $x_4 = b$. Then the solutions may be written

$$x_1 = 7 - 4a + b$$
$$x_2 = 8 + 2a + 3b$$
$$x_3 = a$$
$$x_4 = b$$

The solutions are all 4-tuples of the form $(7 - 4a + b, 8 + 2a + 3b, a, b)$, where a and b can be any numbers. In writing the solutions as 4-tuples we are generalizing the notation (x, y) for points in the xy plane and (x, y, z) for points in xyz space.   ●

Example 3-4

Suppose a system of homogeneous equations has an unaugmented matrix that reduces to

$$\begin{bmatrix} 1 & 3 & 0 & -5 & 0 \\ 0 & 0 & 1 & 4 & 0 \\ 0 & 0 & 0 & 0 & 1 \end{bmatrix}$$

Since the row-leading 1s correspond to x_1, x_3, and x_5, we keep these on the left and write

$$x_1 = -3x_2 + 5x_4$$
$$x_3 = -4x_4$$
$$x_5 = 0$$

We can choose any numbers we want for x_2 and x_4, say a and b, and can write the solutions as

$$
\begin{aligned}
x_1 &= -3a + 5b \\
x_2 &= a \\
x_3 &= -4b \\
x_4 &= b \\
x_5 &= 0
\end{aligned}
$$

or as 5-tuples of the form $(-3a + 5b,\ a,\ -4b,\ b,\ 0)$.   ●

Example 3-5

In solving the system

$$
\begin{aligned}
x + 2y + 4z &= 6 \\
y + z &= 1 \\
x + 3y + 5z &= 10
\end{aligned}
$$

its matrix reduces to

$$
\left[\begin{array}{ccc:c}
1 & 0 & 2 & 0 \\
0 & 1 & 1 & 0 \\
0 & 0 & 0 & 1
\end{array}\right]
$$

That is,

$$
\begin{aligned}
x + 2z &= 0 \\
y + z &= 0 \\
0x + 0y + 0z &= 1
\end{aligned}
$$

This system has no solution since the third "equation" $0x + 0y + 0z = 1$ has no solution. Hence the original system has no solution. Geometrically, the three original equations represent three planes having no point in common, as in Fig. 3-1c.   ●

It was mentioned earlier in this section that the triple $(a,\ 2a,\ a)$ represents infinitely many solutions, one for each value of a. To indicate that we have a whole set of solutions, we frequently use set notation:

$$
\{(a,\ 2a,\ a) : a \in \mathcal{R}\}
$$

The latter is read as "the set of all ordered triples of the form $(a,\ 2a,\ a)$, where a is any number." The symbol $\in$ means "is an element of" and $\mathcal{R}$

denotes the set of all real numbers. To pick out particular points that belong to this set, we take particular real numbers for a. For example $a = 3$ yields $(3, 6, 3)$ as one particular member of this set, and $a = -\sqrt{3}$ yields $(-\sqrt{3}, -2\sqrt{3}, -\sqrt{3})$ as another point in this set.

A set such as $\{(2a - 5b, a, b) : a \in \mathcal{R}, b \in \mathcal{R}\}$ would be read "the set of all ordered triples of the form $(2a - 5b, a, b)$, where a and b are real numbers." It also has infinitely many points in it, one for each choice of two numbers a and b.

It is frequently important to be able to *rewrite* solutions in a different form.† The sets of solutions

$$A = \{(a, 2a, a) : a \in \mathcal{R}\}$$
$$B = \{(2a - 5b, a, b) : a, b \in \mathcal{R}\}$$

are not very revealing in their present form. It would be hard to describe these sets geometrically, for instance. Set A can be rewritten by factoring out the a in the triple $(a, 2a, a)$, obtaining $a(1, 2, 1)$. We are using a special kind of multiplication here. By definition, the product of a number times an ordered triple is obtained by multiplying each entry of the triple by the number. Thus, by definition, set A can be written $\{a(1, 2, 1) : a \in \mathcal{R}\}$. The significance of the ordered triple $(1, 2, 1)$ will be explored in the next section.

Set B can be rewritten by defining an addition of ordered triples as follows:

$$(x, y, z) + (u, v, w) = (x + u, y + v, z + w)$$

that is, we add corresponding entries of the triples. It follows that $(2a - 5b, a, b)$ can be written as the sum of an "a part" and a "b part":

$$(2a, a, 0) + (-5b, 0, b)$$

We do this by writing the a part of each entry of the original triple in one triple, and writing the b part of each entry in a second triple. We can now factor out a and b, obtaining

$$a(2, 1, 0) + b(-5, 0, 1)$$

Thus set B can be written

$$\{a(2, 1, 0) + b(-5, 0, 1) : a, b \in \mathcal{R}\}$$

† The idea of rewriting expressions recurs remarkably often in mathematics. For example, the equation $3x^2 - 6x - 12y - 12 + 3y^2 = 0$ appears—at first—to be nothing but a jumble of symbols. But rewritten as $(x - 1)^2 + (y - 2)^2 = 9$, it is quickly seen to be the equation of a circle with center $(1, 2)$ and radius 3.

The significance of the triples $(2, 1, 0)$ and $(-5, 0, 1)$ will be dealt with in the next section.

In like fashion, the triple $(4 + 2a, 7, a)$ can be split into a numerical part and an a part. Moving from entry to entry, we put the numerical parts of each entry into one triple $(4, 7, 0)$ and the a terms into a second triple $(2a, 0, a)$, thus obtaining

$$(4, 7, 0) + (2a, 0, a)$$

Factoring out a yields

$$(4, 7, 0) + a(2, 0, 1)$$

We can also split the triple $(2 - 3a + 5b, -6 + a, 2 + b)$ into a numerical part, an a part, and a b part:

$$(2, -6, 2) + (-3a, a, 0) + (5b, 0, b)$$

or, after factoring out a and b,

$$(2, -6, 2) + a(-3, 1, 0) + b(5, 0, 1)$$

Ordered pairs, 4-tuples, 5-tuples, etc., can be rewritten in exactly the same way. For example,

$$\begin{aligned}
(2 + 3a - 2b, a + b, 5a, -7 + a, b) &= (2, 0, 0, -7, 0) + (3a, a, 5a, a, 0) \\
&\quad + (-2b, b, 0, 0, b) \\
&= (2, 0, 0, -7, 0) + a(3, 1, 5, 1, 0) \\
&\quad + b(-2, 1, 0, 0, 1)
\end{aligned}$$

We will now solve a system of equations from start to finish, using the above ideas to write the set of solutions as simply as possible.

Example 3-6

Find the solutions of the system

$$\begin{aligned}
x - 2y \quad &= \quad 2 \\
2x + y - 5z &= -1 \\
4x \quad - 8z &= \quad 0
\end{aligned}$$

Its augmented matrix is

$$\begin{bmatrix}
1 & -2 & 0 & \vdots & 2 \\
2 & 1 & -5 & \vdots & -1 \\
4 & 0 & -8 & \vdots & 0
\end{bmatrix}$$

which reduces to

$$\left[\begin{array}{ccc|c} 1 & 0 & -2 & 0 \\ 0 & 1 & -1 & -1 \\ 0 & 0 & 0 & 0 \end{array}\right]$$

We read off

$$\begin{aligned} x &= & 2z \\ y &= -1 + & z \end{aligned}$$

As before, we let $z = a$ and have

$$\begin{aligned} x &= & 2a \\ y &= -1 + & a \\ z &= & a \end{aligned}$$

Thus the set of solutions is

$$\{(2a, -1 + a, a) : a \in \mathcal{R}\}$$

Lastly we rewrite the triple $(2a, -1 + a, a)$ as

$$(0, -1, 0) + a(2, 1, 1)$$

And so, the solution set, rewritten, is

$$\{(0, -1, 0) + a(2, 1, 1) : a \in \mathcal{R}\}$$

●

Exercises

Each of the matrices (already in reduced form) in Exercises 3-1 to 3-9 represents a system of equations with unknowns x_1, x_2, x_3, etc. Find the set of solutions for each system. Rewrite each solution set to exhibit the numerical part, the a part, b part, etc.

3-1 $\left[\begin{array}{cc|c} 1 & 0 & -3 \\ 0 & 1 & 2 \end{array}\right]$

3-2 $\left[\begin{array}{cccc} 1 & 0 & 0 & 0 \\ 0 & 1 & 0 & 0 \\ 0 & 0 & 1 & 0 \\ 0 & 0 & 0 & 1 \end{array}\right]$

3-3 $\begin{bmatrix} 1 & 0 & 0 & 9 & | & 2 \\ 0 & 1 & 0 & -1 & | & 1 \\ 0 & 0 & 1 & 3 & | & 4 \end{bmatrix}$ 3-4 $\begin{bmatrix} 1 & 4 & 0 & 3 \\ 0 & 0 & 1 & -1 \end{bmatrix}$

3-5 $\begin{bmatrix} 1 & 4 & 0 & | & 3 \\ 0 & 0 & 1 & | & -1 \end{bmatrix}$ 3-6 $\begin{bmatrix} 1 & 0 & 6 & | & 0 \\ 0 & 1 & 5 & | & 0 \\ 0 & 0 & 0 & | & 1 \end{bmatrix}$

3-7 $\begin{bmatrix} 1 & 0 & 0 & -1 & 4 & | & 7 \\ 0 & 1 & 0 & 5 & 3 & | & 5 \\ 0 & 0 & 1 & 2 & 3 & | & -9 \end{bmatrix}$

3-8 $\begin{bmatrix} 0 & 0 & 1 & -4 & -2 & 0 \\ 0 & 0 & 0 & 0 & 0 & 1 \\ 0 & 0 & 0 & 0 & 0 & 0 \end{bmatrix}$

(Hint: There are six columns, hence six unknowns $x_1, x_2, \ldots, x_6$. Only x_3 and x_6 have row-leading 1s, thus the other unknowns x_1, x_2, x_4, and x_5 can be any real number.)

3-9 $\begin{bmatrix} 0 & 1 & 2 & 0 & 0 & 2 & | & 1 \\ 0 & 0 & 0 & 1 & 0 & 3 & | & -2 \\ 0 & 0 & 0 & 0 & 1 & 0 & | & 0 \\ 0 & 0 & 0 & 0 & 0 & 0 & | & 0 \end{bmatrix}$

Solve each of the systems in Exercises 3-10 to 3-17. Rewrite each solution set to exhibit the numerical part, a part, b part, etc.

3-10
$$x \qquad + 3z = 1$$
$$2x + 3y + 18z = \tfrac{7}{2}$$
$$2x + 2y + 14z = 3$$

3-11
$$x_1 + x_2 + 2x_3 = 0$$
$$x_1 - x_2 + x_3 = 0$$

3-12
$$x_1 + x_2 + 2x_3 = 4$$
$$x_1 - x_2 + x_3 = 2$$

3-13
$$u + 2v + 4w = 6$$
$$v + w = 1$$
$$u + 3v + 5w = 10$$

3-14
$$x_1 + x_2 + x_3 + x_4 = 0$$
$$x_1 - x_2 + x_3 - x_4 = 0$$
$$-x_1 + x_2 - x_3 - x_4 = 0$$

3-15 $x - y + 2z = 0$

3-16
$$x_1 + 3x_2 + x_3 + 3x_5 \qquad = 14$$
$$x_3 - x_4 + 3x_5 + 3x_7 \quad = 8$$
$$-x_3 + x_4 - 3x_5 - x_6 + x_7 = 1$$

3-17 $x + 3y = c$
$x + 2y = d$

4.   GEOMETRY AND SOLUTIONS

In the last section we obtained solution sets whose members were of the form $a(-, -, -)$, $(-, -, -) + a(-, -, -) + b(-, -, -)$, etc. In this section we will study sets such as these geometrically. As we will see, they represent lines and planes.

We first remind the reader of the general equation for lines in the xy plane. We take as a starting point the idea of slope and the fact that a line is determined by a point on it together with its slope†. From this starting point we can obtain the following theorem describing lines in terms of equations.

Theorem 4-1   Any line in the xy plane is the solution set of a linear equation

$$rx + sy = t$$

The key to the proof is rewriting the slope relationship $(y - q)/(x - p) = m$ in the form $rx + sy = t$. The details are left to the reader as an exercise.

For lines in xyz space we take as a starting point the fact that any line is determined by one point on it and two slope relationships.‡ We are led to the following description of lines in xyz space.

Theorem 4-2   Any line in xyz space is the solution set of a system of two equations of the form

$$px + qy = c$$
$$ry + sz = d$$

The proof depends on the observation that the first equation describes the slope relationship m_1 and the second equation describes m_2.

---

† Given a point (p, q) in the xy plane plus a value m for the slope, there is a unique line through (p, q) having the slope m. In the usual terminology, slope $= \Delta y/\Delta x =$ (change in y)/(change in x).
‡ That is, given a point (p, q, r) in xyz space plus values for $m_1 = \Delta y/\Delta x$ and $m_2 = \Delta z/\Delta y$, there is a unique line through (p, q, r) having the "slopes" m_1 and m_2.

These two theorems will help us to describe solution sets whose members are of the forms $(-, -) + a(-, -)$ or $(-, -, -) + a(-, -, -)$. We begin with two examples.

Example 4-1

The set of points of the form $(\frac{5}{2}, 0) + a(2, 1)$ arises as the set of solutions of the equation $2x - 4y = 5$.

[To see this, consider the one equation as a system with matrix $[2 \quad -4 \vdots 5]$. The matrix reduces to $[\underline{1} \quad -2 \vdots \frac{5}{2}]$, from which we read the solutions $(\frac{5}{2} + 2a, a)$. Lastly, separate the numerical part and the a part to obtain $(\frac{5}{2}, 0) + a(2, 1)$.]

By Theorem 4-1, the set of points is therefore a line $\mathfrak{L}$ in the xy plane. It still remains to uncover the meaning of the points $(\frac{5}{2}, 0)$ and $(2, 1)$ that are displayed. Taking $a = 0$ in the expression $(\frac{5}{2}, 0) + a(2, 1)$ yields $(\frac{5}{2}, 0)$ as one particular point on $\mathfrak{L}$.

Consider now the effect of adding multiples $a(2, 1)$ to the point $(\frac{5}{2}, 0)$. From the entries of $(2, 1)$ we deduce that for each 1 unit moved in the y direction, we move 2 units in the x direction as we move along $\mathfrak{L}$ from the point $(\frac{5}{2}, 0)$. This can be pictured in two ways: (1) $\mathfrak{L}$ has slope $\frac{1}{2}$; (2) $\mathfrak{L}$ is parallel to the segment from $(0, 0)$ to $(2, 1)$. (See Fig. 4-1.) ●

Example 4-2

The set of points $(4, 1, 0) + a(2, 5, 1)$ arises as the set of solutions of the system

$$5x - 2y = 18$$
$$y - 5z = 1$$

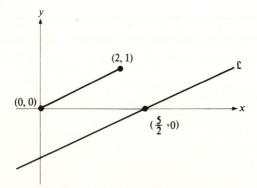

Fig. 4-1   The line $\mathfrak{L}$ is parallel to the segment from the origin to $(2, 1)$.

(The student should verify this.) By Theorem 4-2, the set is therefore a line $\mathcal{M}$ in xyz space. Again, we wish to uncover the meaning of the points $(4, 1, 0)$ and $(2, 5, 1)$. Taking $a = 0$, we find that $(4, 1, 0)$ is a particular point on $\mathcal{M}$. The point $(2, 5, 1)$ indicates the direction of $\mathcal{M}$, in the sense that we obtain points on the line by moving from $(4, 1, 0)$ 2 units in the x direction and 5 units in the y direction for each 1 unit moved in the z direction. A convenient way to sketch $\mathcal{M}$ is to first sketch the segment joining $(0, 0, 0)$ and $(2, 5, 1)$, as in Fig. 4-2. $\mathcal{M}$ is parallel to this segment and passes through the point $(4, 1, 0)$.   ●

In each of these examples the set of points was the set of solutions to a system of one or two equations, and therefore was a line. Each had the same basic form:

$$\begin{pmatrix} \text{Particular point} \\ \text{on the line} \end{pmatrix} + a\begin{pmatrix} \text{point indicating} \\ \text{direction} \end{pmatrix}$$

This is always the case for such sets.

Theorem 4-3

Lines are the sets of points of the form

1 $\{(c, d) + a(e, f) : a \in \mathcal{R}\}$ in the xy plane
2 $\{(c, d, e) + a(f, g, h) : a \in \mathcal{R}\}$ in xyz space

As in the examples, (c, d) or (c, d, e) will be a particular point on the line,

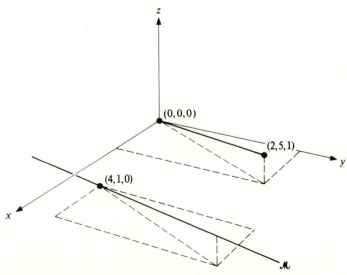

Fig. 4-2   The line $\mathcal{M}$ is parallel to the segment from the origin to $(2, 5, 1)$.

and the line will be parallel to the segment joining the origin with (e, f) or (f, g, h).

In particular, according to Theorem 4-3, a set of points of the form $(0, 0) + a(e, f)$ is a line through the origin $(0, 0)$. Since $(0, 0) + a(e, f) = a(e, f)$, the form $a(e, f)$ describes a line through the origin in the xy plane. Hence we have the following corollary.

Corollary 4-4   Lines that pass through the origin are the sets of points of the form

  1 $\{a(e, f) : a \in \mathcal{R}\}$ in the xy plane

  2 $\{a(f, g, h) : a \in \mathcal{R}\}$ in xyz space

Since (e, f) indicates the direction of the line, we see that the points of the form $a(e, f)$ describe the unique line through the origin and (e, f). Similarly for the form $a(f, g, h)$.

We now consider planes in xyz space. Various definitions are possible for a plane in xyz space. For example, (1) a plane is the set of all points equidistant from two given points; or (2) a plane is the set of all points that lie on perpendiculars dropped to a given line at a given point on that line. Whatever the definition, any plane in xyz space can be shown to satisfy a single linear equation.

Theorem 4-5   Any plane in xyz space is the solution set of a linear equation

$$px + qy + rz = s$$

Using this general equation for planes, we can easily describe the sets of points of the form $(-, -, -) + a(-, -, -) + b(-, -, -)$. First two examples.

Example 4-3

Consider the set of points of the form

$$a(-\tfrac{5}{2}, 1, 0) + b(2, 0, 1)$$

This set arises as the solution set of the equation $2x + 5y - 4z = 0$. (The reader can check this.) From Theorem 4-5 we know that this set of solutions is a plane $\mathcal{P}$, but how is it positioned in xyz space? Taking $a = b = 0$ shows that $\mathcal{P}$ contains the origin $(0, 0, 0)$. Taking $a = 1$ and $b = 0$ shows that $\mathcal{P}$ contains $(-\tfrac{5}{2}, 1, 0)$, and taking $a = 0$ and $b = 1$ shows that $\mathcal{P}$ contains the point $(2, 0, 1)$. Therefore $\mathcal{P}$ is the unique plane containing the origin, $(-\tfrac{5}{2}, 1, 0)$, and $(2, 0, 1)$. (See Fig. 4-3.)   ●

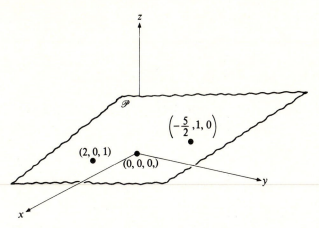

Fig. 4-3   The plane containing $(0, 0, 0)$, $(-\frac{5}{2}, 1, 0)$, $(2, 0, 1)$.

Example 4-4

Consider the set Ω of all points of the form

$$(3, 0, 0) + a(-\tfrac{5}{2}, 1, 0) + b(2, 0, 1)$$

Ω was obtained by adding $(3, 0, 0)$ to each point of the plane $\mathscr{P}$ of the previous example. By Theorem 4-5, Ω is also a plane since it is the solution set of the equation $2x + 5y - 4z = 6$. Since each point of Ω is 3 units in the x direction from a corresponding point on $\mathscr{P}$, it is intuitively clear that Ω is parallel to $\mathscr{P}$ (Fig. 4-4). Taking $a = b = 0$ reveals that $(3, 0, 0)$ is a particular point on Ω.   ●

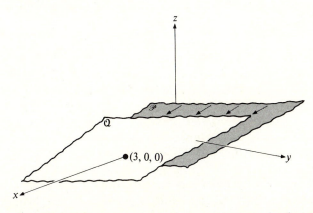

Fig. 4-4   Parallel planes. Ω is obtained by adding $(3, 0, 0)$ to each point of $\mathscr{P}$.

The above examples are typical of all planes in xyz space.

Theorem   Planes in xyz space are the sets of points of the form
4-6

$$\{(c, d, e) + a(f, g, h) + b(i, j, k) : a, b \in \mathcal{R}\}$$

As in the last example, (c, d, e) is a particular point on the plane, and the plane is parallel to the plane through the origin, (f, g, h), and (i, j, k).

Corollary   Planes that pass through the origin are the sets of points of the form
4-7

$$\{a(f, g, h) + b(i, j, k) : a, b \in \mathcal{R}\}$$

The use of our theorems is illustrated by describing some solution sets geometrically.

Example 4-5

Consider the system

$$
\begin{aligned}
5x + 5y - 2z &= 5 \\
10x + 5y - 3z &= 0 \\
5x - z &= -5
\end{aligned}
$$

Each separate equation is the equation of a plane and the set of solutions will be their intersection. The matrix for the system reduces to

$$
\begin{bmatrix}
1 & 0 & -\frac{1}{5} & \vdots & -1 \\
0 & 1 & -\frac{1}{5} & \vdots & 2 \\
0 & 0 & 0 & \vdots & 0
\end{bmatrix}
$$

The set of solutions is

$$\{(-1, 2, 0) + a(\tfrac{1}{5}, \tfrac{1}{5}, 1) : a \in \mathcal{R}\}$$

and thus by Theorem 4-3 is the line through the point $(-1, 2, 0)$ and parallel to the line segment joining $(0, 0, 0)$ and $(\tfrac{1}{5}, \tfrac{1}{5}, 1)$. The three planes intersect in this line, as in Fig. 3-1b.   ●

Example 4-6

The system

$$
\begin{aligned}
2x - 4y + 6z &= 3 \\
-4x + 8y - 12z &= -6 \\
4x - 8y + 12z &= 6
\end{aligned}
$$

has an augmented matrix which reduces to

$$
\left[\begin{array}{ccc:c}
1 & -2 & 3 & \frac{3}{2} \\
0 & 0 & 0 & 0 \\
0 & 0 & 0 & 0
\end{array}\right]
$$

and solutions of the form $(\frac{3}{2} + 2a - 3b, a, b)$, or

$$
(\tfrac{3}{2}, 0, 0) + a(2, 1, 0) + b(-3, 0, 1)
$$

According to Theorem 4-6 this is a plane passing through $(\frac{3}{2}, 0, 0)$, and parallel to the plane through $(0, 0, 0)$, $(2, 1, 0)$, and $(-3, 0, 1)$.   ●

Exercises

4-1 (a) Suppose that a system of equations has the solution set $\{(0, 1, 2) + a(-2, -2, 0) : a \in \mathcal{R}\}$. Describe the set of solutions geometrically.

(b) Suppose that a system of equations has the solution set $\{(1, 3, 1) + a(1, 0, 0) + b(0, 2, -4) : a, b \in \mathcal{R}\}$. Describe the set of solutions geometrically.

4-2 Each of the following is the reduced matrix of a system of equations. For each, write the set of solutions (if any solutions exist) in terms of its basic parts by separating out the numerical part, the a part, etc. Then describe the solution set geometrically.

(a) $\begin{bmatrix} 1 & 0 & 4 \\ 0 & 1 & 3 \end{bmatrix}$ (b) $\begin{bmatrix} 1 & -5 \\ 0 & 0 \end{bmatrix}$

(c) $\left[\begin{array}{cc:c} 1 & -5 & 3 \\ 0 & 0 & 0 \end{array}\right]$ (d) $\left[\begin{array}{ccc:c} 0 & 1 & -1 & 3 \\ 0 & 0 & 0 & 0 \end{array}\right]$

(e) $\left[\begin{array}{ccc:c} 1 & 0 & 2 & 5 \\ 0 & 1 & -3 & 3 \\ 0 & 0 & 0 & 0 \end{array}\right]$ (f) $\left[\begin{array}{ccc:c} 1 & -2 & 5 & 4 \end{array}\right]$

4-3 Describe geometrically the intersection of the two planes in xyz space: $y - 3z = 2$ and $2y - 2z = 5$. (Note that these are planes in xyz space, and hence the matrix for this system needs a column for the unknown x. Supply the term $0x$ for each equation.)

4-4 Describe geometrically the intersection of the two planes $x + 3y - 5z = 2$ and $-2x - 6y + 10z = -4$.

4-5 Describe geometrically the intersection of the three planes $x + y + z = 3$, $x - y = 1$, $y + 2z = 2$.

4-6 Describe geometrically the intersection of the three planes $x - y - 2z = 0$, $x + y - z = 4$, $2x - 3z = 2$.

4-7 Prove Theorem 4-6. (Hint: Start with the equation $px + qy + rz = s$ and write its solution set in the required form.)

5. EXISTENCE AND UNIQUENESS OF SOLUTIONS; RANK

As noted previously, some systems of linear equations have no solutions at all; some have unique solutions; some have infinitely many solutions. In this section we establish tests for determining in advance which of these situations holds in a particular case.

Testing for existence of a solution—to know in advance whether or not a solution to a system of equations exists—is often helpful. If we know that a solution does exist, then we know that it is worthwhile to look for one. For instance, we may then be justified in expending expensive computer time to search for a solution. On the other hand, knowing that no solution exists can spare us a great deal of labor. Historically, almost 3 centuries passed between the discovery of a formula for solving all fourth-degree polynomials and the belated discovery that there was no general formula for solving all fifth-degree polynomials. In the interim, untold hours were spent by mathematicians trying to derive a nonexistent formula.

Uniqueness of a solution is also of practical importance. When a solution is known to be unique, once any solution is found, one can stop searching for others. We use this principle frequently without being fully aware of it. For instance, when using logarithms to solve a problem, if the log of the answer is 0.693 we know that the answer is 2; no other number has that logarithm. On the other hand, if for some other problem we find that the sine of an angle is 0.5, we cannot be sure without further information that the angle is $\pi/6$; it might also be $5\pi/6$.

We have seen that the matrix for a system of equations provides a systematic method for completely solving the system. The matrix also provides a systematic and relatively painless way of determining whether there *are* any solutions, and if so, *how many* there are.

We first develop a matrix test for homogeneous systems.

Every homogeneous system has at least one solution, namely the solution where all the unknowns equal zero.

As for the uniqueness of solutions, consider a homogeneous system (with unknowns x, y, and z) with

$$\begin{bmatrix} 1 & 0 & 3 \\ 0 & 1 & 2 \\ 0 & 0 & 0 \end{bmatrix}$$

as its reduced matrix. To solve the system we keep only the variables corresponding to these row-leading 1s (x and y) on the left, obtaining

$$x = -3z$$
$$y = -2z$$

Letting $z = a$, as usual, we have the solutions

$$x = -3a$$
$$y = -2a$$
$$z = \quad a$$

Thus, this system has an infinite number of solutions, one for each value we choose for z (the unknown with no row-leading 1). This lack of a row-leading 1 for z yields infinitely many solutions. This is always the case; a homogeneous system will have infinitely many solutions when there are *fewer* row-leading 1s than unknowns.

On the other hand, if the reduced matrix for a homogeneous system has the same number of row-leading 1s as there are unknowns, then the reduced matrix will be essentially of the form

$$\begin{bmatrix} 1 & 0 & 0 \\ 0 & 1 & 0 \\ 0 & 0 & 1 \end{bmatrix} \quad \text{or} \quad \begin{bmatrix} 1 & 0 & 0 & 0 \\ 0 & 1 & 0 & 0 \\ 0 & 0 & 1 & 0 \\ 0 & 0 & 0 & 1 \\ 0 & 0 & 0 & 0 \end{bmatrix}$$

etc.

In this case we can immediately read off the value of every unknown and the solution will be unique: $(0, 0, 0)$ or $(0, 0, 0, 0)$, etc.

The preceding discussion shows the importance of the number of row-leading 1s in determining the number of solutions a homogeneous system has. This leads us to make the following definition.

Definition

The **rank** of a matrix is the number of row-leading 1s in its reduced echelon form.

The above discussion essentially proves the following theorem.

Theorem 5-1

Given a homogeneous system of linear equations with n unknowns:

1 It has at least one solution.

2 If the rank of its matrix is n, there is a unique solution.

3 If the rank of its matrix is less than n, there are infinitely many solutions.

We will now develop an analogous theorem for nonhomogeneous systems.

We have seen several examples of nonhomogeneous systems that have no solution. In each case the reduced augmented matrix had a row of the form

$$0 \quad 0 \quad \cdots \quad 0 \mathrel{\vdots} \underline{1}$$

For example, a system with

$$\begin{bmatrix} \underline{1} & 0 & 2 & \vdots & 0 \\ 0 & \underline{1} & 3 & \vdots & 0 \\ 0 & 0 & 0 & \vdots & \underline{1} \end{bmatrix}$$

as a reduced matrix has no solution. The presence of the row

$$0 \quad 0 \quad 0 \mathrel{\vdots} \underline{1}$$

in this matrix can be translated in terms of rank. If we write this matrix as $[A \mathrel{\vdots} B]$, where B is the column to the right of the dotted line, we immediately see that the rank of the unaugmented matrix A and the rank of the augmented matrix $[A \mathrel{\vdots} B]$ are different (the rank of A is 2, the rank of $[A \mathrel{\vdots} B]$ is 3). In general, a nonhomogeneous system will have no solution exactly when the rank of its augmented matrix $[A \vdots B]$ and the rank of its unaugmented matrix A are different, for it is precisely in this case that we have a row of the form

$$0 \quad 0 \quad \cdots \quad 0 \quad \vdots \quad \underline{1}$$

Now suppose we have a nonhomogeneous system with matrix $[A \vdots B]$ where the rank of $[A \vdots B]$ and the rank of A are equal. Then the system *has* a solution, for no row $0 \quad 0 \quad \cdots \quad 0 \quad \vdots \quad \underline{1}$ can exist. Just as in the homogeneous case, the system has a unique solution if the rank of A is equal to the number of unknowns, and there are infinitely many solutions if the rank of A is less than the number of unknowns.

We summarize this discussion in the following theorem.

Theorem 5-2   Given a nonhomogeneous system of linear equations with n unknowns and augmented matrix $[A \vdots B]$:

1 If the rank of A and the rank of $[A \vdots B]$ are unequal, there is no solution.

2 If the rank of A equals the rank of $[A \vdots B]$:
(a) There is a unique solution if the rank of A is n.
(b) There are infinitely many solutions if the rank of A is less than n.

Exercises

5-1 We can often judge the rank of a matrix without simplifying it all the way to its reduced form. We need only go far enough to see how many row-leading 1s there would be if we finished the reduction process. For each of the matrices below, do the following:

(a) Find the rank by underlining the entries that will become row-leading 1s in reducing the matrix.
(b) Determine whether the system of equations the matrix represents has a solution.
(c) Determine whether the solution is unique or whether there are infinitely many solutions.

$$A = \begin{bmatrix} 3 & 4 & 5 & 6 & \vdots & 7 \\ 0 & 5 & 8 & 9 & \vdots & 2 \end{bmatrix} \qquad B = \begin{bmatrix} 1 & 2 & 3 \\ 0 & 5 & 6 \\ 0 & 0 & 9 \end{bmatrix}$$

$$C = \begin{bmatrix} 86.27 & 6.381 & 0 \\ 0 & 19.8 & 0.37 \end{bmatrix} \qquad D = [\tfrac{2}{3} \quad \tfrac{5}{3} \quad 6 \quad 1 \ \vdots \ \tfrac{7}{5}]$$

$$E = \begin{bmatrix} 2 & 5 & -4 & 6 & -3 & 2 \\ 0 & 0 & 2 & 5 & -4 & -2 \\ 0 & 0 & 0 & 4 & 4 & 4 \\ 0 & 0 & 0 & 0 & 0 & 0 \end{bmatrix} \qquad F = \begin{bmatrix} 23 & 23 & 23 \\ -23 & -23 & -23 \\ 46 & 46 & 46 \end{bmatrix}$$

$$G = \begin{bmatrix} 46 & 5.3 & 1.1 & \vdots & 8 \\ 0 & 0 & .2 & \vdots & 0 \\ 0 & 0 & 3 & \vdots & 5 \end{bmatrix}$$

5-2 Find numbers a for which the following systems have solutions:

(a)
$$\begin{aligned} x - 3y &= 2 \\ 3x + 2y &= -4 \\ x + 8y &= a \end{aligned}$$

(b)
$$\begin{aligned} 2x - y + 2z &= 2 \\ 4x \qquad + 2z &= 1 \\ -3y + 3z &= a \end{aligned}$$

5-3 Is it possible for a homogeneous system with 15 equations and 12 unknowns to have:

(a) No solution
(b) A unique solution
(c) Infinitely many solutions
Explain.

5-4 Is it possible for a nonhomogeneous system with 12 equations and 15 unknowns to have:

(a) No solution
(b) A unique solution
(c) Infinitely many solutions
Explain.

6. APPLICATIONS

Example 6-1 Chemical Equations

In this example we show how a system of equations may be used to solve the problem of balancing a chemical equation.

One of the products currently being distilled from seawater is bromine. One of its uses is in fire-retardant chemicals used for treating clothing.

Bromine (Br_2) vapor is obtained from the ocean. When sodium carbonate (Na_2CO_3) is combined with this vapor, sodium bromide (NaBr), sodium bromate ($NaBrO_3$), and carbon dioxide (CO_2) are obtained. That is, we have the reaction

$$Na_2CO_3 + Br_2 \longrightarrow NaBr + NaBrO_3 + CO_2$$

The relative amounts of each chemical are not indicated in the above reaction. To balance this reaction, that is, to get an exact equation, we need to find positive integers x_1, x_2, x_3, x_4, and x_5 so that

$$x_1 Na_2CO_3 + x_2 Br_2 = x_3 NaBr + x_4 NaBrO_3 + x_5 CO_2$$

Equating the number of atoms of each particular kind, we have the system

$$
\begin{aligned}
2x_1 &= x_3 + x_4 &\quad \text{sodium (Na)} \\
2x_2 &= x_3 + x_4 &\quad \text{bromine (Br)} \\
x_1 &= x_5 &\quad \text{carbon (C)} \\
3x_1 &= 3x_4 + 2x_5 &\quad \text{oxygen (O)}
\end{aligned}
$$

Its matrix is

$$
\begin{bmatrix}
2 & 0 & -1 & -1 & 0 \\
0 & 2 & -1 & -1 & 0 \\
1 & 0 & 0 & 0 & -1 \\
3 & 0 & 0 & -3 & -2
\end{bmatrix}
$$

and it reduces to

$$
\begin{bmatrix}
1 & 0 & 0 & 0 & -1 \\
0 & 1 & 0 & 0 & -1 \\
0 & 0 & 1 & 0 & -\frac{5}{3} \\
0 & 0 & 0 & 1 & -\frac{1}{3}
\end{bmatrix}
$$

Hence

$$
\begin{aligned}
x_1 &= x_5 \\
x_2 &= x_5 \\
x_3 &= \tfrac{5}{3}x_5 \\
x_4 &= \tfrac{1}{3}x_5
\end{aligned}
$$

To have a feasible solution, x_5 must be chosen so that all five variables are positive integers. If we choose $x_5 = 3$, we have $x_1 = 3$, $x_2 = 3$, $x_3 = 5$, $x_4 = 1$, $x_5 = 3$. Therefore, the chemical equation becomes

$$3Na_2CO_3 + 3Br_2 = 5NaBr + NaBrO_3 + 3CO_2 \qquad \bullet$$

Example 6-2 Stability of Sample Groups

A certain political scientist studying the political thinking of people classifies everyone as conservative, liberal, or moderate. Based on past experience the following data on how a person's political tendency changes in any 1-year period have been gathered. For example, of the conservatives at any given time, 80 percent will be conservatives 1 year later, 5 percent will have become liberals, and 15 percent will have become moderates. The political scientist wants to pick a sample made up of the three types so that the number of each type in his sample remains constant from year to year.

	One year later		
Now	% conservative	% liberal	% moderate
Conservative	80	5	15
Liberal	5	80	15
Moderate	5	5	90

Let x_1 represent the number of present conservatives, x_2 the number of present liberals, and x_3 the number of present moderates to be put into the sample. Based on the data above, at the end of 1 year the number of conservatives in the sample will be the sum of the following three numbers: $0.80x_1$ (80 percent of the present conservatives), $0.05x_2$ (5 percent of the present liberals), and $0.05x_3$ (5 percent of the present moderates). Since he wants the number of conservatives in his sample after 1 year to be equal to the present number of conservatives in the sample, he must have

$$0.80x_1 + 0.05x_2 + 0.05x_3 = x_1$$

Similarly, he has the following equations for the liberal group and the moderate group:

$$0.05x_1 + 0.80x_2 + 0.05x_3 = x_2$$
$$0.15x_1 + 0.15x_2 + 0.90x_3 = x_3$$

These three equations give the system

$$-0.20x_1 + 0.05x_2 + 0.05x_3 = 0$$
$$0.05x_1 - 0.20x_2 + 0.05x_3 = 0$$
$$0.15x_1 + 0.15x_2 - 0.10x_3 = 0$$

Solving, we obtain the solution set

$$\{(a,\ a,\ 3a) : a \in \mathcal{R}\}$$

Taking $a = 100$, for example, we have $(100, 100, 300)$ as a solution. That is, beginning with 100 conservatives, 100 liberals, and 300 moderates now, at the end of 1 year (and hence at the end of each future year) the number in each group will remain the same. Thus to have the three groups remain stable, they must be in the proportion of one conservative and one liberal for every three moderates.

This stability of group sizes does not hold for samples having the three groups in different proportions. For example, a sample of 500 with 200 conservatives, 200 liberals, and 100 moderates would change in a year's time to 175 conservatives, 175 liberals, and 150 moderates, according to the given data.   ●

Example 6-3   An Input-Output Model of the Economy

Systems of equations play a large role in models set up to describe the behavior of the economy. We present here a simplified version of a Leontief input-output model.

We divide the economy into a consumer sector and three basic production sectors (or industries)—manufacturing, agriculture, and services. Each of these three industries requires inputs from all production sectors in order to produce its output. The level of activity in each of the industries depends on and affects all the industries. For example, a rise in farm output requires larger inputs from the manufacturing sector (more farm equipment and fertilizer will be needed), from the services sector (new farm equipment sales will rise and there will be a need for more repairs on existing farm equipment), and from the agricultural sector itself (more seed and breeding cattle will be needed). The level of activity of each industry also depends on consumer demand for finished products from that industry.

For a hypothetical economy Table A below shows the total amount of output from each production sector and how this amount is used by the various production sectors and the consumer sector (all entries are in

millions of dollars). For example, the manufacturing sector has a total
output worth $100 million. Of this, $20 million is used by the manufactur-
ing sector itself, $12 million is used by each of the agricultural and ser-
vices sectors, and $56 million goes into the consumer sector in the form of
products made by the manufacturing sector.

TABLE A

	Manufacturing	Agriculture	Services	Consumer demand	Total output
Manufacturing	20	12	12	56	100
Agriculture	20	3	4	33	60
Services	25	6	8	41	80

We are interested in knowing how much each industry must have as
input in order to produce $1 of output. On the basis of the figures above,
in order to produce $60 million output, the agricultural sector uses inputs
of $12 million, $3 million, and $6 million from manufacturing, agricul-
ture, and services, respectively. Therefore, we assume that to produce $1
of output the agricultural sector requires inputs of $0.20 ($= \frac{12}{60}$), $0.05
($= \frac{3}{60}$), and $0.10 ($= \frac{6}{60}$) from the three sectors.

Input-output analysis assumes that these ratios remain the same regardless of the total output of each industry. If we make the analogous
assumptions for each of the production sectors, we obtain Table B below
showing the amounts that each sector requires as input in order to produce $1 of output. The entries in each column are obtained by dividing
Table A's entries by the total output (100, 60, 80, respectively) of the
appropriate sector of the economy. We can use these data to study the
effects of fluctuations in consumer demands on the levels of activity for
each industry. That is, we are able to determine how much total output
each industry must have to satisfy different levels of consumer demands.

TABLE B

	Inputs per $1.00 of output		
	Manufacturing	Agriculture	Services
Manufacturing	$0.20	$0.20	$0.15
Agriculture	0.20	0.05	0.05
Services	0.25	0.10	0.10

We let x_1, x_2, and x_3 be the total outputs (in millions of dollars) from manufacturing, agriculture, and services respectively, and we let c_1, c_2, and c_3 be the consumer demands (in millions of dollars) from each of the three industries. We have the following input-output equation for each of the three production sectors:

$$\begin{pmatrix}\text{Total output of} \\ \text{that sector}\end{pmatrix} = \begin{pmatrix}\text{sum of outputs} \\ \text{from that sector} \\ \text{used by the 3} \\ \text{production sectors}\end{pmatrix} + \begin{pmatrix}\text{consumer demand} \\ \text{from that sector}\end{pmatrix}$$

We obtain three equations:

$$\begin{aligned}x_1 &= 0.20x_1 + 0.20x_2 + 0.15x_3 + c_1 \quad \text{(manufacturing)} \\ x_2 &= 0.20x_1 + 0.05x_2 + 0.05x_3 + c_2 \quad \text{(agriculture)} \\ x_3 &= 0.25x_1 + 0.10x_2 + 0.10x_3 + c_3 \quad \text{(services)}\end{aligned}$$

Thus, given particular consumer demand figures, a solution to this system (provided all parts of the solution are nonnegative) will give the level of output each sector must have in order to satisfy the demands of the production sectors and the consumer sector.

The matrix for this system is

$$\begin{bmatrix} 0.80 & -0.20 & -0.15 & \vdots & c_1 \\ -0.20 & 0.95 & -0.05 & \vdots & c_2 \\ -0.25 & -0.10 & 0.90 & \vdots & c_3 \end{bmatrix}$$

If for example $c_1 = \$69$ million, $c_2 = \$28$ million, and $c_3 = \$54$ million, the solution is

$$\begin{aligned}x_1 &= \$120 \text{ million} \\ x_2 &= \$60 \text{ million} \\ x_3 &= \$100 \text{ million}\end{aligned}$$

and these outputs will exactly satisfy all production and consumer demands. If the consumer demands change to $c_1 = \$83.5$ million, $c_2 = \$23.5$ million, and $c_3 = \$58$ million, then the outputs must be

$$\begin{aligned}x_1 &= \$140 \text{ million} \\ x_2 &= \$60 \text{ million} \\ x_3 &= \$110 \text{ million}\end{aligned}$$

Interestingly enough, a 16 percent decrease in consumer demand for agricultural products caused no drop in agricultural production, because of the needs of other sectors in the economy.   ●

Exercises

6-1 Assume that every automobile is categorized as either "economy" or "full-size." Suppose the following data have been gathered. In a 1-year period 90 percent of the economy-car owners still own economy cars and 10 percent become full-size-car owners. Also, 70 percent of the full-size-car owners remain full-size-car owners and 30 percent become economy-car owners. What percentage must now be economy-car owners and what percentage must now be full-size-car owners in order that these two groups remain stable?

6-2 Census figures collected now and 10 years ago in a certain county yield the following data in which the entries indicate percentages of people who changed their location during that 10-year period. For example, of urban dwellers 10 years ago, 80 percent remain urban dwellers now and 20 percent became suburban dwellers.

	% urban dwellers	% suburban dwellers
Urban dwellers	80	20
Suburban dwellers	10	90

(a) Assuming that these percentages remain the same over the next 10-year period, if a sample of each of the two groups is to be chosen, in what proportion should the groups be chosen so that they remain stable?
(b) At the present time the county has 7200 urban dwellers and 2500 suburban dwellers. How many of each type were there 10 years ago? How many of each type will there be 10 years from now if the above figures govern the coming 10 years as well?

6-3 Balance the chemical equation

$$Cl_2 + KOH \longrightarrow KCl + KClO_3 + H_2O$$

6-4 Balance the chemical equation

$$C_3H_5(NO_3)_3 \longrightarrow H_2O + CO + N_2 + O_2$$

6-5 Suppose a hypothetical economy is divided into two production sectors A and B whose input-output table is shown below. Entries are in millions of dollars.

	A	B	Consumer demand	Total output
A	2	3	5	10
B	4	6	5	15

Find the total outputs of A and B if:

(a) The consumer demand for A drops to 4 and the consumer demand for B remains the same
(b) The consumer demand for A and for B each drop to 4

Review of Chapter 1†

1 To solve a system of equations by gaussian elimination:

(a) Write the _____ for the system.

(b) Put the matrix into _____ form.
(c) Read off the solutions.

Unknowns with row-leading 1s can be kept to the left; unknowns with no row-leading 1s can be any real number.

2 In reducing a matrix we use only these three operations:

(a) _____

(b) _____

(c) _____

3 Solutions of equations with n unknowns are ordered n-tuples. If there are infinitely many solutions, we use notation like $\{(2a, a, -4a) : a \in \mathcal{R}\}$. If we factor out the a, this is the set of all triples of the form $a(\_\_, \_\_, \_\_)$. Similarly the solution set $\{(a, 2a - b, 3b, b) : a, b \in \mathcal{R}\}$ can be expressed as the set of all 4-tuples of the form $a(\_\_, \_\_, \_\_, \_\_) + b(\_\_, \_\_, \_\_, \_\_)$. The solution set $\{(3 + 2a - b, a, 2 + a - b) : a, b \in \mathcal{R}\}$ can be written as $(\_\_, \_\_, \_\_) + a(\_\_, \_\_, \_\_) + b(\_\_, \_\_, \_\_)$.

† We recommend that the student first reread the chapter quickly in search of the main definitions and theorems, and then return to this review section to fill in the blanks and work on the review exercises.

4   Geometrically, the set $\{(2a, a, -4a) : a \in \mathcal{R}\}$ is a _____ through the origin and the point _____. The set $\{(3 + 2a - b, a, 2 + a - b) : a, b \in \mathcal{R}\}$ is a _____ through the point _____ and parallel to the _____ through the origin and the points _____ and _____. More generally, sets whose points are of the form $(-, -) + a(-, -)$ or $(-, -, -) + a(-, -, -)$ are _____. In particular, if they pass through the origin, they have the form _____ or _____. Sets whose points are of the form $(-, -, -) + a(-, -, -) + b(-, -, -)$ are _____. If a plane passes through the origin, it has the form _____.

5   The rank of a matrix is the number of _____ 1s in its _____ form.

6   Given a homogeneous system with n unknowns:

(a)   The system always has at least one solution, namely, _____.

(b)   If the rank of its matrix is n, the solution is _____.

(c)   If the rank of its matrix is less than n, then _____.

7   Given a nonhomogeneous system with n unknowns and matrix $[A \vdots B]$:

(a)   There is no solution if the rank of _____ and the rank of _____ are _____.

(b)   Assuming the rank of A equals the rank of $[A \vdots B]$:

   (i)   The solution is _____ provided the rank of $[A \vdots B]$ is n.

   (ii)   There are infinitely many solutions provided the rank of $[A \vdots B]$ is _____.

Review Exercises

1   Solve the following systems of equations by gaussian elimination. Express the solutions in the form $a(\quad)$ or $a(\quad) + b(\quad)$, etc., wherever possible.

(a) $\begin{aligned} x - y + 4z &= 4 \\ 2x \quad\;\; + 6z &= -2 \end{aligned}$ (b) $\begin{aligned} x - y + 4z &= 0 \\ 2x \quad\;\; + 6z &= 0 \end{aligned}$

(c)
$$3x - 2y + z = 1$$
$$y - 2z = -3$$
$$x - 4y = 2$$

(d)
$$2x - 3y + 2z + t = 0$$
$$-4x + 6y + 2t = 0$$
$$4x - 6y + 8z + 6t = 0$$
$$5z + 5t = 0$$

(e)
$$x_1 + 2x_2 + x_3 - x_4 + x_5 = 0$$
$$-2x_1 - 4x_2 - 2x_3 + 3x_4 = 0$$
$$3x_1 + 6x_2 + 3x_3 + 9x_5 = 0$$

(f)
$$2x - 4y + 5z = 1$$
$$x - 3z = 2$$
$$5x - 8y + 7z = 6$$
$$3x - 4y + 2z = 3$$
$$x - 4y + 8z = -1$$

(g)
$$\tfrac{1}{2}x + y = 0$$
$$3x - \tfrac{5}{6}y + 4z = 0$$
$$\tfrac{1}{4}x - \tfrac{1}{6}y + z = 0$$

2 Find numbers a for which the following systems have solutions:

(a)
$$x - 3y = 2$$
$$3x + 2y = 4$$
$$x + 8y = a$$

(b)
$$2x - y + 2z = 2$$
$$4x + 2z = -1$$
$$-3y + 3z = a$$

3 (a) Let $p(x) = a_2 x^2 + a_1 x + a_0$. Suppose we want to pick a_2, a_1, a_0 such that $p(1) = b_1$, $p(2) = b_2$, $p(3) = b_3$. Show that this yields three linear equations in the three unknowns a_2, a_1, a_0 that always have a *unique* solution. This says that given three points, there is a unique polynomial of degree 2 whose graph passes through them. (Hint: Find the rank of the unaugmented matrix.)
(b) Find a polynomial $p(x) = a_2 x^2 + a_1 x + a_0$ that pauses through the points $(1, 4)$, $(0, -3)$, $(-2, -5)$.

4 Evaluate
$$\int \frac{2x + 4}{(x - 1)(x^2 + 1)} \, dx$$

by first finding A, B, and C in the equation

$$\frac{2x + 4}{(x - 1)(x^2 + 1)} = \frac{A}{x - 1} + \frac{Bx + C}{x^2 + 1}$$

(Hint: Clear the fractions on both sides and compare like powers of x; this will yield a system of equations with unknowns A, B, and C.)

5 Find the intersection of the planes

$$4x - z = 5$$
$$2x + y - 2z = 3$$

Write the solution in suitable form and describe their intersection.

6 Show that the planes $x = 1$, $2x + y + z = 8$, $y + z = 4$ have no point of intersection.

7 A railroad has three types of tank cars (types I, II, III), each able to move chemicals A, B, and C according to the table below. A company wants exactly 12 units of A, 10 units of B, and 16 units of C. How can the railroad transport these amounts most efficiently?

	Capacity by type of car		
Chemicals	I	II	III
A	1	1	1
B	0	1	2
C	2	1	1

8 From a pile of nickels, dimes, and quarters, I choose 17 coins. I have twice as many dimes as quarters. "Aha," I announce, "the total value of the coins I've picked is $1.85." Prove that I lied!

9 Find the points of intersection of the circle $x^2 + y^2 = 13$ and the ellipse $3x^2 + 4y^2 = 48$. (Hint: First solve the equations treating x^2 and y^2 as the two unknowns, then solve for x and y.)

10 Answer these three questions for each of the following types of systems of equations:

(a) Is it possible for the system to have no solution?
(b) Is it possible for the system to have a unique solution?
(c) Is it possible for the system to have infinitely many solutions?

  (i)   A homogeneous system of 17 equations with 23 unknowns.
  (ii) A nonhomogeneous system of 17 equations with 23 unknowns.
  (iii) A homogeneous system of 23 equations with 17 unknowns.
  (iv) A nonhomogeneous system of 23 equations with 17 unknowns.

CHAPTER 2

VECTOR SPACES AND SUBSPACES

Mathematics is used to solve problems. One of the key ingredients in the problem-solving process is the *abstraction* of the original problem into mathematical symbols and ideas.

This abstraction can occur in two ways. The first is the creation of abstract *models* of concrete physical situations. For instance, euclidean geometry is a model of the space we live in, and a system of equations relating various quantities is a model of the complex relations among real-life objects.

The second way, perhaps less familiar, is the creation of one abstract *theory* to cover several similar models that are governed by the same mathematical laws. In calculus, for example, instantaneous velocity, the slope of a curve, and the rate of change of a function are all dealt with by similar models. The abstract theory that covers all of these is the theory of differentiation. In like manner area, volume, and arc length are all covered by the theory of integration.

The virtue of this second type of abstraction is that by studying the one abstract theory we learn simultaneously about *all* the models governed by these same laws.

The theory we are about to develop, that of vector spaces, is an abstraction of this second type. It is an abstraction from many similar mathematical models.

7. DEFINITION OF VECTOR SPACE

Consider the following four mathematical objects:

The xy plane

The set of all 2×2 matrices†

The set of all polynomials

The set of all solutions to a given system of homogeneous equations

These four have similarities that may not be immediately obvious (and they all turn out to be vector spaces, as we will soon see):

(a)   Each is a *set* of objects.

(b)   Each has the property that any two objects in the set can be *added* and the result is an object in the same set.

(c)   Each has the property that any object in the set can be *multiplied* by a number, and the result is an object in the set.

For example, the xy plane is the set $\{(x, y): x, y \in \mathcal{R}\}$. We add two objects (a, b) and (c, d) according to the rule $(a, b) + (c, d) = (a + c, b + d)$. We multiply an object (a, b) by a number r according to the rule $r(a, b) = (ra, rb)$.

Similarly, the set of all 2×2 matrices may be written as the set

$$\left\{ \begin{bmatrix} a & b \\ c & d \end{bmatrix} : a, b, c, d \in \mathcal{R} \right\}$$

We add two matrices by adding entries in the corresponding positions,

that is, $\begin{bmatrix} a & b \\ c & d \end{bmatrix} + \begin{bmatrix} e & f \\ g & h \end{bmatrix} = \begin{bmatrix} a + e & b + f \\ c + g & d + h \end{bmatrix}$

We multiply $\begin{bmatrix} a & b \\ c & d \end{bmatrix}$ by the number r by multiplying each entry of the

matrix by r, that is, $r\begin{bmatrix} a & b \\ c & d \end{bmatrix} = \begin{bmatrix} ra & rb \\ rc & rd \end{bmatrix}$

---

† The symbol 2×2, read "two by two," refers to the size of the matrices: two rows and two columns. Thus 2×2 matrices are of the form $\begin{bmatrix} a & b \\ c & d \end{bmatrix}$

The three properties just mentioned form the nucleus of the definition of vector space. We must have a set of objects (called *vectors*), a rule for adding the vectors (called *vector addition*), and a rule for multiplying vectors by real numbers (called *scalar multiplication*, since real numbers are called *scalars*).

To complete the definition of vector space around the above nucleus, we need to list the mathematical laws that must hold in every vector space. In addition to knowing how to add vectors, we will also insist that our definition of addition obeys certain reasonable laws. Similarly we will insist that our definition of scalar multiplication also obeys certain reasonable laws. A precise definition of vector space will now be given.

Definition

A **vector space** is a set $\mathcal{V}$ of elements (called *vectors*) together with two operations, vector addition and scalar multiplication, satisfying the following laws (for all vectors $\mathbf{X}$, $\mathbf{Y}$, and $\mathbf{Z}$ in $\mathcal{V}$ and all (real) scalars r and s).

Laws for addition

A1.   If $\mathbf{X}$ and $\mathbf{Y}$ are any two vectors in $\mathcal{V}$, then $\mathbf{X} + \mathbf{Y} \in \mathcal{V}$.   Closure under addition

A2. $(\mathbf{X} + \mathbf{Y}) + \mathbf{Z} = \mathbf{X} + (\mathbf{Y} + \mathbf{Z})$.   Associative law

A3. $\mathbf{X} + \mathbf{Y} = \mathbf{Y} + \mathbf{X}$.   Commutative law

A4.   There is a vector in $\mathcal{V}$, denoted $\mathbf{0}$, such that $\mathbf{X} + \mathbf{0} = \mathbf{X}$ ($\mathbf{0}$ is called the *zero vector*).   Additive identity law

A5.   For every $\mathbf{X} \in \mathcal{V}$ there is a vector $-\mathbf{X}$ such that $\mathbf{X} + (-\mathbf{X}) = \mathbf{0}$ ($-\mathbf{X}$ is called the *additive inverse* of $\mathbf{X}$).   Additive inverse law

Laws for scalar multiplication

S1.   If $\mathbf{X} \in \mathcal{V}$ and r is any real number, then $r\mathbf{X} \in \mathcal{V}$.   Closure under scalar multiplication

S2. $(rs)\mathbf{X} = r(s\mathbf{X})$.   Associative law

S3. $r(\mathbf{X} + \mathbf{Y}) = r\mathbf{X} + r\mathbf{Y}$.   Left distributive law

S4. $(r + s)\mathbf{X} = r\mathbf{X} + s\mathbf{X}$.   Right distributive law

S5. $1\mathbf{X} = \mathbf{X}$.   Unit law

The definition of vector space given above appears rather involved, but it can be easily remembered, since every law given is a law that also governs the adding and multiplying of real numbers.

Let us consider briefly the meaning of the laws listed in the definition of vector space.

A1 and S1.   There is an analogy here with football. When the ball goes out of bounds, the play stops. In a vector space, the set $\mathcal{V}$ is the playing field; A1 and S1 ensure that the "play" stays within bounds.

A2. $(X + Y) + Z$ means "add X and Y first, then add Z to the result," while $X + (Y + Z)$ means "add Y and Z first, then add X to the result." A2 states that these two must always be equal. Without this law, $X + Y + Z$ is ambiguous: which of the additions does one perform first? The associative law says that it does not matter.

A3.   The analogous law for numbers is familiar—for instance, $2 + 3 = 3 + 2$. Vectors can be strange mathematical objects and the vector addition can also be strange, but we still always require $X + Y = Y + X$.

A4.   This law, to be verified, requires that we find a vector in $\mathcal{V}$ with a special property—namely, that if it is added to any vector $X \in \mathcal{V}$ we get X as the result. This special vector behaves like the number zero, and hence is denoted $\mathbf{0}$.

A5.   This law says that every vector X has an "opposite" vector, denoted $-X$.

The other laws have similar interpretations. Note that in the ten laws, parentheses indicate that the terms inside them are to be combined first.

We now give examples of some specific vector spaces of two types: those whose elements are ordered pairs, triples, 4-tuples, etc., and those whose elements are matrices. Vector spaces of these two types will play an important role in the remainder of this book.

To describe these vector spaces we must specify the set of vectors, how vector addition is defined, and how scalar multiplication is defined. In what follows, $\mathcal{R}$ denotes the set of real numbers.

Example 7-1   The xy Plane

Set: $\{(a, b) : a, b \in \mathcal{R}\}$

Definition of addition: $(a, b) + (c, d) = (a + c, b + d)$

Definition of scalar multiplication: $r(a, b) = (ra, rb)$

Later on we will verify that the ten laws hold. From now on we will denote this vector space by $\mathcal{R}^2$. The 2 is a reminder that the vectors are ordered *pairs*; the $\mathcal{R}$ is a reminder that the entries in each vector are real numbers.   ●

Example 7-2 *xyz* Space

Set: $\{(a,\, b,\, c) : a,\, b,\, c \in \mathcal{R}\}$

Definition of addition: $(a,\, b,\, c) + (d,\, e,\, f) = (a + d,\, b + e,\, c + f)$

Definition of scalar multiplication: $r(a,\, b,\, c) = (ra,\, rb,\, rc)$

We will denote this vector space by $\mathcal{R}^3$.   ●

$\mathcal{R}^2$ and $\mathcal{R}^3$ are the vector spaces with which the student is probably most familiar. Vectors in $\mathcal{R}^2$ can be visualized as points in the *xy* plane; similarly, vectors in $\mathcal{R}^3$ can be visualized as points in *xyz* space. We can also picture vectors in $\mathcal{R}^2$ and $\mathcal{R}^3$ by arrows. In particular, the vector $(a,\, b)$ in $\mathcal{R}^2$ can be pictured by an arrow drawn from the origin $(0,\, 0)$ to the point $(a,\, b)$.

Using arrows to represent vectors gives us a geometric picture of vector addition. By definition, $(a,\, b) + (c,\, d) = (a + c,\, b + d)$. If we let **A** be the arrow representing $(a,\, b)$ and **B** the arrow representing $(c,\, d)$, as in Fig. 7-1, the arrow representing the sum $(a + c,\, b + d)$ is the arrow from the origin to the opposite corner of the parallelogram determined by **A** and **B**. (See Exercise 7-10.) We denote this arrow by **A + B**.

Arrows also give us a picture of scalar multiplication. Let **A** again be the arrow for $(a,\, b)$. All points of the form $r(a,\, b)$ lie on the line through

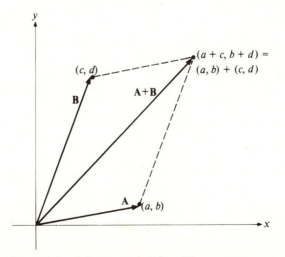

Fig. 7-1 Parallelogram method for adding vectors.

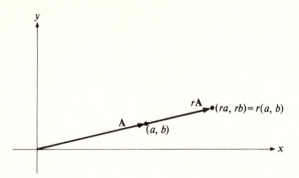

Fig. 7-2

$(0, 0)$ and (a, b). Thus the arrow $r\mathbf{A}$ representing the scalar multiple $r(a, b)$ lies on the same line as $\mathbf{A}$ (Fig. 7-2). Furthermore, $r\mathbf{A}$ is $|r|$ times as long as $\mathbf{A}$ (Exercise 7-11). If $r < 0$, $r\mathbf{A}$ points in the opposite direction from $\mathbf{A}$.

Similar interpretations can be made for vectors in $\mathcal{R}^3$. Throughout the book we will make frequent use of this geometric interpretation of vectors in $\mathcal{R}^2$ and $\mathcal{R}^3$. In particular we will apply these ideas in Sec. 11 to describe vector quantities in physics.

Example 7-3 $\mathcal{R}^n$ (n any positive integer)

Set: $\{(a_1, a_2, \ldots, a_n) : a_1, a_2, \ldots, a_n \in \mathcal{R}\}$

Definition of addition:

$$(a_1, a_2, \ldots, a_n) + (b_1, b_2, \ldots, b_n) = (a_1 + b_1, a_2 + b_2, \ldots, a_n + b_n)$$

Definition of scalar multiplication:

$$r(a_1, a_2, \ldots, a_n) = (ra_1, ra_2, \ldots, ra_n)$$

If $n = 2$ or $n = 3$, we have the vector spaces $\mathcal{R}^2$ and $\mathcal{R}^3$ of Examples 7-1 and 7-2.   ●

Example 7-4   The Vector Space of All 2×2 Matrices

Set: $\left\{ \begin{bmatrix} a & b \\ c & d \end{bmatrix} : a, b, c, d \in \mathcal{R} \right\}$

Definition of addition: $\begin{bmatrix} a & b \\ c & d \end{bmatrix} + \begin{bmatrix} e & f \\ g & h \end{bmatrix} = \begin{bmatrix} a + e & b + f \\ c + g & d + h \end{bmatrix}$

Definition of scalar multiplication: $r\begin{bmatrix} a & b \\ c & d \end{bmatrix} = \begin{bmatrix} ra & rb \\ rc & rd \end{bmatrix}$

From now on we will denote this vector space by $\mathcal{M}^{2 \times 2}$.   ●

Example 7-5

The vector space of all $m \times n$ matrices (m and n stand for fixed positive integers; $m \times n$ means that the matrices have m rows and n columns).

Set: $\left\{ \begin{bmatrix} a_{11} & a_{12} & \cdots & a_{1n} \\ a_{21} & a_{22} & \cdots & a_{2n} \\ \hline a_{m1} & a_{m2} & \cdots & a_{mn} \end{bmatrix} : \text{every entry a real number} \right\}$

The entry a_{ij} is in the ith row and jth column. For example, a_{35} would be in the third row and fifth column, a_{m2} would be in the mth row and second column. Two $m \times n$ matrices are added by adding their corresponding entries. An $m \times n$ matrix is scalar multiplied by r by multiplying each entry of the matrix by r. This vector space will be denoted $\mathcal{M}^{m \times n}$. In particular, if $m = 2$ and $n = 2$, we have the vector space $\mathcal{M}^{2 \times 2}$ of Example 7-4.   ●

Example 7-6 The x Axis in xyz Space

Set: $\{(x, 0, 0) : x \in \mathcal{R}\}$

Definition of addition: $(a, 0, 0) + (b, 0, 0) = (a + b, 0\ 0)$

Definition of scalar multiplication: $r(a, 0, 0) = (ra, 0, 0)$

Note that addition and scalar multiplication are defined just as they are for $\mathcal{R}^3$. The x axis is an example of a vector space contained within a larger vector space, namely $\mathcal{R}^3$. In Sec. 10 we will study in more detail examples of vector spaces contained in larger ones.   ●

To prove that each of the above six examples is a vector space, all ten laws listed in the definition would have to be verified. Note that these laws must hold for *all* vectors in the vector space. Because of this it is important to know what *all* the vectors in each example look like; that is, we must know what an *arbitrary* vector looks like in each example. For instance, in $\mathcal{R}^2$ (Example 7-1) we can use notation such as (a, b), (c, d),

(u, v), or (x, y) to denote an arbitrary vector. An arbitrary vector in $\mathcal{M}^{2 \times 2}$ can be written as

$$\begin{bmatrix} q & r \\ s & t \end{bmatrix} \quad \text{or} \quad \begin{bmatrix} a & b \\ c & d \end{bmatrix}$$

An arbitrary vector in Example 7-6, the x axis in $\mathcal{R}^3$, can be written as $(a, 0, 0)$ or $(b, 0, 0)$, for example.

In order to check laws A4 and A5, we must also know specifically what the zero vector $\mathbf{0}$ looks like and what additive inverses $-\mathbf{X}$ look like. For instance, in $\mathcal{R}^2$ our intuition tells us that the zero vector is $(0, 0)$ and the additive inverse of an arbitrary vector (a, b) is $(-a, -b)$. We will verify that this is the case in the next section. In $\mathcal{M}^{2 \times 2}$, the zero vector is

$$\begin{bmatrix} 0 & 0 \\ 0 & 0 \end{bmatrix}$$

and the additive inverse of

$$\begin{bmatrix} a & b \\ c & d \end{bmatrix}$$

is

$$\begin{bmatrix} -a & -b \\ -c & -d \end{bmatrix}$$

Exercises

For each of the vector spaces in Exercises 7-1 through 7-9:

(a)   Write three arbitrary vectors.
(b)   Write what you would expect the zero vector to be.
(c)   Write what you would expect to have as the additive inverses of the arbitrary vectors from part (a).

7-1. $\mathcal{R}^3$ **7-2.** $\mathcal{R}^5$

7-3. $\mathcal{M}^{2 \times 2}$ **7-4.** $\mathcal{M}^{2 \times 3}$

7-5. $\mathcal{M}^{m \times n}$ **7-6.** The x axis in $\mathcal{R}^3$

7-7.   The y axis $\{(0, y) : y \in \mathcal{R}\}$ in $\mathcal{R}^2$

7-8.   The plane through the origin of $\mathcal{R}^3$ consisting of all points $a(2, 1, 0) + b(-3, 0, 1)$

7-9.   The vector space consisting of all solutions of the homogeneous system

$$
\begin{aligned}
x - y + 4z &= 0 \\
2x \quad\;\; + 6z &= 0
\end{aligned}
$$

7-10.   Show that the points $(0, 0), (a, b), (c, d), (a + c, b + d)$ determine a parallelogram (as in Fig. 7-1). Thus the arrow $\mathbf{A} + \mathbf{B}$ is indeed the diagonal of the parallelogram determined by $\mathbf{A}$ and $\mathbf{B}$. [Hint: Compute the slopes of the opposite sides of the quadrilateral whose vertices are the four points.]

7-11.   Show that the arrow $r\mathbf{A}$ from $(0, 0)$ to $r(a, b)$ is $|r|$ times as long as the arrow $\mathbf{A}$ from $(0, 0)$ to (a, b). [Hint: Write $r(a, b)$ as (ra, rb) and use the distance formula for points in the xy plane.]

8.   PROVING VECTOR SPACE LAWS

Since we now know how to write arbitrary vectors and have a good idea of what the zero vector and additive inverses are in the examples of the previous section, we are now in a position to prove that each of our examples is indeed a vector space. We will carry out several parts of the proof for $\mathcal{R}^2$. Proofs for the other examples follow the same pattern and are left as exercises.

Example 8-1   Proof That $\mathcal{R}^2$ Is a Vector Space

A1. Closure under addition   We must show that if $\mathbf{X}$ and $\mathbf{Y}$ are two arbitrary vectors in $\mathcal{R}^2$, then $\mathbf{X} + \mathbf{Y}$ is still in $\mathcal{R}^2$. Thus, let (a, b) and (c, d) be arbitrary vectors in $\mathcal{R}^2$. Their sum $(a, b) + (c, d) = (a + c, b + d)$. Since $a + c$ and $b + d$ are still real numbers, the sum is an ordered pair of real numbers, and thus lies in $\mathcal{R}^2$, as desired.

A2. Associative law for addition   For laws such as this, whose statement is an equation, we adopt the following format. We begin with the general statement of the law, introduce notation suitable for $\mathcal{R}^2$, and then work on both sides until equality is evident. Reasons for our steps are given at the right.

STATEMENTS

$$(\mathbf{X} + \mathbf{Y}) + \mathbf{Z} \stackrel{?}{=} \mathbf{X} + (\mathbf{Y} + \mathbf{Z})$$

REASONS

Statement of law to be proved

$((a, b) + (c, d)) + (e, f)$	$(a, b) + ((c, d) + (e, f))$	Appropriate notation for $\mathcal{R}^2$
$(a + c, b + d) + (e, f)$	$(a, b) + (c + e, d + f)$	Definition of addition in $\mathcal{R}^2$
$((a + c) + e, (b + d) + f)$	$(a + (c + e), b + (d + f))$	Definition of addition in $\mathcal{R}^2$
$(a + (c + e), b + (d + f))$		Associative law for addition of real numbers†

At the last step we have the desired equality. One point is especially worth noting about this proof. Expressions inside parentheses had to be evaluated first before we could continue. For instance $((a, b) + (c, d))$ had to be evaluated as $(a + c, b + d)$ before proceeding. This guides us to some extent in the order in which we do the steps of the proof.

A3. Commutative law for addition   We follow the same format as that for A2.

STATEMENTS

$$\mathbf{X} + \mathbf{Y} \stackrel{?}{=} \mathbf{Y} + \mathbf{X}$$

REASONS

Statement of law to be proved

$(a, b) + (c, d)$	$(c, d) + (a, b)$	Appropriate notation for $\mathcal{R}^2$
$(a + c, b + d)$	$(c + a, d + b)$	Definition of addition in $\mathcal{R}^2$
$(c + a, d + b)$		Commutative law for addition of real numbers

† It is assumed the reader is familiar with the usual laws governing addition and multiplication of numbers. These laws are needed in proving the analogous laws for vector spaces such as $\mathcal{R}^2$.

A4. Identity law for addition   This law is somewhat different from the previous ones. It asserts the existence of a vector in our vector space with a certain property. Our first step, therefore, is to find a candidate for the zero vector; next we will verify its property. (Whatever the vector space, this will be the mode of attack.)

We expect that $(0, 0)$ is the zero vector of $\mathcal{R}^2$. To prove this, let (a, b) be an arbitrary vector in $\mathcal{R}^2$. Then

$$(0, 0) + (a, b) = (0 + a, 0 + b) \qquad \text{By definition of addition in } \mathcal{R}^2$$
$$= (a, b) \qquad\qquad\quad \text{Since zero is the additive identity for } \mathcal{R}$$

A5. Additive inverse law   Let (a, b) be an arbitrary vector in $\mathcal{R}^2$. As in the proof of A4, our first task is to find a candidate for the additive inverse of (a, b). We claim that $(-a, -b)$ is the additive inverse of (a, b). To show this, note that

$$(a, b) + (-a, -b) = (a + (-a), b + (-b)) = (0, 0)$$

by the additive inverse law for $\mathcal{R}$. Since $(0, 0)$ is the zero vector for $\mathcal{R}^2$, $(-a, -b)$ has the required property.

S1. Closure under scalar multiplication   Let (a, b) be an arbitrary vector in $\mathcal{R}^2$ and let r be an arbitrary scalar. We must show that the scalar multiple $r(a, b)$ is still in $\mathcal{R}^2$. But $r(a, b) = (ra, rb)$, by definition of scalar multiplication. Since ra and rb are still real numbers, the scalar multiple is still an ordered pair of real numbers, and therefore remains in $\mathcal{R}^2$, as desired.

S2. Associative law for scalar multiplication   Again we can use the format of A2.

STATEMENTS		REASONS
$(rs)\mathbf{X} \overset{?}{=} r(s\mathbf{X})$		Statement of law to be proved
$(rs)(a, b)$	$r(s(a, b))$	Appropriate notation for $\mathcal{R}^2$
$((rs)a, (rs)b)$	$r(sa, sb)$	Definition of scalar multiplication for $\mathcal{R}^2$
	$(r(sa), r(sb))$	Definition of scalar multiplication for $\mathcal{R}^2$
	$((rs)a, (rs)b)$	Associative law for multiplication of real numbers

S3, S4, and S5 can also be verified using the same format as in A2. In general, that format is appropriate for all laws except the closure laws (A1 and S1) and the additive identity and inverse laws (A4 and A5).   ●

Example 8-2

We will prove the left distributive law S3 for $\mathcal{M}^{3 \times 2}$.

$$r(\mathbf{X} + \mathbf{Y}) \stackrel{?}{=} r\mathbf{X} + r\mathbf{Y}$$
          Statement of law to be proved

$$r\left(\begin{bmatrix} a & b \\ c & d \\ e & f \end{bmatrix} + \begin{bmatrix} g & h \\ i & j \\ k & l \end{bmatrix}\right) \quad \middle| \quad r\begin{bmatrix} a & b \\ c & d \\ e & f \end{bmatrix} + r\begin{bmatrix} g & h \\ i & j \\ k & l \end{bmatrix}$$
          Appropriate notation for $\mathcal{M}^{3 \times 2}$

$$\begin{bmatrix} ra & rb \\ rc & rd \\ re & rf \end{bmatrix} + \begin{bmatrix} rg & rh \\ ri & rj \\ rk & rl \end{bmatrix}$$
          Definition of scalar multiplication for $\mathcal{M}^{3 \times 2}$

$$r\begin{bmatrix} a+g & b+h \\ c+i & d+j \\ e+k & f+l \end{bmatrix} \quad \middle| \quad \begin{bmatrix} ra+rg & rb+rh \\ rc+ri & rd+rj \\ re+rk & rf+rl \end{bmatrix}$$
          Definition of addition for $\mathcal{M}^{3 \times 2}$

$$\begin{bmatrix} r(a+g) & r(b+h) \\ r(c+i) & r(d+j) \\ r(e+k) & r(f+l) \end{bmatrix}$$
          Definition of scalar multiplication for $\mathcal{M}^{3 \times 2}$

$$\begin{bmatrix} r(a+g) & r(b+h) \\ r(c+i) & r(d+j) \\ r(e+k) & r(f+l) \end{bmatrix}$$
          Left distributive law for real numbers

    ●

Once one has proved the ten basic laws for a particular vector space, many other laws, analogous to familiar facts about ordinary numbers, will be automatically true in that vector space. The following theorem lists several of these laws. They are true in any vector space, for any vectors **X**, **Y**, and **Z**, and any scalar r.

Theorem 8-1

1   If $\mathbf{X} + \mathbf{Y} = \mathbf{X} + \mathbf{Z}$, then $\mathbf{Y} = \mathbf{Z}$ (the left cancellation law)

2   If $\mathbf{Y} + \mathbf{X} = \mathbf{Z} + \mathbf{X}$, then $\mathbf{Y} = \mathbf{Z}$ (the right cancellation law)

3   $0\mathbf{X} = \mathbf{0}$

$$4 \quad r\mathbf{0} = \mathbf{0}$$

$$5 \quad (-1)\mathbf{X} = -\mathbf{X}$$

$$6 \quad -(-\mathbf{X}) = \mathbf{X}$$

Proof   (1) Let

$$\mathbf{X} + \mathbf{Y} = \mathbf{X} + \mathbf{Z}$$

Then

$-\mathbf{X} + (\mathbf{X} + \mathbf{Y}) = -\mathbf{X} + (\mathbf{X} + \mathbf{Z})$	Adding $-\mathbf{X}$ to the left side of both sides of the equation
$(-\mathbf{X} + \mathbf{X}) + \mathbf{Y} = (-\mathbf{X} + \mathbf{X}) + \mathbf{Z}$	Associative law for addition (A2)
$\mathbf{0} + \mathbf{Y} = \mathbf{0} + \mathbf{Z}$	Additive inverse law (A5)
$\mathbf{Y} = \mathbf{Z}$	Additive identity law (A4)

The proofs of parts (2), (3), (4), (5), and (6) are left as exercises. ∎

Exercises

8-1 Verify the left distributive, right distributive, and unit laws (S3, S4, and S5) for $\mathcal{R}^2$.

8-2 Verify the commutative, identity, inverse, left distributive, and right distributive laws (A3, A4, A5, S3, and S4) for $\mathcal{M}^{2 \times 2}$.

8-3 Prove the right cancellation law. This is part (2) of Theorem 8-1.

8-4 Prove that $0\mathbf{X} = \mathbf{0}$. This is part (3) of Theorem 8-1. (Hint: In words, we are trying to show that $0\mathbf{X}$ is the zero vector. Begin by adding $0\mathbf{X}$ to $1\mathbf{X}$ and proceed from there.)

8-5 Prove that $r\mathbf{0} = \mathbf{0}$. This is part (4) of Theorem 8-1. (Hint: Begin by adding $r\mathbf{0}$ to $r\mathbf{X}$.)

8-6 Prove that $(-1)\mathbf{X} = -\mathbf{X}$. This is part (5) of Theorem 8-1. [Hint: First state in words the meaning of $(-1)\mathbf{X} = -\mathbf{X}$. Then add $(-1)\mathbf{X}$ to $1\mathbf{X}$ and see what happens.]

8-7 Verify the equation $-(-X) = X$ and explain in words its meaning. This is part (6) of Theorem 8-1.

8-8 Suppose that $\mathcal{V}$ is a vector space. Prove that there is only one vector in $\mathcal{V}$ with the property of law A4. In other words, prove that the zero vector is unique. (Hint: Let $\mathbf{0}$ and $\mathbf{0}'$ be two such vectors, and add them to the same vector X. Use this to show that $\mathbf{0}' = \mathbf{0}$.)

8-9 Given X, prove that there is only one vector with the property of law A5. In other words, prove that additive inverses are unique. (Hint: Let $-X$ and Y both be additive inverses of X, and show that $Y = -X$.)

8-10 We define subtraction of vectors by the rule $X - Y = X + (-Y)$. Prove that $X - (-Y) = X + Y$.

9. OTHER EXAMPLES OF VECTOR SPACES

For the balance of the text, the most frequently encountered vector spaces will be $\mathcal{R}^2$, $\mathcal{R}^3$, other $\mathcal{R}^n$s, and $\mathcal{M}^{m \times n}$s. But other examples are of great importance in many other branches of mathematics and its applications. As we will see, various collections of functions form vector spaces, and hence the abstract theory of vector spaces can be used to study and give new information about functions.

Example 9-1 The Vector Space of All Polynomials of Degree ≤ 2 (The *degree* of a polynomial is the highest power of x that has a nonzero coefficient in the polynomial)

Set: $\{ax^2 + bx + c : a, b, c \in \mathcal{R}\}$

Definition of addition: $(ax^2 + bx + c) + (dx^2 + ex + f) = (a + d)x^2 + (b + e)x + (c + f)$

Definition of scalar multiplication: $r(ax^2 + bx + c) = (ra)x^2 + (rb)x + (rc)$

We will denote this vector space $\mathcal{P}^2$.   ●

Example 9-2 The Vector Space of All Polynomials of Degree $\leq n$ (n being any positive integer)

Set: $\{a_n x^n + a_{n-1} x^{n-1} + \cdots + a_1 x + a_0 : a_n, \ldots, a_0 \in \mathcal{R}\}$

Definition of addition: $(a_n x^n + \cdots + a_1 x + a_0) +$

$$(b_n x^n + \cdots + b_1 x + b_0) = (a_n + b_n)x^n + \cdots + (a_1 + b_1)x + (a_0 + b_0)$$

Definition of scalar multiplication: $r(a_n x^n + \cdots + a_1 x + a_0)$
$= (r a_n)x^n + \cdots + (r a_1)x + (r a_0)$

This vector space is denoted $\mathscr{P}^n$. In particular, if $n = 2$, we have the vector space $\mathscr{P}$ of Example 9-1.   ●

Example 9-3   The Vector Space of All Solutions to the Homogeneous System

$$x + y + 3z = 0$$
$$x - y - \ z = 0$$

The solutions here are all triples of the form $(-a, -2a, a)$, and thus the set of vectors is

$$\{(-a, -2a, a) : a \in \mathscr{R}\}$$

Addition and scalar multiplication are defined as in $\mathscr{R}^3$: to add two vectors in this set, add corresponding entries; to scalar multiply a vector by r, multiply each entry by r. We know that this set is a line through the origin in $\mathscr{R}^3$. As in Example 7-6, this is an example of a vector space contained within a larger vector space.   ●

Example 9-4   The Vector Space of All Solutions to the Homogeneous System

$$x - y - 3z = 0$$

Since the solution set consists of all triples of the form $(a + 3b, a, b)$, the set of vectors is

$$\{(a + 3b, a, b) : a, b \in \mathscr{R}\}$$

Addition and scalar multiplication are defined as in $\mathscr{R}^3$. This vector space is a plane through the origin in $\mathscr{R}^3$.   ●

It should be noted that Examples 9-3 and 9-4 are special cases of a general type of vector space: the set of solutions of *any* homogeneous system of equations is a vector space.

Example 9-5   The Vector Space of All Functions Whose Domain is $\mathcal{R}$ and Whose Values are Real Numbers

Set: $\{f : f$ is a function whose domain is $\mathcal{R}$ and whose values are in $\mathcal{R}\}$

Definition of addition: $f + g$ is the function defined by the rule $(f + g)(x) = f(x) + g(x)$

Definition of scalar multiplication: rf is the function defined by the rule $(rf)(x) = r(f(x))$ ●

Example 9-6   The Vector Space of All Continuous Functions Whose Domain is $[0, 1]$

Set: $\{f : f$ is a continuous function whose domain is $[0, 1]\}$

Addition and scalar multiplication are defined as in Example 9-5.   ●

Again, verifying that each of these six examples is a vector space requires checking the ten laws for a vector space. (Section 10 will give a shortcut applicable to Examples 9-1, 9-3, and 9-4, each of which is contained in a larger vector space.)

We will verify a few of the laws for these examples.

Example 9-7   The Right Distributive Law for $\mathscr{P}^2$

We must verify that $(r + s)\mathbf{X} = r\mathbf{X} + s\mathbf{X}$. We will work on each side of this equation separately and see that equality holds. It will be left to the reader to supply the reasons for each step.

$$\begin{aligned}
(r + s)\mathbf{X} &= (r + s)(ax^2 + bx + c) \\
&= ((r + s)a)x^2 + ((r + s)b)x + (r + s)c \\
&= (ra + sa)x^2 + (rb + sb)x + (rc + sc)
\end{aligned}$$

Also

$$\begin{aligned}
r\mathbf{X} + s\mathbf{X} &= r(ax^2 + bx + c) + s(ax^2 + bx + c) \\
&= ((ra)x^2 + (rb)x + (rc)) + ((sa)x^2 + (sb)x + (sc)) \\
&= (ra + sa)x^2 + (rb + sb)x + (rc + sc)
\end{aligned}$$

Thus it is clear that $(r + s)\mathbf{X} = r\mathbf{X} + s\mathbf{X}$.   ●

Example 9-8   The Commutative Law for the Vector Space in Example 9-5

We wish to show that $X + Y = Y + X$, that is, that $f + g = g + f$ for all functions f and g. To show that the two functions $f + g$ and $g + f$ are equal, we must show that they produce the same values when applied to an arbitrary number x. That is to say, we must show that $(f + g)(x) = (g + f)(x)$ for all real numbers x. By definition, $(f + g)(x) = f(x) + g(x)$ and $(g + f)(x) = g(x) + f(x)$. Since $f(x)$ and $g(x)$ are numbers they commute—that is, $f(x) + g(x) = g(x) + f(x)$. Therefore $(f + g)(x) = (g + f)(x)$, as desired.   ●

Example 9-9   The Additive Identity Law for the Vector Space in Example 9-6

In this case we must find a function Z such that $f + Z = f$ for all continuous functions whose domain is $[0, 1]$. The obvious choice for Z is the function whose rule is $Z(x) = 0$ for all numbers x in $[0, 1]$. This function is continuous, as are all constant functions. To show that $f + Z = f$ we must show that $(f + Z)(x) = f(x)$. But this is clear since $(f + Z)(x) = f(x) + Z(x) = f(x) + 0 = f(x)$.   ●

Exercises

For each of the following vector spaces in Exercises 9-1 to 9-7:

(a)   Write three arbitrary vectors.
(b)   Write what you would expect the zero vector to be.
(c)   Write what you would expect to have as the additive inverses of the arbitrary vectors from part (a).

9-1 $\mathscr{P}^2$.

9-2 $\mathscr{P}^n$.

9-3   The vector space of Example 9-3.

9-4   The vector space of Example 9-4.

9-5   The vector space of Example 9-5.

9-6   The set of solutions to the equation $x - 3y = 0$.

9-7 The set of solutions to the system

$$x - 2y + z = 0$$
$$2x - 4y + 2z = 0$$

9-8 Prove laws A4, A5, and S3 for $\mathscr{P}^2$.

9-9 Prove laws A1, A5, and S1 for the vector space of Example 9-3.

9-10 Prove laws A1, S1, and S2 for the vector space of Example 9-4.

9-11 Explain why the set $\{(2 + a, a, -a) : a \in \mathscr{R}\}$ is not a vector space (using the same addition and scalar multiplication as for $\mathscr{R}^3$).

9-12 Let $\mathscr{P}$ be the set of all polynomials with real coefficients, that is,

$$\mathscr{P} = \{a_k x^k + a_{k-1} x^{k-1} + \cdots + a_1 x + a_0 : a_k, \ldots, a_0 \in R, k = 0, 1, 2, \ldots\}$$

Addition and scalar multiplication are defined as for the vector spaces $\mathscr{P}^n$.

(a) Verify that $\mathscr{P}$ is closed under addition (law A1) and closed under scalar multiplication (law S1).
(b) Verify the additive identity law (A4) and the left distributive law (S3) for $\mathscr{P}$. (Note: $\mathscr{P}$ satisfies all ten laws, and hence is a vector space.)

10. SUBSPACES

We have seen several examples of vector spaces contained in larger vector spaces: for example, the x axis of Example 7-6, the line $(-a, -2a, a)$ of Example 9-3, and the plane $(a + 3b, a, b)$ of Example 9-4, all contained in $\mathscr{R}^3$. Vector spaces within vector spaces arise frequently in many branches of mathematics. Furthermore, they have a geometric significance: in $\mathscr{R}^2$ and $\mathscr{R}^3$ any line or plane through the origin is itself a vector space. We are thus led to the following definition.

Definition

Let $\mathscr{V}$ be a vector space. A subset $\mathscr{W}$ of $\mathscr{V}$ is said to be a **subspace** of $\mathscr{V}$ provided $\mathscr{W}$ is itself a vector space (where vectors are added and multiplied by scalars in the *same way* as they are in $\mathscr{V}$).

Example 10-1

The set of points $\{(-a, -2a, a) : a \in \mathscr{R}\}$, from Example 9-3, is a subspace of $\mathscr{R}^3$. The points $(-a, -2a, a)$ all belong to $\mathscr{R}^3$, and addition and scalar multiplication are defined in precisely the same way as in $\mathscr{R}^3$. ●

Example 10-2

The set $\{(a, 0) : a \in \mathcal{R}\}$ is a subspace of $\mathcal{R}^2$. Geometrically, this set is the x axis in the xy plane.   ●

If we wanted to prove that the above two examples were subspaces, we would have to verify that they satisfy all ten laws for a vector space. But there is a quicker and easier way, as the following theorem shows.

Theorem 10-1

Let $\mathcal{W}$ be a nonempty subset of a vector space $\mathcal{V}$. $\mathcal{W}$ is a subspace of $\mathcal{V} \Leftrightarrow \mathcal{W}$ is closed under vector addition and scalar multiplication.†

Proof‡

(Note that the left side says "all ten laws are true," while the right side says "A1 and S1 are true.")

$\Rightarrow$: This is clear, for if all ten laws are true, then A1 and S1, which are two of the ten, are certainly true.

$\Leftarrow$: By assumption, laws A1 and S1 hold. We must verify the other eight laws for vectors in $\mathcal{W}$. Laws A2, A3, and S2 to S5 (associative, commutative, distributive, and unit laws) pose no problem since we know they hold for all vectors in $\mathcal{V}$, and consequently also for those in $\mathcal{W}$, which is part of $\mathcal{V}$.

Verification of A4: We must show that $\mathbf{0} \in \mathcal{W}$ (of course $\mathbf{0}$ is already known to be in $\mathcal{V}$). To do this, pick any vector $\mathbf{X} \in \mathcal{W}$. By law S1, true by assumption, $0\mathbf{X} \in \mathcal{W}$. But $\mathbf{0} = 0\mathbf{X}$ as noted in Theorem 8-1. Hence $\mathbf{0} \in \mathcal{W}$.

Verification of A5: We must show that if $\mathbf{X} \in \mathcal{W}$, then $-\mathbf{X} \in \mathcal{W}$. But if $\mathbf{X} \in \mathcal{W}$, then so is $(-1)\mathbf{X}$ by law S1, again true by assumption. But $(-1)\mathbf{X} = -\mathbf{X}$ by Theorem 8-1. Hence $-\mathbf{X} \in \mathcal{W}$, as desired.   ■

By virtue of this theorem, to verify that a nonempty¶ subset $\mathcal{W}$ of a vector space is a subspace, we need only show that $\mathcal{W}$ is closed under addition and scalar multiplication.

Example 10-3

$\mathcal{U} = \{(a, 0, 0) : a \in \mathcal{R}\}$ is a subspace of $\mathcal{R}^3$.

$\mathcal{U}$ *is closed under addition*   We choose two arbitrary vectors in $\mathcal{U}$, say $(a, 0, 0)$ and $(b, 0, 0)$. Then their sum $(a, 0, 0) + (b, 0, 0) = (a + b, 0, 0)$ is

† The symbol $\Leftrightarrow$ is read "if and only if."
‡ To prove a theorem of the form $A \Leftrightarrow B$, we must do two things: (1) Assume that the statement A is true and prove that the statement B is true. (2) Assume that B is true and prove that A is true.
¶ Technically, the first step in verifying that $\mathcal{W}$ is a subspace is to verify that it is nonempty (that is, it contains at least one vector). In all our examples, the fact that the subset is nonempty is evident.

still an element of $\mathcal{U}$, since the first entry, $a + b$, is a real number and since the second and third entries are zero. The vector $(a + b, 0, 0)$ has the required form to belong to $\mathcal{U}$.

$\mathcal{U}$ *is closed under scalar multiplication*   We need only note that for an arbitrary scalar r and an arbitrary vector $(a, 0, 0)$ in $\mathcal{U}$, $r(a, 0, 0) = (ra, 0, 0)$. The latter is in the proper form—(*real number*, 0, 0)—to belong to $\mathcal{U}$.

Geometrically, $\mathcal{U}$ is the x axis in $\mathcal{R}^3$.   ●

Example 10-4

$\mathcal{W} = \{(a, b, b) : a, b \in \mathcal{R}\}$ is a subspace of $\mathcal{R}^3$. Note that to belong to $\mathcal{W}$, an ordered triple must have its second and third entries equal.

$\mathcal{W}$ *is closed under addition*   We choose two arbitrary vectors of the proper form, say (a, b, b) and (c, d, d). Their sum $(a, b, b) + (c, d, d) = (a + c, b + d, b + d)$ has the proper form to belong to $\mathcal{W}$. That is, its entries are all real numbers, and the second and third entries are equal.

$\mathcal{W}$ *is closed under scalar multiplication*   We note that for an arbitrary scalar r, $r(a, b, b) = (ra, rb, rb)$, which again has the proper form to belong to $\mathcal{W}$. Geometrically, $\mathcal{W}$ is a plane through the origin in $\mathcal{R}^3$, as can be seen by rewriting the points of $\mathcal{W}$ in the form $a(1, 0, 0) + b(0, 1, 1)$.●

Example 10-5

Consider the homogeneous system

$$x + y + 3z = 0$$
$$x - y - \ z = 0$$

If $\mathcal{X}$ is the set of solutions to this system, then $\mathcal{X} = \{(-a, -2a, a) : a \in \mathcal{R}\}$ and is a subspace of $\mathcal{R}^3$.

$\mathcal{X}$ *is closed under addition*   Let $(-a, -2a, a)$ and let $(-b, -2b, b)$ be two arbitrary vectors in $\mathcal{X}$. Then $(-a, -2a, a) + (-b, -2b, b) = (-a - b, -2a - 2b, a + b) = (-(a + b), -2(a + b), a + b)$, which is in the required form to belong to $\mathcal{X}$—the first entry is the negative of the third entry and the second entry is -2 times the third entry—and hence their sum is in $\mathcal{X}$. (Note that we had to rewrite $-a - b$ as $-(a + b)$ and $-2a - 2b$ as $-2(a + b)$ to see clearly that the first two entries had the required form.)

$\mathfrak{X}$ *is closed under scalar multiplication*   If $(-a, -2a, a)$ is in $\mathfrak{X}$, and if r is an arbitrary scalar, then $r(-a, -2a, a) = (r(-a), r(-2a), ra) = (-(ra), -2(ra), ra)$ is also in $\mathfrak{X}$. Again we had to rewrite $r(-a)$ as $-(ra)$ and $r(-2a)$ as $-2(ra)$ to confirm the form required for membership in $\mathfrak{X}$.

Therefore, the set of solutions to this system of equations is a subspace of $\mathcal{R}^3$.

More generally, the set of solutions to any homogeneous system with n unknowns is a subspace of $\mathcal{R}^n$.   ●

Example 10-6

$$\mathcal{Y} = \left\{ \begin{bmatrix} a & 0 \\ 0 & b \end{bmatrix} : a, b \in \mathcal{R} \right\}$$

is a subspace of $\mathcal{M}^{2 \times 2}$.

$\mathcal{Y}$ is closed under addition since for any two elements

$$\begin{bmatrix} a & 0 \\ 0 & b \end{bmatrix} \quad \text{and} \quad \begin{bmatrix} c & 0 \\ 0 & d \end{bmatrix} \quad \text{of } \mathcal{Y}$$

their sum

$$\begin{bmatrix} a & 0 \\ 0 & b \end{bmatrix} + \begin{bmatrix} c & 0 \\ 0 & d \end{bmatrix} = \begin{bmatrix} a+c & 0 \\ 0 & b+d \end{bmatrix}$$

has the proper form to belong to $\mathcal{Y}$ since its 1, 2 entry and 2, 1 entry are both zero.

$\mathcal{Y}$ is closed under scalar multiplication since

$$r \begin{bmatrix} a & 0 \\ 0 & b \end{bmatrix} = \begin{bmatrix} ra & 0 \\ 0 & rb \end{bmatrix}$$

again has the proper form for membership in $\mathcal{Y}$.   ●

Example 10-7

$\mathcal{Z} = \{ax^3 + bx : a, b \in \mathcal{R}\}$ is a subspace of $\mathcal{P}^3$. $\mathcal{Z}$ is closed under vector addition since $(ax^3 + bx) + (cx^3 + dx) = (a + c)x^3 + (b + d)x$, which has the required form to belong to $\mathcal{Z}$ (that is, the x^2 term and the constant term are both zero).

$\mathcal{Z}$ is closed under scalar multiplication since $r(ax^3 + bx) = (ra)x^3 + (rb)x$, which, by its form, is a vector in $\mathcal{Z}$.   ●

Example 10-8

If $\mathcal{V}$ is any vector space, it is a subspace of itself, since it is closed under vector addition and scalar multiplication.   ●

Example 10-9

If $\mathcal{V}$ is a vector space, then $\{\mathbf{0}\}$ is a subspace of $\mathcal{V}$. $\{\mathbf{0}\}$ is closed under vector addition, since the only "arbitrary" vector here is $\mathbf{0}$, and $\mathbf{0} + \mathbf{0} = \mathbf{0}$. $\{\mathbf{0}\}$ is closed under scalar multiplication since $r\mathbf{0} = \mathbf{0}$ for any scalar r.   ●

Example 10-10

$\{(a, 1, 0) : a \in \mathcal{R}\}$ is *not* a subspace of $\mathcal{R}^3$. This set is not closed under addition: $(a, 1, 0) + (b, 1, 0) = (a + b, 2, 0)$, which does not have the required 1 as its second entry. This set also is not closed under scalar multiplication.   ●

Example 10-11

$\{(a^2, 0) : a \in \mathcal{R}\}$ is *not* a subspace of $\mathcal{R}^2$. This set is not closed under scalar multiplication: for example, $-1(3^2, 0) = (-9, 0)$, but the entry -9 cannot be written as a^2 since a^2 is never negative. Note, however, that this set *is* closed under addition. The reader should verify this.   ●

In these examples the vectors in the subsets have a certain form; verification that we actually have a subspace amounts to showing that their sums and scalar multiples have the proper form to again belong to the set. Thus, as in Sec. 7, we see that it is important to know what the vectors look like when dealing with a specific vector space.

The following theorem and corollary spell out two types of sets that always turn out to be subspaces.

Theorem 10-2   The set of solutions of a homogeneous system of linear equations with n unknowns is a subspace of $\mathcal{R}^n$.

Proof   Let $\mathcal{S}$ be the set of solutions. $\mathcal{S}$ is not empty since it contains at least the zero vector $(0, 0, \ldots, 0)$. By Theorem 10-1, to show that $\mathcal{S}$ is a subspace we need only verify that $\mathcal{S}$ is closed under vector addition and scalar multiplication.

S *is closed under vector addition*   Let $(u_1, u_2, \ldots, u_n)$ and $(v_1, v_2, \ldots, v_n)$ be two vectors in S (that is, they are solutions to the system). We must show that their sum $(u_1 + v_1, u_2 + v_2, \ldots, u_n + v_n)$ is also a solution. To that end, suppose that

$$a_1 x_1 + a_2 x_2 + \cdots + a_n x_n = 0 \tag{1}$$

is any one of the equations. Substituting the $u + v$'s for the x's in the left side, we obtain $a_1(u_1 + v_1) + \cdots + a_n(u_n + v_n)$, which equals $(a_1 u_1 + \cdots + a_n u_n) + (a_1 v_1 + \cdots + a_n v_n)$. Since we already know that $a_1 u_1 + \cdots + a_n u_n = 0$ and $a_1 v_1 + \cdots + a_n v_n = 0$, we have their sum equal to zero, as desired.

S *is closed under scalar multiplication*   Let $(u_1, u_2, \ldots, u_n)$ be a solution to the system and let r be a scalar. We must show that the scalar multiple $(ru_1, ru_2, \ldots, ru_n)$ is also a solution. Substituting the ru's for the x's in the left side of (1), we obtain $a_1(ru_1) + \cdots + a_n(ru_n)$, which equals $r(a_1 u_1 + \cdots + a_n u_n)$. Since we already know that $a_1 u_1 + \cdots + a_n u_n = 0$, this product equals zero, as desired.   ∎

Corollary 10-3   Lines and planes through the origin in $\mathcal{R}^2$ or $\mathcal{R}^3$ are subspaces.

Proof   Lines and planes through the origin are sets of solutions of homogeneous systems (see Sec. 4). By Theorem 10-2 they must be subspaces.   ∎

The set of solutions S of a nonhomogeneous system with n unknowns is *never* a subspace of $\mathcal{R}^n$. If S were a subspace, it would have the zero vector $(0, 0, \ldots, 0)$ in it (Exercise 10-19), but the zero vector is never a solution to a nonhomogeneous system.

Exercises

In Exercises 10-1 to 10-11 determine whether the sets are subspaces of $\mathcal{R}^3$. If the set is a subspace, prove it. If not, explain why.

10-1 $\{(0, a, b) : a, b \in \mathcal{R}\}$

10-2 $\{(a, 0, 2) : a \in \mathcal{R}\}$

10-3 $\{(a^2, 0, 0) : a \in \mathcal{R}\}$

10-4 $\{(a, b, c) : a, b, c \in \mathcal{R}\}$

10-5 $\{(2a, -5a, b) : a, b \in \mathcal{R}\}$

10-6 $\{(a, b, a + b) : a, b \in \mathcal{R}\}$

10-7 $\{a(2, 0, 1) : a \in \mathcal{R}\}$

10-8 $\{(a^3, 0, 0) : a \in \mathcal{R}\}$

10-9 $\{a(1, -1, 2) + b(0, 0, 3) : a, b \in \mathcal{R}\}$

10-10 $\{(0, a, b + 1) : a, b \in \mathcal{R}\}$

10-11 $\{(2, 0, 1) + a(4, 1, 3) : a \in \mathcal{R}\}$

In Exercises 10-12 to 10-14 determine whether the sets are subspaces of $\mathcal{M}^{2 \times 2}$.

10-12 $\left\{ \begin{bmatrix} a & -a \\ 0 & 0 \end{bmatrix} : a \in \mathcal{R} \right\}$

10-13 $\left\{ \begin{bmatrix} a & 0 \\ b & a+b \end{bmatrix} : a, b \in \mathcal{R} \right\}$

10-14 $\left\{ \begin{bmatrix} 1 & a \\ a & 1 \end{bmatrix} : a \in \mathcal{R} \right\}$

In Exercises 10-15 to 10-18 determine whether the sets are subspaces of $\mathcal{P}^3$.

10-15   The set of all polynomials with zero as the constant term.

10-16 $\{ax^2 + x : a \in \mathcal{R}\}$.

10-17   All polynomials whose degree is exactly 2.

10-18   All polynomials whose derivative is constant.

10-19   Let $\mathcal{W}$ be a subspace of a vector space $\mathcal{V}$. Explain why we must have $\mathbf{0} \in \mathcal{W}$.

10-20   Let $\mathcal{V}$ be a vector space and let $\mathbf{X}$ be a particular vector in $\mathcal{V}$. Let $\mathcal{W} = \{r\mathbf{X} : r \in \mathcal{R}\}$.

(a) Prove that $\mathcal{W}$ is a subspace of $\mathcal{V}$.
(b) Suppose $\mathcal{V} = \mathcal{R}^2$ and $\mathbf{X} = (3, 1)$. Describe $\mathcal{W}$ geometrically.

10-21 Let $\mathcal{V}$ be a vector space and let $\mathbf{X}_1$ and $\mathbf{X}_2$ be two particular vectors in $\mathcal{V}$. Let $\mathcal{W} = \{r\mathbf{X}_1 + s\mathbf{X}_2 : r, s \in \mathcal{R}\}$.

(a) Prove that $\mathcal{W}$ is a subspace of $\mathcal{V}$.
(b) Suppose that $\mathcal{V} = \mathcal{R}^3$, $\mathbf{X}_1 = (1, 0, 1)$, and $\mathbf{X}_2 = (1, 1, 0)$. Describe $\mathcal{W}$ geometrically.
(c) Suppose that $\mathcal{V} = \mathcal{R}^3$, $\mathbf{X}_1 = (1, 0, 1)$, and $\mathbf{X}_2 = (3, 0, 3)$. Describe $\mathcal{W}$ geometrically.

11. APPLICATIONS

Example 11-1 Vectors in Physics

The reader may be acquainted with the terms *vector* and *scalar* from physics. (In fact, these terms first arose there.) A *vector quantity* is a physical quantity, such as force or velocity, that has both a magnitude and a direction. A *scalar quantity*, such as speed or time, involves only magnitude. Vector quantities are often pictured by arrows; see the comments following Example 7-2. An arrow is a very natural model for a vector quantity since an arrow has both a magnitude (its length) and a direction associated with it.

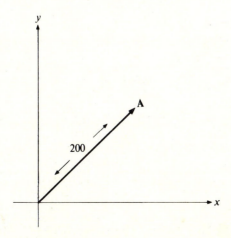

Fig. 11-1

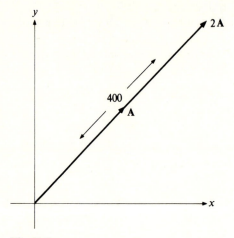

Fig. 11-2

For example, suppose an airplane is flying northeast with an airspeed of 200 mi/h. We can represent the velocity by an arrow **A** in the plane starting at the origin, 200 units long, and pointing in the northeast direction (Fig. 11-1). If the airplane keeps on the same course (that is, its direction is unchanged) but increases its airspeed to 400 mi/h, we can represent the velocity by the arrow **2A** (twice as long as **A** and in the same direction) as in Fig. 11-2.

Many problems in physics involving vector quantities require that these quantities be "added." For example, suppose that an airplane is flying with an airspeed of 120 mi/h due north and the wind is blowing

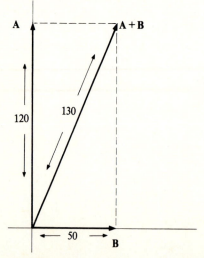

Fig. 11-3 Adding vectors by the parallelogram method.

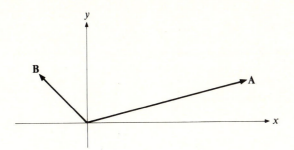

Fig. 11-4

toward the east at 50 mi/h. We let **A** be the arrow representing the plane's velocity and **B** be the arrow representing the wind velocity (Fig. 11-3). To find the velocity of the plane relative to the ground, that is, the total effect of both velocities, we add **A** and **B** by completing the parallelogram determined by **A** and **B** (by drawing the two dotted lines in Fig. 11-3). This is reasonable to do, since in 1 h the airplane would be moved 120 mi north by its engines and 50 mi east by the wind; the net effect would be to reach the end of the diagonal. The resulting diagonal arrow **A** + **B** gives the total effect of the two velocities. (In this case **A** + **B** has magnitude 130, by the Pythagorean theorem. The wind has boosted the plane's speed by 10 mi/h. Its direction is approximately NNE.) Experimental evidence supports this parallelogram method of addition as the correct one for vector addition.

The addition of vector quantities can be described in a second way, called the polygon method (or head-to-tail method). Suppose **A** and **B** are the two arrows in Fig. 11-4. To find **A** + **B** we redraw **B**, keeping its original direction and length, so that its tail coincides with the head of the arrow **A**. The sum is the arrow from the origin to the tip of **B** (Fig. 11-5). This method yields the same result as the parallelogram method, and is especially efficient when several arrows are to be added.

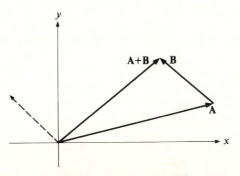

Fig. 11-5   Head-to-tail method for adding vectors.

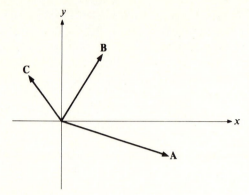

Fig. 11-6

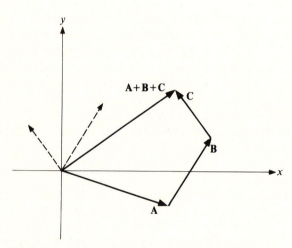

Fig. 11-7

For example, suppose we wish to determine the total gravitational force exerted on a satellite by the sun, moon, and earth. Figure 11-6 shows the three respective forces represented by arrows **A**, **B**, and **C**. In Fig. 11-7 we use the head-to-tail method to find the total force **A** + **B** + **C**. We move **B** parallel so that its tail touches the tip of **A**, and then we move **C** so that its tail touches the tip of the moved-over arrow **B**. The resultant force is given by the arrow **A** + **B** + **C** from the origin to the tip of **C**.

●

Example 11-2   Vectors as Bookkeeping Devices

In this example we will have a brief look at several ways in which large amounts of data can be handled using vectors and the operations of vector addition and scalar multiplication.

Suppose that a warehouse stocks 800 different types of items. We can describe this by an "inventory vector" $\mathbf{A} \in \mathcal{R}^{800}$, where the first entry is the number of items of type 1 on hand, at the beginning of the month, the second entry equals the number of items of type 2, etc. Let $\mathbf{B} \in \mathcal{R}^{800}$ be the vector whose entries equal the numbers of each type of item received by the warehouse during this month and let $\mathbf{C} \in \mathcal{R}^{800}$ be the vector whose entries give the number of items of each type shipped from the warehouse during this month. Then $\mathbf{A} + \mathbf{B} - \mathbf{C}$ gives the inventory at the end of the month. Further, suppose that the entries of a vector $\mathbf{D} \in \mathcal{R}^{800}$ give the minimum number of items that should be kept in stock at the warehouse. Then each negative entry of $\mathbf{D} - (\mathbf{A} + \mathbf{B} - \mathbf{C})$ indicates that more of that item should be ordered, and each nonnegative entry of $\mathbf{D} - (\mathbf{A} + \mathbf{B} - \mathbf{C})$ tells that that item is in sufficient supply.

A bank managing the stockholders' records for a corporation might use vectors in the following way: Suppose the corporation has 5000 stockholders. If the stockholders are listed alphabetically, we construct a vector $\mathbf{A} \in \mathcal{R}^{5000}$ whose first entry is the number of shares owned by the first stockholder, whose second entry is the number of shares owned by the second stockholder, etc. If the corporation wishes to pay a 5 percent stock dividend and a \$0.27 per share cash dividend, then the bank need determine only the scalar multiple $0.05\mathbf{A}$ to compute the stock dividend amounts and $0.27\mathbf{A}$ to compute the cash dividend amounts.

A meteorologist might use vectors to keep records of temperature. On each day in May he records the daily high and low temperature at a certain location. At the end of the month he can form two vectors $\mathbf{H}$ and $\mathbf{L}$ in $\mathcal{R}^{31}$, where the entries of $\mathbf{H}$ are the 31 daily high temperatures and the entries of $\mathbf{L}$ are the 31 daily low temperatures. Various combinations of these vectors give useful temperature data. For example, the 31 entries of $\mathbf{H} - \mathbf{L}$ give the daily temperature ranges, and the 31 entries of $\frac{1}{2}(\mathbf{H} + \mathbf{L})$ give the average daily temperatures.

These three illustrations show how vectors can work neatly with large amounts of data and also show the applicability of vectors in $\mathcal{R}^n$, for large values of n. In practice, computer programs would be written to carry out these calculations (especially if the amount of data is very large); several different programming languages contain special statements to work with vectors.   ●

Exercises

11-1 Let $\mathbf{A}$, $\mathbf{B}$, $\mathbf{C}$, and $\mathbf{D}$ be the arrows from the origin to the points $(5, 2)$, $(1, 3)$, $(-3, 4)$, and $(-2, -6)$ respectively. (Note: In the following, use either the parallelogram law or the head-to-tail method to compute sums.)

(a) Draw **A** + 3**B**.
(b) Draw **A** + (**B** + **C**).
(c) Draw (**A** + **B**) + **C**, and compare this with the answer to (b).
(d) Verify that **D** = − 2**B**.
(e) Verify that 2**B** − **A** = **C**.
(f) Find **A** + **B** + **C** + **D** using the head-to-tail method.

11-2 Suppose that a box has two cords attached to it. One man pulls with a force of 30 lb along the positive x axis. A second man pulls with a force of 40 lb along the positive y axis. Find the magnitude and direction of the combined forces.

11-3 The wind blows due east at 50 mi/h. An airplane travels with an airspeed (speed produced by its engines alone) of 130 mi/h. Sketch the direction the plane must head in order to actually fly directly north. What will its ground speed (speed taking into account the wind effect also) be then?

11-4 Suppose that the meteorologist in Example 11-2 records all his temperatures in degrees Celsius (centigrade). Write a vector equation that changes the temperatures in the vector **H** to degrees Fahrenheit.

11-5 Suppose that a company measures 100 different machine parts in inches. Write a vector equation that would be used to translate the measurements into centimeters.

Review of Chapter 2

1 A vector space $\mathcal{V}$ consists of a set of elements, called vectors, and two operations: _____ and _____.
These two operations satisfy ten laws:
A1. Closure under addition: If **X** and **Y** are in the vector space $\mathcal{V}$, then

_____.

A2. Associative law: (**X** + **Y**) + **Z** = _____.

A3. _____ law: **X** + **Y** = **Y** + **X**.

A4. Additive identity law: There is a vector, denoted **0**, such that ____

_____.

A5. Additive inverse law: _____.

S1. Closure under scalar multiplication: If **X** is any vector and *r* is any scalar, then _____.

S2. Associative law:_____.

S3. Left distributive law: $r(\mathbf{X} + \mathbf{Y}) =$ _____.

S4. Right distributive law: _____.

S5. Unit law: _____.

2   To verify that a given set with two given operations is a vector space, we must verify all ten laws. The chief examples of vector spaces in this text are the $\mathscr{R}^n$s and the $\mathscr{M}^{m \times n}$s. To prove any of the ten laws for a particular vector space, we must translate the law into notation appropriate for that vector space. For example, to prove the commutative law for $\mathscr{M}^{2 \times 3}$, we would begin with

$$\begin{bmatrix} a & b & c \\ d & e & f \end{bmatrix} + \begin{bmatrix} & & \end{bmatrix} \overset{?}{=} \begin{bmatrix} & & \end{bmatrix} + \begin{bmatrix} & & \end{bmatrix}$$

To prove the left distributive law for $\mathscr{R}^4$, we would begin with

_____ $\overset{?}{=}$ _____.

3   To check that a nonempty subset $\mathscr{W}$ of a vector space $\mathscr{U}$ is a subspace of $\mathscr{V}$, we do not need to verify all ten laws, but only laws _____ and _____. That is, (1) starting with arbitrary elements **X** and **Y** in ____, show that _____; and (2) starting with an arbitrary vector **X** in ____ and an arbitrary scalar *r*, show that _____.

Review Exercises

1   The following are *not* vector spaces. Determine which of the ten laws hold and which fail.

(a)   The set of all ordered pairs of real numbers, with addition as usual: $(a, b) + (c, d) = (a + c, b + d)$, but scalar multiplication defined by the rule $r(a, b) = (ra, 0)$

(b)   The same set as in (a), scalar multiplication defined as usual, but addition defined by the rule $(a, b) + (c, d) = (a + 2c, b + 2d)$

2   Determine if the following are subspaces of the given vector spaces:

(a)   The subset of $\mathcal{M}^{2 \times 3}$ consisting of all matrices of the form

$$\begin{bmatrix} a & b & 0 \\ 0 & 0 & 0 \end{bmatrix}$$

(b)   The subset of $\mathcal{R}^2$: $\{(a, b) : a^2 + b^2 \leq 1\}$
(c)   The set of points on the graph of the curve $y = x^2$ in $\mathcal{R}^2$

3   Let $\mathcal{V}$ be an arbitrary vector space. Prove that for all vectors $\mathbf{X}$ in $\mathcal{V}$ and all scalars r:

(a) $r(-\mathbf{X}) = -(r\mathbf{X})$
(b) $r(-\mathbf{X}) = (-r)\mathbf{X}$

4   For each of the vector spaces $\mathcal{R}^3$, $\mathcal{M}^{2 \times 3}$, $\mathcal{R}^n$, $\mathcal{M}^{m \times n}$, and $\mathcal{P}^3$:

(a)   What does the zero vector look like?
(b)   What do additive inverses look like?

5(a)   Verify the commutative, identity, and inverse laws for addition and the associative and left distributive laws for scalar multiplication (A3, A4, A5, S2, and S3) for $\mathcal{R}^4$.
(b)   Verify the associative and inverse laws for addition and the right distributive and unit laws for multiplication (A2, A5, S4, and S5) for $\mathcal{M}^{3 \times 2}$.
(c)   Verify the commutative, identity, and inverse laws for addition and the right distributive law for multiplication (A3, A4, A5, and S4) for $\mathcal{P}^3$.

6   Let $\mathbf{X}$ be any vector in an arbitrary vector space $\mathcal{V}$. Prove that $\mathbf{X} + \mathbf{X} = 2\mathbf{X}$.

7   Prove that $\{(2a, b, 3b) : a, b \in \mathcal{R}\}$ is a subspace of $\mathcal{R}^3$.

8   Prove that

$$\left\{ a \begin{bmatrix} 1 & 0 & 0 \\ 0 & 1 & 0 \end{bmatrix} + b \begin{bmatrix} 0 & 2 & 0 \\ 0 & 0 & 0 \end{bmatrix} : a, b \in \mathcal{R} \right\}$$

is a subspace of $\mathcal{M}^{2 \times 3}$.

9   Prove that $\{2ax^2 + a : a \in \mathcal{R}\}$ is a subspace of $\mathcal{P}^2$.

10   Determine whether or not the point $(7, 1, 3)$ is in the subspace $\{(3a - 2b, \ b + a, \ b + a) : a, \ b \in \mathcal{R}\}$ of $\mathcal{R}^3$. [Hint: Put $(7, 1, 3) =$

$(3a - 2b, b + a, b + a)$. Comparing corresponding entries yields three equations with unknowns a and b. Solve for a and b, if possible.]

11(a)   Prove that the set of points on the line $5x - y = 0$ is a subspace of $\mathcal{R}^2$.

(b)   Is the set of points on the line $5x - y = 1$ a subspace of $\mathcal{R}^2$? Explain.

(c)   Explain why lines and planes in $\mathcal{R}^3$ that are not through the origin are not subspaces of $\mathcal{R}^3$.

12   Which of the following are subspaces of the vector space defined in Example 9-6? (Call this vector space $\mathcal{F}$.)

(a) $\{f : f \in \mathcal{F} \text{ and } f(x) \geq 0 \text{ for all } x \in [0, 1]\}$

(b) $\{f : f \in \mathcal{F} \text{ and } f(0) = f(1)\}$

(c) $\{f : f \in \mathcal{F} \text{ and } f \text{ has a derivative for all } x \text{ in } [0, 1]\}$

(d) $\{f : f \in \mathcal{F} \text{ and } f(0) = 0\}$

(e) $\{f : f \in \mathcal{F} \text{ and } f(0) = 1\}$

*13   Write a computer program that:

(a)   Adds vectors in $\mathcal{R}^3$ (or, more generally, $\mathcal{R}^N$, where one inputs the value of N).

(b)   Scalar multiplies vectors in $\mathcal{R}^3$ (or $\mathcal{R}^N$ for input N).

CHAPTER 3

VECTOR SPACES AND DIMENSION

We have an intuitive idea of dimension. For instance, we think of lines as one-dimensional, planes as two-dimensional, and xyz space as three-dimensional. Our intuitive idea is reflected in the typical dictionary definition, which uses words like *length*, *breadth*, and *thickness*. We think of one-dimensional objects as those having just length, two-dimensional objects as those having length and breadth, etc. Expressing it another way, we say that a line is one-dimensional because we can move from point to point on the line by motion in the one direction along the line; that the xy plane is two-dimensional because we can move from point to point by motion parallel to the x axis followed by motion parallel to the y axis; and that xyz space is three-dimensional because motion from point to point can be achieved by motion in three directions: parallel to the x axis, parallel to the y axis, and parallel to the z axis. Summarizing, our intuitive idea of dimension can be expressed as the *minimum number of directions* necessary to allow movement between any two points in the vector space.

In order to make the idea of dimension precise, rather than intuitive, we need to introduce two concepts—*spanning* and *independence*. When certain vectors *span* a vector space $\mho$, moving from point to point in $\mho$ can be accomplished by motions parallel to those vectors. (For example, three vectors lying along the x, y, and z axes respectively will span xyz space.) This concept will be discussed at length in Sec. 12. On the other hand, if certain vectors are *independent*, all these directions are needed for spanning the vector space. (For example, three vectors along the x, y, and z axes are independent; leave out the vector on the z axis, say, and we could not move "up and down" in xyz space.) Independence is the subject of Sec. 13.

If certain vectors both span a vector space $\mathcal{V}$ and are independent, then they provide a minimum collection of directions for movement in $\mathcal{V}$, and their number is $\mathcal{V}$'s dimension. (Vectors chosen along the x, y, and z axes have both properties, as mentioned above, and therefore show that xyz space is three-dimensional.) The subject of dimension is covered in Sec. 14.

12. SPANNING

Let $\mathcal{S}$ be the set of points on the plane with equation

$$x + 3y + 2z = 0$$

It is easily seen that

$$\mathcal{S} = \{a(-3, 1, 0) + b(-2, 0, 1) : a, b \in \mathcal{R}\}$$

By Theorem 10-2, $\mathcal{S}$ is a subspace of $\mathcal{R}^3$. Every vector in $\mathcal{S}$ is written in terms of two specific vectors, $(-3, 1, 0)$ and $(-2, 0, 1)$. We express this fact by saying that $\mathcal{S}$ is *spanned* by $(-3, 1, 0)$ and $(-2, 0, 1)$. And as a shorthand notation we write

$$\mathcal{S} = \langle(-3, 1, 0), (-2, 0, 1)\rangle$$

The intuitive interpretation of the above expression is that motion from point to point in the plane $\mathcal{S}$ can be achieved by moving parallel to the arrows that represent $(-3, 1, 0)$ and $(-2, 0, 1)$. We illustrate this in Fig. 12-1, where the dotted lines indicate the motions parallel to $(-3, 1, 0)$ and $(-2, 0, 1)$. For example, to move from A to B we first move parallel to $(-2, 0, 1)$ and then move parallel to $(-3, 1, 0)$. Similarly for motion between C and D.

In like fashion, the solution set $\mathcal{C}$ of the system

$$x + 2y - 10z = 0$$
$$y - 5z = 0$$

is the line

$$\mathcal{C} = \{a(0, 5, 1) : a \in \mathcal{R}\}$$

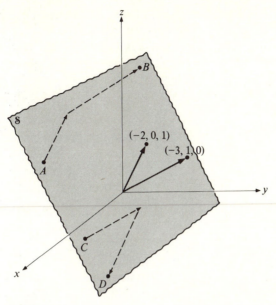

Fig. 12-1

We say that this subspace $\mathcal{C}$ is *spanned* by $(0, 5, 1)$, and we write

$$\mathcal{C} = \langle (0, 5, 1) \rangle$$

as a shorthand notation. Motion from point to point on this line $\mathcal{C}$ requires only one direction.

To generalize these examples, we have the following definition.

Definition

A vector space $\mathcal{U}$ is **spanned** by the vectors $\mathbf{X}_1, \mathbf{X}_2, \ldots, \mathbf{X}_k$, and we write $\mathcal{U} = \langle \mathbf{X}_1, \mathbf{X}_2, \ldots, \mathbf{X}_k \rangle$, provided every vector in $\mathcal{U}$ can be written in terms of $\mathbf{X}_1, \mathbf{X}_2, \ldots, \mathbf{X}_k$; that is, provided that every vector in $\mathcal{U}$ is of the form

$$a_1 \mathbf{X}_1 + a_2 \mathbf{X}_2 + \cdots + a_k \mathbf{X}_k$$

for scalars $a_1, a_2, \ldots, a_k$.

A vector of the form $a_1 \mathbf{X}_1 + a_2 \mathbf{X}_2 + \cdots + a_k \mathbf{X}_k$ is called a **linear combination** of the vectors $\mathbf{X}_1, \mathbf{X}_2, \ldots, \mathbf{X}_k$.

The vector space $\mathcal{R}^2$ is spanned by the vectors $(1, 0)$, $(0, 1)$ since every vector (a, b) in $\mathcal{R}^2$ is of the form $a(1, 0) + b(0, 1)$. We can write, using shorthand notation, $\mathcal{R}^2 = \langle (1, 0),\ (0, 1) \rangle$. Similarly, $\mathcal{R}^3 = \langle (1, 0, 0),$ $(0, 1, 0),\ (0, 0, 1) \rangle$. Also $\mathcal{M}^{2 \times 2}$ is spanned by

$$\begin{bmatrix} 1 & 0 \\ 0 & 0 \end{bmatrix},\ \begin{bmatrix} 0 & 1 \\ 0 & 0 \end{bmatrix},\ \begin{bmatrix} 0 & 0 \\ 1 & 0 \end{bmatrix},\ \begin{bmatrix} 0 & 0 \\ 0 & 1 \end{bmatrix},$$

since an arbitrary vector $\begin{bmatrix} a & b \\ c & d \end{bmatrix}$ equals the linear combination

$$a\begin{bmatrix} 1 & 0 \\ 0 & 0 \end{bmatrix} + b\begin{bmatrix} 0 & 1 \\ 0 & 0 \end{bmatrix} + c\begin{bmatrix} 0 & 0 \\ 1 & 0 \end{bmatrix} + d\begin{bmatrix} 0 & 0 \\ 0 & 1 \end{bmatrix}$$

The importance of spanning is that if vectors $\mathbf{X}_1, \ldots, \mathbf{X}_k$ span $\mathcal{V}$, then we know that it takes *at most* these k "directions" to move from point to point in $\mathcal{V}$ (see Exercise 12-20), and so the dimension of $\mathcal{V}$ will be at most k. Therefore, an important problem is that of determining whether a given vector space is spanned by some given vectors. The following examples illustrate the method of solution.

Example 12-1

$\mathcal{R}^2$ is spanned by $(1, 1)$ and $(3, 2)$. Verification entails showing that an arbitrary vector (a, b) in $\mathcal{R}^2$ is of the form $x(1, 1) + y(3, 2)$. That is, we must show that the equation

$$(a, b) = x(1, 1) + y(3, 2) \tag{1}$$

can be solved for x and y. Carrying out the computations on the right side, we have

$$(a, b) = (x + 3y,\ x + 2y)$$

The only way these two vectors can be equal is that their corresponding entries be equal. This amounts to two equations in the two unknowns x and y:

$$x + 3y = a$$
$$x + 2y = b$$

We solve by our usual matrix method:

$$\begin{bmatrix} 1 & 3 & \vdots & a \\ 1 & 2 & \vdots & b \end{bmatrix} \longrightarrow \begin{bmatrix} 1 & 3 & \vdots & a \\ 0 & -1 & \vdots & b-a \end{bmatrix}$$

$$\longrightarrow \begin{bmatrix} 1 & 3 & \vdots & a \\ 0 & 1 & \vdots & a-b \end{bmatrix} \longrightarrow$$

$$\begin{bmatrix} 1 & 0 & \vdots & a-3(a-b) \\ 0 & 1 & \vdots & a-b \end{bmatrix} = \begin{bmatrix} 1 & 0 & \vdots & -2a+3b \\ 0 & 1 & \vdots & a-b \end{bmatrix}$$

Therefore, there *are* numbers x and y such that equation (1) holds, namely $x = -2a + 3b$ and $y = a - b$. ●

Example 12-2

$\mathcal{R}^3$ is spanned by $(1, 0, 2)$, $(0, 1, 0)$, and $(1, -1, 1)$. As above, for any (a, b, c) we must show that we can solve

$$(a, b, c) = x(1, 0, 2) + y(0, 1, 0) + z(1, -1, 1)$$

for x, y, and z.

Rewriting yields the system

$$\begin{aligned} x \quad\quad + z &= a \\ y - z &= b \\ 2x \quad\quad + z &= c \end{aligned}$$

Solving by reducing the matrix, we obtain $x = -a + c$, $y = 2a + b - c$, $z = 2a - c$. Thus any vector in $\mathcal{R}^3$ can be written as a linear combination of the three given vectors:

$$(a, b, c) = (-a + c)(1, 0, 2) + (2a + b - c)(0, 1, 0) + (2a - c)(1, -1, 1)$$

●

Example 12-3

$\mathcal{R}^2$ is *not* spanned by $(4, -1)$ and $(-8, 2)$. In this case we will show that there is at least one vector (a, b) that cannot be written as

$$(a, b) = x(4, -1) + y(-8, 2)$$

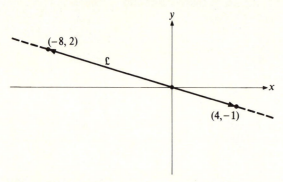

Fig. 12-2   Two vectors on the same line through the origin cannot span $\mathcal{R}^2$.

Our matrix method for solution yields

$$\begin{bmatrix} 4 & -8 & \vdots & a \\ -1 & 2 & \vdots & b \end{bmatrix} \longrightarrow \begin{bmatrix} 1 & -2 & \vdots & a/4 \\ -1 & 2 & \vdots & b \end{bmatrix}$$

$$\longrightarrow \begin{bmatrix} 1 & -2 & \vdots & a/4 \\ 0 & 0 & \vdots & (a/4) + b \end{bmatrix}$$

From the last row of this matrix we must have $(a/4) + b = 0$ in order to have a solution to the original system. That is, we must have $a = -4b$. Hence *only* the points of the form $(-4b, b)$ can be written as linear combinations of $(4, -1)$ and $(-8, 2)$. Since not every vector in $\mathcal{R}^2$ is of this form, the two given vectors do not span $\mathcal{R}^2$. Geometrically $(4, -1)$ and $(-8, 2)$ both lie on the same line $\mathcal{L}$ through the origin, and only points on this line $\mathcal{L}$ can be obtained as linear combinations of $(4, -1)$ and $(-8, 2)$. (See Fig. 12-2.)   ●

In summary, to show that given vectors span a vector space, we form the basic equation

Arbitrary vector in vector space = linear combination of given vectors

or, equivalently.

$$\text{Arbitrary vector} = a_1 \mathbf{X}_1 + a_2 \mathbf{X}_2 + \cdots + a_k \mathbf{X}_k \tag{2}$$

We then rewrite (2) using appropriate notation for the arbitrary vector and the given vectors $\mathbf{X}_i$. To show spanning, we must show that a solution for the unknown scalars $a_1, a_2, \ldots, a_k$ exists.

Example 12-4

Show that $\mathcal{M}^{2 \times 2}$ is spanned by

$$\begin{bmatrix} 1 & 0 \\ 0 & 0 \end{bmatrix}, \begin{bmatrix} 0 & 3 \\ 0 & 0 \end{bmatrix}, \begin{bmatrix} 0 & 0 \\ 1 & 1 \end{bmatrix}, \begin{bmatrix} 0 & -3 \\ 0 & 1 \end{bmatrix}$$

Our basic equation (2) is

$$\text{Arbitrary vector in } \mathcal{M}^{2 \times 2} = a_1 \mathbf{X}_1 + a_2 \mathbf{X}_2 + a_3 \mathbf{X}_3 + a_4 \mathbf{X}_4$$

that is,

$$\begin{bmatrix} r & s \\ t & u \end{bmatrix} = a_1 \begin{bmatrix} 1 & 0 \\ 0 & 0 \end{bmatrix} + a_2 \begin{bmatrix} 0 & 3 \\ 0 & 0 \end{bmatrix} + a_3 \begin{bmatrix} 0 & 0 \\ 1 & 1 \end{bmatrix} + a_4 \begin{bmatrix} 0 & -3 \\ 0 & 1 \end{bmatrix}$$

$$= \begin{bmatrix} a_1 & 3a_2 - 3a_4 \\ a_3 & a_3 + a_4 \end{bmatrix}$$

Comparing corresponding entries, we have

$$a_1 = r$$
$$3a_2 - 3a_4 = s$$
$$a_3 = t$$
$$a_3 + a_4 = u$$

This system of equations in the four unknowns a_1, a_2, a_3, a_4 has the solution $a_1 = r$, $a_2 = (s/3) + u - t$, $a_3 = t$, $a_4 = u - t$. Hence the arbitrary vector $\begin{bmatrix} r & s \\ t & u \end{bmatrix}$ can be written as a linear combination of the four given vectors. $\qquad \bullet$

One other type of problem arises concerning spanning: given a vector space $\mathcal{V}$, to find vetors that span $\mathcal{V}$. This problem can be difficult in some cases. But for lines and planes through the origin and other solution sets of homogeneous systems, we already have a method of determining a spanning set: our process of writing the set of solutions in the form $a(\quad) + b(\quad) + \cdots$.

Example 12-5

We wish to find vectors that span the plane $\mathcal{U}$ whose equation is $x + 4y - 5z = 0$. We know that the solutions are all of the form $(-4a + 5b, a, b)$. Rewriting, we have

$$\underbrace{(-4a + 5b, a, b)}_{\substack{\text{arbitrary vector} \\ \text{in } \mathcal{U}}} = \underbrace{a(-4, 1, 0) + b(5, 0, 1)}_{\substack{\text{linear combination of} \\ (-4, 1, 0) \text{ and } (5, 0, 1)}}$$

Since an arbitrary vector in $\mathcal{U}$ can be written as a linear combination of $(-4, 1, 0)$ and $(5, 0, 1)$, these two vectors span the plane $\mathcal{U}$.   ●

Example 12-6

We wish to find vectors that span the solution set $\mathcal{S}$ of the system

$$\begin{aligned} x_1 - 2x_2 + 4x_3 - x_4 - x_5 &= 0 \\ x_1 \quad\quad + 2x_3 - x_4 + 2x_5 &= 0 \end{aligned}$$

As the reader may verify, the set $\mathcal{S}$ consists of all 5-tuples of the form $(-2a + b - 2c, a - \frac{3}{2}c, a, b, c)$. Rewriting, we have

$$\underbrace{(-2a + b - 2c, a - \tfrac{3}{2}c, a, b, c)}_{\substack{\text{arbitrary vector} \\ \text{in } \mathcal{S}}} = \underbrace{\begin{aligned} &a(-2, 1, 1, 0, 0) \\ &+ b(1, 0, 0, 1, 0) \\ &+ c(-2, -\tfrac{3}{2}, 0, 0, 1) \end{aligned}}_{\substack{\text{linear combination of } (-2, 1, 1, 0, 0), \\ (1, 0, 0, 1, 0), (-2, -\frac{3}{2}, 0, 0, 1)}}$$

Thus $\mathcal{S}$ is spanned by $(-2, 1, 1, 0, 0)$, $(1, 0, 0, 1, 0)$, and $(-2, -\frac{3}{2}, 0, 0, 1)$.   ●

Exercises

12-1 Show that $\mathcal{R}^2$ is spanned by $(3, 0)$, $(0, 5)$.

12-2 Show that $\mathcal{R}^2 = \langle (1, -2), (1, -1) \rangle$.

12-3 Show that $\mathcal{R}^2$ is not spanned by (2, 7), (4, 14).

12-4 Show that $\mathcal{R}^3 = \langle (1, 0, -1), (0, 2, 0), (0, -2, 1) \rangle$.

12-5 Show that $\mathcal{R}^3$ is not spanned by $(-1, 3, 1), (2, -6, 0), (0, 0, 3)$.

12-6 Find vectors that span $\{(a, 3a + 2b, b) : a, b \in \mathcal{R}\}$.

12-7 Find vectors that span the set of solutions of the system

$$\begin{aligned} x + y &= 0 \\ -3x - 3y &= 0 \end{aligned}$$

12-8 Find vectors that span the set of solutions of

$$\begin{aligned} x_1 + 2x_3 - x_4 &= 0 \\ x_2 - 2x_3 &= 0 \\ x_1 + 2x_2 - 2x_3 - x_4 &= 0 \end{aligned}$$

12-9 Find vectors that span $\mathcal{R}^4$.

12-10 Do (1, 1, 0), (2, 3, 0) span $\{(a, b, 0) : a, b \in \mathcal{R}\}$? Explain.

12-11 Do (1, 1, 1), (2, 2, 3) span $\{(a, a, b) : a, b \in \mathcal{R}\}$? Explain.

12-12 Show that $\mathcal{M}^{2 \times 2}$ is spanned by

$$\begin{bmatrix} 1 & 0 \\ 0 & 0 \end{bmatrix}, \begin{bmatrix} 0 & 2 \\ 0 & 0 \end{bmatrix}, \begin{bmatrix} 0 & 1 \\ 1 & 0 \end{bmatrix}, \begin{bmatrix} 0 & 0 \\ -3 & 1 \end{bmatrix}.$$

12-13 Is $\mathcal{M}^{2 \times 2}$ spanned by

$$\begin{bmatrix} 1 & 0 \\ 1 & -1 \end{bmatrix}, \begin{bmatrix} 0 & 2 \\ 1 & 1 \end{bmatrix}, \begin{bmatrix} 1 & 0 \\ 0 & 0 \end{bmatrix}, \begin{bmatrix} 3 & 3 \\ 3 & 0 \end{bmatrix}?$$

Explain.

12-14 Let $X_1, X_2, \ldots, X_k$ be k vectors in a vector space $\mathcal{V}$. Let $\mathcal{S} = \langle X_1, X_2, \ldots, X_k \rangle$, that is, let $\mathcal{S}$ be the set of all linear combinations of $X_1, \ldots, X_k$. Prove that $\mathcal{S}$ is a subspace of $\mathcal{V}$. (Note: This generalizes Exercises 10-20 and 10-21.)

12-15 Describe geometrically each of these subspaces of $\mathcal{R}^3$:

(a) $\langle (1, 1, 0), (-1, 1, 3) \rangle$ (b) $\langle (2, 5, 5) \rangle$
(c) $\langle (1, 0, 1), (-3, 0, -3) \rangle$
(d) $\langle (2, 0, 0), (0, 3, 1), (1, 2, 5) \rangle$

12-16 Show that $\langle(2, 3, 0)\rangle = \langle(-4, -6, 0)\rangle$. (Hint: Show that every vector in the left set is also in the right set, and vice versa.)

12-17 Show that $\langle(1, 2)\rangle = \langle(1, 2), (2, 4)\rangle$.

12-18 Show that $\mathscr{P}^2$ is spanned by the polynomials 1, x, x^2.

12-19 Show that $\mathscr{P}^2 = \langle1 + x, 2x, x^2 + 3\rangle$.

12-20 The vectors $\mathbf{X} = (8, 0, -11)$ and $\mathbf{Y} = (-4, 1, 8)$ are both in $\langle(4, 1, -3), (0, 2, 5)\rangle$. Show that $\mathbf{Y}$ can be reached from $\mathbf{X}$ by motions parallel to the vectors $(4, 1, -3)$ and $(0, 2, 5)$. [Hint: Solve the equation $(-4, 1, 8) = (8, 0, -11) + a(4, 1, -3) + b(0, 2, 5)$ for a and b.] This is a particular illustration of the fact that if vectors span $\mathscr{V}$, then motion from point to point in $\mathscr{V}$ can be achieved by motions parallel to the spanning vectors.

13. INDEPENDENCE

Section 12 introduced the concept of spanning. We noted that intuitively, if k vectors span a vector space $\mathscr{V}$, then to move from any point in $\mathscr{V}$ to any other point in $\mathscr{V}$ requires motion in *at most* the k directions given by those k vectors. We say "at most" because conceivably the k spanning "directions" are not all needed. For instance, let $\mathscr{V}$ be the subspace of $\mathscr{R}^2$ spanned by $(1, 2)$ and $(2, 4)$. The two given vectors lie on the same line through the origin; therefore they give the same direction (Fig. 13-1). To

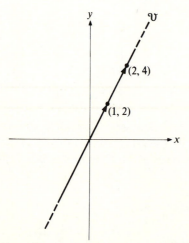

Fig. 13-1 The two vectors lie on the same line through the origin; therefore, only one is needed to span the line.

move from point to point in $\mathcal{V}$ requires motion along either one of these vectors; the other is not needed.

Argued another way, $\mathcal{V}$ consists of all vectors $a(1, 2) + b(2, 4)$, which can be rewritten as $(a + 2b)(1, 2)$ since $(2, 4) = 2(1, 2)$. Thus any vector in $\mathcal{V}$ is a scalar multiple of $(1, 2)$. Only the one vector $(1, 2)$ is needed to span $\mathcal{V}$; $(2, 4)$ can be discarded. That is, $\langle(1, 2), (2, 4)\rangle = \langle(1, 2)\rangle$.

As a second example, let $\mathcal{W} = \langle(1, 0, 0), (0, 2, 0), (2, 6, 0)\rangle$. As the reader may notice, $(2, 6, 0)$ is a linear combination of the first two vectors: $(2, 6, 0) = 2(1, 0, 0) + 3(0, 2, 0)$. Because of this, it turns out that $(2, 6, 0)$ is not needed in this spanning set: any linear combination of the three given vectors will reduce to a linear combination of the first two. That is, $\mathcal{W} = \langle(1, 0, 0), (0, 2, 0)\rangle$. The details are left as an exercise.

Although the above discussion is intuitive, it does contain the germ of the idea of what makes a direction unnecessary for motion from point to point. In both examples one vector was not needed, essentially because it could be expressed in terms of the other vectors. The following definition generalizes and makes precise this idea.

Definition

Vectors $X_1, X_2, \ldots, X_k$ are said to be **dependent** provided that at least one of the vectors equals a linear combination of the others.†

Example 13-1

$(0, 2)$ and $(0, 6)$ are dependent since $(0, 2) = \frac{1}{3}(0, 6)$. (The two vectors lie on the same line through the origin; only one is needed to span that line.)
●

Example 13-2

$(1, 2, 2)$, $(0, 1, 1)$, and $(4, 0, 0)$ are dependent since $(1, 2, 2) = 2(0, 1, 1) + \frac{1}{4}(4, 0, 0)$. The three vectors lie in the same plane, and motion from point to point on that plane needs only two directions. We have $\langle(1, 2, 2), (0, 1, 1), (4, 0, 0)\rangle = \langle(0, 1, 1), (4, 0, 0)\rangle$.
●

Example 13-3

$(1, 5, 2)$, $(0, 0, 0)$, $(0, 0, 3)$ are dependent since $(0, 0, 0) = 0(1, 5, 2) + 0(0, 0, 3)$.
●

† The zero vector by itself is considered, by convention, to be dependent.

Example 13-4

$x^2 + 3x$, x, and $2x^2 + x$ are dependent since $2x^2 + x = 2(x^2 + 3x) - 5(x)$. That is, one of the given vectors equals a linear combination of the others. (We have gone beyond intuitive meanings here. The purpose of formal definitions is to generalize ideas and make them usable in any vector space.)   ●

Definition

Vectors $X_1, X_2, \ldots, X_k$ (not all 0) are said to be **independent** provided that none of the vectors equals a linear combination of the others.

The intuitive meaning of vectors' independence is, of course, that they are *all* needed for motion from point to point in the vector space.

Example 13-5

$(1, 0)$ and $(0, 1)$ are independent, since neither is a multiple of the other.
                                                                                 ●

Example 13-6

$(1, 0, 0)$, $(0, 1, 0)$, $(0, 0, 1)$ are independent since $(1, 0, 0)$ is not equal to $a(0, 1, 0) + b(0, 0, 1)$ for any a and b; $(0, 1, 0)$ is not equal to a linear combination of $(1, 0, 0)$ and $(0, 0, 1)$; nor is $(0, 0, 1)$ equal to a linear combination of $(1, 0, 0)$ and $(0, 1, 0)$.   ●

Example 13-7

Any one nonzero vector is independent. This follows at once from the definition; there are no other vectors in the list that the one vector could be a multiple of.   ●

The definition of independence given above expresses a *conceptual* meaning: vectors are independent provided none of them can be expressed in terms of the others. The following theorem provides a strictly *numerical* test for independence. Since it is numerical, the test can be applied by a straightforward computation and therefore is frequently more useful than the definition in testing for independence.

Theorem 13-1 $X_1, X_2, \ldots, X_k$ are independent $\Leftrightarrow$ the only solution to the equation

$$a_1 X_1 + a_2 X_2 + \cdots + a_k X_k = 0$$

is the solution $a_1 = 0,\ a_2 = 0,\ \ldots,\ a_k = 0$.

Proof $\Rightarrow$ We are given that $X_1, X_2, \ldots, X_k$ are independent. We carry out the proof by contradiction. To this end, assume that the conclusion is false. Then there is a solution to

$$a_1 X_1 + a_2 X_2 + \cdots + a_k X_k = 0$$

in which one of the a_i's, say a_1, is not zero. Solving for X_1, we have

$$X_1 = \left(-\frac{a_2}{a_1}\right)X_2 + \left(-\frac{a_3}{a_1}\right)X_3 + \cdots + \left(-\frac{a_k}{a_1}\right)X_k$$

This expression is permissible since $a_1 \neq 0$. But it says that X_1 is a linear combination of the other vectors, and hence the vectors are dependent. This is the desired contradiction, since by hypothesis the vectors were independent.

   $\Leftarrow$ We are given that the only solution to the equation

$$a_1 X_1 + a_2 X_2 + \cdots + a_k X_k = 0 \qquad (3)$$

is $a_1 = 0,\ a_2 = 0, \ldots, a_k = 0$. Again, to get a contradiction, we assume that the conclusion is false. Then some vector, say X_1, equals a linear combination of the others:

$$X_1 = b_2 X_2 + \cdots + b_k X_k$$

Adding $-X_1$ to both sides yields

$$0 = (-1)X_1 + b_2 X_2 + \cdots + b_k X_k$$

which gives us a solution to (3) where not all the coefficients are zero, since we have $a_1 = -1$. This is the desired contradiction, since by hypothesis the only solution has all the a_i's equal to zero. ∎

   In view of this theorem, to show that given vectors are independent we form the basic equation

Linear combination of given vectors = zero vector

or, equivalently,

$$a_1 \mathbf{X}_1 + a_2 \mathbf{X}_2 + \cdots + a_k \mathbf{X}_k = \mathbf{0} \tag{4}$$

We then rewrite (4) using appropriate notation for the zero vector and the given vectors $\mathbf{X}_i$. If (4) has the *unique* solution $a_1 = 0, \ldots, a_k = 0$, the vectors are independent; if (4) has a nonzero solution, the vectors are dependent.

Example 13-8

$\mathbf{X}_1 = (1, 1)$, $\mathbf{X}_2 = (3, 2)$ are independent. The basic equation (4) in this case is $a_1 \mathbf{X}_1 + a_2 \mathbf{X}_2 = \mathbf{0}$, and using the actual $\mathbf{X}_1$, $\mathbf{X}_2$, and $\mathbf{0}$, this becomes

$$a_1(1, 1) + a_2(3, 2) = (0, 0)$$

Combining the left side, we obtain

$$(a_1 + 3a_2, a_1 + 2a_2) = (0, 0)$$

That is, we must have

$$a_1 + 3a_2 = 0$$
$$a_1 + 2a_2 = 0$$

(a system of two equations with the two unknowns a_1 and a_2). The matrix solution is easy:

$$\begin{bmatrix} 1 & 3 \\ 1 & 2 \end{bmatrix} \longrightarrow \begin{bmatrix} 1 & 3 \\ 0 & -1 \end{bmatrix} \longrightarrow \begin{bmatrix} 1 & 0 \\ 0 & 1 \end{bmatrix}$$

and we read off the unique solution $a_1 = 0$, $a_2 = 0$. Therefore, $\mathbf{X}_1$ and $\mathbf{X}_2$ are independent. ●

Example 13-9

$\mathbf{X}_1 = (1, 3, 1)$, $\mathbf{X}_2 = (-2, -4, 2)$ are independent. The fundamental equation is $a_1(1, 3, 1) + a_2(-2, -4, 2) = (0, 0, 0)$. Combining the terms on the left yields $(a_1 - 2a_2, 3a_1 - 4a_2, a_1 + 2a_2) = (0, 0, 0)$, or

$$a_1 - 2a_2 = 0$$
$$3a_1 - 4a_2 = 0$$
$$a_1 + 2a_2 = 0$$

The matrix

$$\begin{bmatrix} 1 & -2 \\ 3 & -4 \\ 1 & 2 \end{bmatrix}$$

reduces to

$$\begin{bmatrix} 1 & 0 \\ 0 & 1 \\ 0 & 0 \end{bmatrix}$$

The rank is 2, there are two unknowns, and so, by Theorem 5-1, there is the unique solution $a_1 = 0$, $a_2 = 0$. Therefore, X_1 and X_2 are independent.

●

Example 13-10

$X_1 = (1, 2, 1)$, $X_2 = (3, 0, 1)$, $X_3 = (2, 1, 1)$ are dependent. The basic equation becomes $a_1(1, 2, 1) + a_2(3, 0, 1) + a_3(2, 1, 1) = (0, 0, 0)$. This yields the system of equations

$$\begin{aligned} a_1 + 3a_2 + 2a_3 &= 0 \\ 2a_1 \qquad\quad + a_3 &= 0 \\ a_1 + a_2 + a_3 &= 0 \end{aligned}$$

Its matrix

$$\begin{bmatrix} 1 & 3 & 2 \\ 2 & 0 & 1 \\ 1 & 1 & 1 \end{bmatrix}$$

reduces to

$$\begin{bmatrix} 1 & 0 & \frac{1}{2} \\ 0 & 1 & \frac{1}{2} \\ 0 & 0 & 0 \end{bmatrix}$$

Since the rank is 2 and since there are three unknowns, there are infinitely many solutions, by Theorem 5-1. Since there are infinitely many solutions, there is certainly a solution where a_1, a_2, and a_3 are not all zero. Hence X_1, X_2, X_3 are dependent.

●

Example 13-11

Consider the vectors $\mathbf{X}_1 = 2x$, $\mathbf{X}_2 = x^2 - 3x$, $\mathbf{X}_3 = 2x^2$ in $\mathscr{P}^2$. The equation $a_1\mathbf{X}_1 + a_2\mathbf{X}_2 + a_3\mathbf{X}_3 = \mathbf{0}$ becomes

$$a_1(2x) + a_2(x^2 - 3x) + a_3(2x^2) = \text{zero polynomial}$$

Gathering like powers of x and writing the zero polynomial as $0x^2 + 0x$, we have

$$(a_2 + 2a_3)x^2 + (2a_1 - 3a_2)x = 0x^2 + 0x$$

Comparing like coefficients gives us

$$
\begin{aligned}
a_2 + 2a_3 &= 0 \\
2a_1 - 3a_2 \quad\;\; &= 0
\end{aligned}
$$

This system of homogeneous equations with three unknowns has a matrix of rank 2, and therefore there are infinitely many solutions. The vectors are dependent.   ●

The reader may have noticed that in each of the above examples, the test for independence required solving a homogeneous system of linear equations. This is usually the case. In particular, the examples testing vectors in $\mathscr{R}^n$ led to an interesting phenomenon. In Example 13-8, the vectors being tested were $(1, 1)$ and $(3, 2)$, and the unaugmented matrix of the system turned out to be $\begin{bmatrix} 1 & 3 \\ 1 & 2 \end{bmatrix}$. In Example 13-9, the vectors being tested were $(1, 3, 1)$ and $(-2, -4, 2)$, and the matrix obtained was

$$
\begin{bmatrix}
1 & -2 \\
3 & -4 \\
1 & 2
\end{bmatrix}
$$

In each case the given vectors became the *columns* of the matrix. This gives us a shortcut in testing vectors in $\mathscr{R}^n$ for independence: form the matrix whose columns are the given vectors. Then find its rank. If the rank equals the number of unknowns (which equals the number of given vectors), then the vectors are independent. Thus we have the following test for independence.

Theorem   Given k vectors in $\mathcal{R}^n$, let A be the matrix whose columns are the given
13-2   vectors.

     1   If the rank of A equals k, then the vectors are independent.

     2   If the rank of A is less than k, then the vectors are dependent.

Exercises

Test the vectors in Exercises 13-1 to 13-3 for independence by beginning
with the basic equation for independence and rewriting it in the form of a
system of equations.

13-1 $X_1 = (1, 3)$, $X_2 = (8, 5)$

13-2 $X_1 = (1, 2, 4)$, $X_2 = (3, 0, 1)$, $X_3 = (1, -1, 0)$

13-3 $X = (1, 0, -1, -1)$, $Y = (2, 1, -2, 0)$, $Z = (-1, -2, 1, -3)$

Test the vectors in Exercises 13-4 and 13-5 for independence by using the
matrix test of Theorem 13-2.

13-4 $X_1 = (2, 0, 5)$, $X_2 = (1, 3, 4)$, $X_3 = (2, -6, 2)$.

13-5 $X = (2, 0, 1, 3)$, $Y = (1, 3, -2, 6)$, $Z = (1, 2, -1, 5)$.

13-6   Determine whether

$$\begin{bmatrix} 1 & 2 \\ 0 & 1 \end{bmatrix}, \begin{bmatrix} 1 & 4 \\ 1 & -1 \end{bmatrix}, \begin{bmatrix} 1 & 6 \\ 2 & -3 \end{bmatrix}$$

are independent.

13-7   Determine whether $X = 2x^2 - x$, $Y = 3x^2 + 2x$ are independent.

13-8   Suppose that X_1, X_2 are independent. Let X_3 be a third vector.
(a)   Prove that if X_3 does not belong to $\langle X_1, X_2 \rangle$, then X_1, X_2, X_3 are
independent.
(b)   Give a geometric interpretation if the vectors are in $\mathcal{R}^3$.
(c)   Find a vector X such that $(1, 2, 0)$, $(2, 1, 0)$, X are independent.

13-9   Suppose that X_1, X_2, ..., X_k are independent vectors in a vector space $\mathcal{V}$. Let Y be a vector in $\mathcal{V}$. Prove that Y is not an element of $\langle X_1, X_2, ..., X_k \rangle \Leftrightarrow X_1, X_2, ..., X_k, Y$ are independent.

13-10   Prove that if 0 is among the vectors X_1, X_2, ..., X_k, then X_1, X_2, ..., X_k are dependent.

13-11   Test the vectors in the following lists for independence. If the vectors are dependent, find one that depends on other vectors in the list and discard that vector. Test the remaining vectors for independence. Continue discarding dependent vectors until only independent vectors remain.

(a) $(2, 1)$, $(3, 0)$, $(10, 5)$
(b) $(4, 0, 1)$, $(0, 0, 2)$, $(3, 3, 5)$
(c) $(1, 0, -3)$, $(5, 2, 1)$, $(1, 2, 13)$, $(10, 4, 2)$
(d) $(2, 0, 4)$, $(1, 0, 2)$, $(6, 0, 12)$, $(3, 1, 3)$, $(6, 2, 6)$

(Note: In each case we end up with independent vectors that span the same subspace as the original vectors.)

13-12   (a)   Prove that four vectors in $\mathcal{R}^3$ must be dependent. (b) More generally, prove that if $k > n$ then k vectors in $\mathcal{R}^n$ must be dependent. [Hint: Use Theorem 13-2.]

14.   BASES AND DIMENSION

If n vectors span a vector space $\mathcal{V}$, then we know intuitively that getting from point to point requires no more than n different directions; if these n vectors are also independent, then we know intuitively that the n directions are all needed. Therefore it would be reasonable, in such a case, to say that $\mathcal{V}$ is n-dimensional. These ideas are the source of the precise definition of dimension.

Definition

Vectors X_1, X_2, ..., X_n constitute a **basis** for $\mathcal{V}$ provided:

(a)   They span $\mathcal{V}$.

(b)   They are independent.

If $\mathcal{V}$ has a basis consisting of n vectors, then the **dimension** of $\mathcal{V}$ is n.

Example 14-1

$(1, 0)$, $(0, 1)$ form a basis for $\mathcal{R}^2$. We saw in Sec. 12 that these vectors span $\mathcal{R}^2$, and in Example 13-5 that they are independent. Thus $\mathcal{R}^2$, as expected, is two-dimensional.

Since $(1, 0)$, $(0, 1)$ is the most natural basis to look at and the simplest to work with, we call this basis the *usual basis* (or *natural basis*) for $\mathcal{R}^2$.

$\bullet$

Example 14-2

$(1, 0, 0)$, $(0, 1, 0)$, $(0, 0, 1)$ form a basis for $\mathcal{R}^3$. Spanning was noted in Sec. 12 and independence was shown in Example 13-6. Thus $\mathcal{R}^3$ is three-dimensional. The given vectors form the usual basis (or natural basis) for $\mathcal{R}^3$.

$\bullet$

Example 14-3

The n n-tuples $(1, 0, 0, \ldots, 0)$, $(0, 1, 0, \ldots, 0)$, $\ldots$, $(0, \ldots, 0, 0, 1)$ form the usual basis for $\mathcal{R}^n$. Therefore, $\mathcal{R}^n$ is n-dimensional.

$\bullet$

Example 14-4

The vectors

$$\begin{bmatrix} 1 & 0 & 0 \\ 0 & 0 & 0 \end{bmatrix}, \begin{bmatrix} 0 & 1 & 0 \\ 0 & 0 & 0 \end{bmatrix}, \begin{bmatrix} 0 & 0 & 1 \\ 0 & 0 & 0 \end{bmatrix}, \begin{bmatrix} 0 & 0 & 0 \\ 1 & 0 & 0 \end{bmatrix}, \begin{bmatrix} 0 & 0 & 0 \\ 0 & 1 & 0 \end{bmatrix}, \begin{bmatrix} 0 & 0 & 0 \\ 0 & 0 & 1 \end{bmatrix}$$

form a basis for $\mathcal{M}^{2 \times 3}$. Thus $\mathcal{M}^{2 \times 3}$ is six-dimensional.

$\bullet$

Example 14-5

The most natural basis for $\mathcal{M}^{m \times n}$ consists of the mn matrices (each with m rows and n columns)

$$\begin{bmatrix} 1 & 0 & 0 & \cdots & 0 \\ 0 & 0 & 0 & \cdots & 0 \\ \hline 0 & 0 & 0 & \cdots & 0 \end{bmatrix}, \begin{bmatrix} 0 & 1 & 0 & \cdots & 0 \\ 0 & 0 & 0 & \cdots & 0 \\ \hline 0 & 0 & 0 & \cdots & 0 \end{bmatrix}, \begin{bmatrix} 0 & 0 & 1 & \cdots & 0 \\ 0 & 0 & 0 & \cdots & 0 \\ \hline 0 & 0 & 0 & \cdots & 0 \end{bmatrix}, \ldots,$$

$$\begin{bmatrix} 0 & 0 & 0 & \cdots & 0 \\ 0 & 0 & 0 & \cdots & 0 \\ \hline 0 & 0 & 0 & \cdots & 1 \end{bmatrix}$$

Hence the dimension of $\mathcal{M}^{m \times n}$ is mn.

$\bullet$

Example 14-6

The three polynomials x^2, x, 1 form a basis for $\mathscr{P}^2$. They span $\mathscr{P}^2$, since any polynomial in $\mathscr{P}^2$ is of the form $ax^2 + bx + c1$, a linear combination of the three given vectors. They are independent since the fundamental equation $a_1 x^2 + a_2 x + a_3 1 = 0x^2 + 0x + 0 \cdot 1$ has (as we see by equating like coefficients) the unique solution $a_1 = 0$, $a_2 = 0$, $a_3 = 0$.   ●

Example 14-7

(1, 1), (3, 2) is a basis for $\mathscr{R}^2$. These two vectors span $\mathscr{R}^2$ (Example 12-1), and they are independent (Example 13-8). This example together with Example 14-1 shows that a vector space can have more than one basis.   ●

Example 14-8

Recall that the zero vector alone constitutes a subspace of any vector space (Example 10-9). Since it is a subspace, it is therefore a vector space, and it should have a dimension. We decree that its dimension is zero.   ●

Example 14-9

We wish to find a basis for the set of solutions $\mathscr{S}$ to the system

$$
\begin{aligned}
x_1 + x_2 + 5x_3 + 2x_4 &= 0 \\
x_1 \qquad\;\; - 2x_3 + 4x_4 &= 0
\end{aligned}
$$

As the reader can verify, $\mathscr{S}$ consists of all 4-tuples of the form

$$(2a - 4b,\ -7a + 2b,\ a,\ b) = a(2,\ -7,\ 1,\ 0) + b(-4,\ 2,\ 0,\ 1) \qquad (5)$$

We claim that $(2, -7, 1, 0)$, $(-4, 2, 0, 1)$ form a basis for $\mathscr{S}$.

Spanning   Since every vector in $\mathscr{S}$ is of the form $(2a - 4b,\ -7a + 2b,\ a,\ b)$, equation (5) shows that an arbitrary vector in $\mathscr{S}$ is a linear combination of $(2, -7, 1, 0)$, $(-4, 2, 0, 1)$. Therefore these two vectors span $\mathscr{S}$.

Independence   We use the matrix test of Theorem 13-2. Writing the two vectors as columns, we see that the matrix

$$\begin{bmatrix} 2 & -4 \\ -7 & 2 \\ 1 & 0 \\ 0 & 1 \end{bmatrix}$$

has rank 2. Therefore, $(2, -7, 1, 0)$, $(-4, 2, 0, 1)$ are independent.   ●

Theorem 13-2 gives a matrix test for independence of vectors in $\mathcal{R}^n$. A similar test for spanning $\mathcal{R}^n$ can be developed as well. Putting the two tests together then gives a simple matrix test for bases of $\mathcal{R}^n$. We first illustrate the method with an example.

Example 14-10

$(1, 2)$, $(4, 3)$ is a basis for $\mathcal{R}^2$.

Independence   By Theorem 13-2, these vectors are independent if and only if the rank of $\begin{bmatrix} 1 & 4 \\ 2 & 3 \end{bmatrix}$ is 2. Clearly the rank of this matrix *is* 2.

Spanning   We must show that for any (a, b), the equation

$$(a, b) = x(1, 2) + y(4, 3)$$

can be solved for x and y. This equation yields the system

$$x + 4y = a$$
$$2x + 3y = b$$

whose matrix is

$$\begin{bmatrix} 1 & 4 & \vdots & a \\ 2 & 3 & \vdots & b \end{bmatrix}$$

For the system of equations to have a solution, we must show that

$$\text{Rank of } \begin{bmatrix} 1 & 4 & \vdots & a \\ 2 & 3 & \vdots & b \end{bmatrix} = \text{rank of } \begin{bmatrix} 1 & 4 \\ 2 & 3 \end{bmatrix}$$

by Theorem 5-2. Again, this is clear since both matrices have rank 2.

Since $(1, 2)$, $(4, 3)$ are independent and span $\mathcal{R}^2$, they form a basis for $\mathcal{R}^2$.   ●

In verifying both spanning and independence it was crucial that the matrix $\begin{bmatrix} 1 & 4 \\ 2 & 3 \end{bmatrix}$ be of rank 2 (equal to the dimension of $\mathcal{R}^2$). All problems of this type, in any $\mathcal{R}^n$, would be attacked in exactly the same way; thus we have the following theorem.

Theorem 14-1   Given n vectors in $\mathcal{R}^n$, let A be the matrix whose columns are the given vectors.

1   If rank of $A = n$, the vectors are a basis for $\mathcal{R}^n$.

2   If rank of $A < n$, the vectors are not a basis for $\mathcal{R}^n$.

We omit the proof.

Corollary 14-2   If we have n independent vectors in $\mathcal{R}^n$, then they form a basis for $\mathcal{R}^n$.

Proof   We form the matrix A whose columns are the n vectors. Since the vectors are independent, the rank of A is n, by Theorem 13-2. Then by Theorem 14-1, the vectors must be a basis for $\mathcal{R}^n$.   ∎

This corollary says that n independent vectors in $\mathcal{R}^n$ automatically span $\mathcal{R}^n$, as well.

In this section we have shown that to determine the dimension of a vector space $\mathcal{V}$ we simply count the number of vectors in a basis for $\mathcal{V}$. The following theorem assures us that we can always carry out this process.

Theorem 14-3   **1**   Every vector space $\mathcal{V}$ has a basis.

2   Any two bases for $\mathcal{V}$ have the same number of vectors in them.

We omit the proof of part (1).† We will discuss part (2) further in Sec. 18.

Part (1) of the theorem assures us that we always have something to count. In many of our examples, such as the $\mathcal{R}^n$s and $\mathcal{M}^{m \times n}$s, we not only know that a basis exists, but also have actually found specific bases.

† A proof of this part of the theorem can be found in George F. Simmons, *Introduction to Topology and Modern Analysis*, McGraw-Hill Book Company, New York, 1963, pp. 198–199.

However, in other examples, such as the vector space of all functions defined on the real numbers (Example 9-5), it can be difficult to find a basis. In such cases we must rely on part (1) of the theorem to be assured that a basis exists.

Part (2) guarantees that any vector space has a *unique* dimension; $\mathcal{V}$ will not get different dimensions depending on which basis is being looked at. For example, every basis for $\mathcal{R}^3$ must consist of exactly three vectors.

Exercises

(Reminder: To show that vectors form a basis, show that (1) they span the space and (2) they are independent. The matrix test of Theorem 14-1 applies only to Exercises 14-1, 14-2, and 14-3.)

14-1 Show that $(1, 2)$, $(3, 5)$ is a basis for $\mathcal{R}^2$.

14-2 Show that $(1, 0, 1)$, $(2, 1, 0)$, $(0, 1, -3)$ is a basis for $\mathcal{R}^3$.

14-3 Determine whether the following are bases:
(a) $(1, 2, 4)$, $(3, 0, 1)$, $(1, 1, 5)$ for $\mathcal{R}^3$
(b) $(-1, 1)$, $(4, -4)$ for $\mathcal{R}^2$
(c) $(1, 1, 1)$, $(2, 2, 3)$, $(0, 0, 4)$ for $\mathcal{R}^3$
(d) $(1, 2, 0, 0, 1)$, $(2, -3, 1, 0, 1)$, $(0, 4, 0, 1, 1)$, $(2, -1, 0, 1, 2)$, $(0, 3, 2, -4, 0)$ for $\mathcal{R}^5$

14-4 Show that $(1, 0, -1)$, $(0, 1, 0)$ is a basis for $\{(a, b, -a) : a, b \in \mathcal{R}\}$.

14-5 Show that $(-4, 2, 1, 0)$, $(1, 3, 0, 1)$ is a basis for $\{(-4a + b, 2a + 3b, a, b) : a, b \in \mathcal{R}\}$.

14-6 Show that $(1, 0, 1)$, $(0, 1, -2)$, $(4, 2, 0)$ is not a basis for $\{(a + 4c, b + 2c, a - 2b) : a, b, c \in \mathcal{R}\}$.

14-7 Verify that $(1, 0, 1)$ is a basis for $\{(a, 0, a) : a \in \mathcal{R}\}$.

14-8 Verify that $\begin{bmatrix} 3 & 0 \\ 0 & 0 \end{bmatrix}$, $\begin{bmatrix} 0 & 1 \\ 0 & 0 \end{bmatrix}$, $\begin{bmatrix} 0 & 0 \\ 1 & 0 \end{bmatrix}$, $\begin{bmatrix} 0 & 0 \\ 2 & 5 \end{bmatrix}$ is a basis for $\mathcal{M}^{2 \times 2}$.

14-9   Find the dimension of the set of solutions of the system

$$\begin{aligned} x \quad\;\; + 4z - \;\; t &= 0 \\ x + y + 2z - 4t &= 0 \end{aligned}$$

14-10   Find the dimension of the set of solutions of the system

$$\begin{aligned} x + 2y \quad\;\;\; &= 0 \\ y - z &= 0 \\ x + \;\; y + z &= 0 \end{aligned}$$

14-11   Explain why $(1, 0)$, $(0, 0)$, $(3, 1)$ is not a basis for $\mathcal{R}^2$.

14-12   Explain why $(4, 3, 1)$, $(2, 0, 1)$ is not a basis for $\mathcal{R}^3$.

14-13   (a)   Why is it impossible for five vectors in $\mathcal{R}^4$ to be a basis for $\mathcal{R}^4$?
(b)   Why is it impossible for three vectors in $\mathcal{R}^4$ to be a basis for $\mathcal{R}^4$?

14-14   Show that x, $3x^2$, $5 + x$ is a basis for $\mathcal{P}^2$.

14-15   Prove that the dimension of $\mathcal{P}^n$ is $n + 1$. (Hint: Guess at a natural basis for $\mathcal{P}^n$ and then prove that it is a basis.)

14-16   Let $\mathcal{P}$ be the vector space of all polynomials (Exercise 9-12).

(a)   Prove that $\mathcal{P}$ does not have a basis with finitely many vectors in it. (Hint: Assume $\mathcal{P}$ *does* have a basis with finitely many vectors in it. Since each vector in this supposed basis is a polynomial, it has a certain degree. Let m be the largest of these degrees. Show that the vector x^{m+1} is not a linear combination of the supposed basis vectors.)
(b)   Prove that the vectors $1, x, x^2, x^3, \ldots$ form a basis for $\mathcal{P}$. Thus $\mathcal{P}$ is an infinite-dimensional vector space. (Hint: We must show that these vectors span $\mathcal{P}$ and are independent. Our definitions of spanning and independence assumed that the number of vectors in the list was finite. To deal with an infinite number of vectors, our definitions need slight modification. To show that $1, x, x^2, \ldots$ *span* $\mathcal{P}$, show that every vector in $\mathcal{P}$ is a linear combination of *finitely many* of these vectors. Similarly, to show that these vectors are *independent*, show that if a *finite* linear combination of them gives the zero vector, then each coefficient must be zero.)

15.   SYSTEMS OF EQUATIONS AND DIMENSION

In this section we see how dimension relates to the theory of systems of linear equations.

Recall that the set of solutions of a homogeneous system with n unknowns is a subspace of $\mathcal{R}^n$ (see Theorem 10-2). Hence, like all vector spaces, it has a dimension. If the rank of the matrix of the system is equal to n, there is a unique solution; if the rank of the matrix is less than n, there are infinitely many solutions (Theorem 5-1). Just as we used the concept of rank to study the *number* of solutions, we can also use rank to determine the *dimension* of the subspace of solutions.

Consider the system

$$\begin{aligned}
x_1 & + 2x_3 + 3x_4 - 5x_5 = 0 \\
x_1 + x_2 & \quad\quad + 4x_4 - 9x_5 = 0
\end{aligned}$$

Its unaugmented matrix reduces to

$$A = \begin{bmatrix} 1 & 0 & 2 & 3 & -5 \\ 0 & 1 & -2 & 1 & -4 \end{bmatrix}$$

Note that A has rank 2. We use the two row-leading 1s to single out the unknowns x_1 and x_2:

$$\begin{aligned}
x_1 &= -2x_3 - 3x_4 + 5x_5 \\
x_2 &= \quad 2x_3 - \quad x_4 + 4x_5
\end{aligned}$$

The remaining three unknowns can be anything—say $x_3 = a$, $x_4 = b$, $x_5 = c$. Thus the solution set $\mathcal{S}$ consists of all vectors of the form

$$a(-2, 2, 1, 0, 0) + b(-3, -1, 0, 1, 0) + c(5, 4, 0, 0, 1)$$

We claim that $(-2, 2, 1, 0, 0)$, $(-3, -1, 0, 1, 0)$, $(5, 4, 0, 0, 1)$ form a basis for $\mathcal{S}$. Since every vector in $\mathcal{S}$ is a linear combination of these three vectors, they span $\mathcal{S}$. Only the independence of these three vectors remains to be checked. The presence of the entries 1, 0, 0 in the first vector, of entries 0, 1, 0 in the second, and of entries 0, 0, 1 in the third, in corresponding places, suggests their independence. To be certain, we use our matrix test for independence and form the matrix B with these three vectors as columns:

$$B = \begin{bmatrix} -2 & -3 & 5 \\ 2 & -1 & 4 \\ 1 & 0 & 0 \\ 0 & 1 & 0 \\ 0 & 0 & 1 \end{bmatrix}$$

The rank of this matrix is 3, and so the three vectors are independent.

Summarizing, the rank of A was 2, therefore three of the five unknowns could be any real number, and it followed that the set of solutions was three-dimensional. For any homogeneous system a similar analysis could be carried out.

Theorem 15-1

Let A be the matrix for a homogeneous system of equations with n unknowns. If the rank of A is r, then the set of solutions (which is a subspace of $\mathcal{R}^n$) has dimension $n - r$.

Proof

We are given that the rank of A is r, that is, there are r unknowns that have row-leading 1s in their columns of the reduced matrix. This leaves $n - r$ unknowns with no row-leading 1s, and each of these unknowns can be any real number, say a_1 for the first of these unknowns, a_2 for the second, etc. Thus the set of solutions consists of all n-tuples of the form

$$a_1(-, -, \ldots, -) + a_2(-, -, \ldots, -) + \cdots + a_{n-r}(-, -, \ldots, -) \qquad (6)$$

(just as in the example above where we obtained

$$a(-2, 2, 1, 0, 0) + b(-3, -1, 0, 1, 0) + c(5, 4, 0, 0, 1)$$

as the set of solutions).

We will show that the $n - r$ vectors $(-, -, \ldots, -)$ in (6) must form a basis for S. They obviously span S. They are also independent. To see this, we use the matrix test for independence and write the $n - r$ vectors as columns of a matrix B. Each of the $n - r$ vectors has a 1 in a position where all the others have zeros, as in the above example. When we reduce B, these 1s become row-leading 1s. Therefore, B has rank $n - r$; thus the $n - r$ vectors in (6) are independent.   ∎

Theorem 15-1 has an analog for nonhomogeneous systems. We illustrate with an example.

Example 15-1

Consider the nonhomogeneous system

$$\begin{aligned} x_1 + 3x_2 \qquad\quad + 2x_4 &= 5 \\ x_1 + 3x_2 + x_3 - 2x_4 &= 3 \end{aligned}$$

Its augmented matrix reduces to

$$[A \mathbin{\vdots} B] = \begin{bmatrix} 1 & 3 & 0 & 2 & \vdots & 5 \\ 0 & 0 & 1 & -4 & \vdots & -2 \end{bmatrix}$$

The set of solutions S consists of all 4-tuples of the form

$$(5, 0, -2, 0) + a(-3, 1, 0, 0) + b(-2, 0, 4, 1)$$

S consists of one *particular solution* $(5, 0, -2, 0)$ and a *two-dimensional subspace part* $a(-3, 1, 0, 0) + b(-2, 0, 4, 1)$. As the reader can verify, the subspace part is the set of solutions of the corresponding homogeneous system

$$\begin{aligned} x_1 + 3x_2 \quad\quad + 2x_4 &= 0 \\ x_1 + 3x_2 + x_3 - 2x_4 &= 0 \end{aligned}$$

whose matrix reduces to A.   ●

The following theorem describes the general case for nonhomogeneous systems.

Theorem 15-2

Let $[A \mid B]$ be the matrix for a nonhomogeneous system of equations with n unknowns. Assume that the system has at least one solution. If the rank of $A = r$, then the solutions are all of the form

$$\underbrace{(c_1, \ldots, c_n)}_{\substack{\text{particular} \\ \text{solution}}} + \underbrace{a_1(-, -, \ldots, -) + a_2(-, -, \ldots, -) + \cdots + a_{n-r}(-, -, \ldots, -)}_{(n-r)\text{-dimensional subspace part}}$$

The subspace part is the set of solutions of the corresponding homogeneous system that has matrix A.

These theorems have some bearing on the geometry of lines and planes. In Sec. 4 we saw that any line in $\mathcal{R}^2$ is the solution set of an equation $sx + ty = u$. The matrix for this system will have rank $r = 1$. Since the number of unknowns $n = 2$, the dimension $n - r$ of the subspace part of the set of solutions is $2 - 1 = 1$. We also saw in Sec. 4 that any line in $\mathcal{R}^3$ is the solution set for a system of two equations

$$\begin{aligned} px + qy &= c \\ sy + tz &= d \end{aligned}$$

with rank $r = 2$. Since $n = 3$, again the subspace part has dimension $n - r = 1$. Thus lines always consist of a constant part plus a one-dimensional subspace part. In view of this, we can regard all lines as one-dimensional. (Caution: Lines not through the origin are not subspaces. Strictly speaking, they do not have a dimension as defined in Sec. 14.)

Lastly, any plane in $\mathcal{R}^3$ is the solution set of one equation $sx + ty + uz = v$. Here $n = 3$ and the rank $r = 1$; so $n - r = 2$. Planes therefore consist of a constant part plus a two-dimensional subspace part, and in that sense are always two-dimensional. (Again, only planes through the origin are subspaces with dimension 2 as defined in Sec. 14.)

These theorems thus justify our intuitive feeling that lines are one-dimensional and planes are two-dimensional.

Example 15-2

Suppose a system of equations has the matrix

$$\begin{bmatrix} 4 & 3 & 1 & 1 & \vdots & 5 \\ 0 & 2 & 3 & -1 & \vdots & -7 \\ 0 & 2 & 3 & -1 & \vdots & -7 \end{bmatrix}$$

This matrix has rank $r = 2$. (This can be seen by inspection, for it is clear that the 1,1 entry and the 2,2 entry would become the row-leading 1s if the matrix were reduced.) By Theorem 15-2, the solution set S can be written as a particular solution plus an $(n - r)$-dimensional subspace part. Since $n - r = 4 - 2 = 2$, the subspace part is two-dimensional. Therefore, the elements of S have the form (   ) $+ a($ $) + b($).   ●

Exercises

15-1 Each of the following matrices represents a system of equations. Determine their ranks by underlining the entries that would become row-leading 1s (do not reduce fully). Then describe the solution sets as completely as possible. Follow the pattern of Example 15-2.

(a) $\begin{bmatrix} 4 & 0 & 23 & 78 \\ 0 & 1 & 2 & 10 \\ 0 & 0 & 0 & 7 \end{bmatrix}$

(b) $\begin{bmatrix} 6 & 0 & 2 & \vdots & 1 \\ 0 & 9 & 5 & \vdots & 5 \end{bmatrix}$

(c) $\begin{bmatrix} 3 & 3 & -\frac{6}{7} & -\frac{5}{2} & \vdots & 3 \end{bmatrix}$

(d) $\begin{bmatrix} 2 & 5 & -30 & 15 & 12 & 41 \\ 0 & 0 & 17 & -3 & 95 & 10 \\ 0 & 0 & 0 & 20 & 40 & -57 \end{bmatrix}$

(e) $\begin{bmatrix} 4 & 89 & 65 \\ 0 & 11.3 & 501 \\ 0 & 0 & 7.1 \end{bmatrix}$

(f) $\begin{bmatrix} 43.6 & 10.7\sqrt{3} & -5\sqrt{2} & \vdots & 87.2 \\ -43.6 & -10.7\sqrt{3} & 5\sqrt{2} & \vdots & -87.2 \end{bmatrix}$

15-2 The system

$$\begin{aligned} x + y + w &= 2 \\ y - 2z &= 1 \\ 2x + 3y - 2z + 2w &= 5 \end{aligned}$$

has a solution $(1, 1, 0, 0)$. Describe all solutions by reducing the matrix of the corresponding homogeneous system

$$\begin{aligned} x + y + w &= 0 \\ y - 2z &= 0 \\ 2x + 3y - 2z + 2w &= 0 \end{aligned}$$

and using Theorem 15-2.

15-3 For each of these homogeneous systems of equations, (1) find the dimension of the subspace of solutions, using Theorem 15-1, and (2) find a basis for the subspace and check that it has the correct number of vectors in it:

(a) $\begin{aligned} x + 3y - 8z &= 0 \\ -x - 3y + 8z &= 0 \end{aligned}$ (b) $\begin{aligned} x + y - 3z &= 0 \\ 2x + 9y + 6z &= 0 \end{aligned}$

(c) $\begin{aligned} 8x + y + t &= 0 \\ x - z - t &= 0 \\ 9x + y - z &= 0 \end{aligned}$ (d) $x + 5y - z - t + u = 0$

16.   COORDINATES

In the previous sections we used the notion of a basis to determine the dimensions of various vector spaces. Another equally important use of a basis is in determining *coordinates* of vectors.

We use coordinates to attach numerical labels to vectors, just as we use latitude and longitude to attach numerical labels to points on the earth for greater ease in locating them. The following theorem and definition show the role a basis plays in this procedure.

Theorem 16-1   Let $X_1, X_2, \ldots, X_n$ be a basis for $\mathcal{V}$. Then every vector in $\mathcal{V}$ can be written in *exactly one way* as a linear combination of $X_1, X_2, \ldots, X_n$.

Proof†

At least one way. Let $X \in \mathcal{V}$. Since $X_1, \ldots, X_n$ span $\mathcal{V}$, we have $X = a_1 X_1 + a_2 X_2 + \cdots + a_n X_n$ for some scalars $a_1, a_2, \ldots, a_n$.

At most one way. Suppose that $X = b_1 X_1 + b_2 X_2 + \cdots + b_n X_n$ also. Then, subtracting the two expressions for X, we have

$$0 = X - X = (a_1 - b_1)X_1 + (a_2 - b_2)X_2 + \cdots + (a_n - b_n)X_n$$

Since $X_1, \ldots, X_n$ are independent, by Theorem 13-1 the only solution to the last equation is $(a_1 - b_1) = 0, (a_2 - b_2) = 0, \ldots, (a_n - b_n) = 0$. Thus $a_1 = b_1, a_2 = b_2, \ldots, a_n = b_n$. The b's were really the a's again. ∎

Thus, given a basis $X_1, \ldots, X_n$, with every vector X we can associate a unique list of n numbers.

Definition

Let $\mathcal{V}$ be a vector space and let $X_1, \ldots, X_n$ be a basis for $\mathcal{V}$. Let $X \in \mathcal{V}$. The unique numbers $a_1, a_2, \ldots, a_n$ such that $X = a_1 X_1 + \cdots + a_n X_n$ are called the **coordinates of X with respect to the basis $X_1, \ldots, X_n$.**

Theorem 16-1 shows that the coordinates of X with respect to a given basis exist and are indeed unique. If we are given a second basis, the coordinates of X with respect to this second basis will be different, in general. For example, the vector $(8, 11)$ has coordinates 3, 5 with respect to the basis $(1, 2), (1, 1)$ for $\mathcal{R}^2$, since

$$(8, 11) = 3(1, 2) + 5(1, 1)$$

Changing to the second basis $(2, 0), (4, 22)$ changes the coordinates of $(8, 11)$ to $3, \frac{1}{2}$ since

$$(8, 11) = 3(2, 0) + \tfrac{1}{2}(4, 22)$$

With respect to the usual basis $(1, 0), (0, 1)$, the vector $(8, 11)$ has coordinates 8, 11. In general, the vector (a, b) has coordinates a, b with respect to the usual basis.

Here are three examples showing how to find coordinates of vectors. The problem, like so many others, reduces to solving a system of linear equations.

---

† To show that something happens in *exactly one way*, it is often easiest to show separately *at most one way* and *at least one way*. Together these yield *exactly one way*.

Example 16-1

Find the coordinates of an arbitrary vector (a, b, c) in $\mathcal{R}^3$ with respect to the basis $(1, 1, 0)$, $(0, 1, 0)$, $(1, 0, \frac{1}{2})$. We seek the unique numbers x, y, z such that

$$(a, b, c) = x(1, 1, 0) + y(0, 1, 0) + z(1, 0, \tfrac{1}{2})$$

This reduces to the system

$$
\begin{aligned}
x \quad\;\; + z &= a \\
x + y \quad\;\; &= b \\
\tfrac{1}{2}z &= c
\end{aligned}
$$

whose matrix is

$$
\begin{bmatrix}
1 & 0 & 1 & \vdots & a \\
1 & 1 & 0 & \vdots & b \\
0 & 0 & \tfrac{1}{2} & \vdots & c
\end{bmatrix}
$$

The solution is $x = a - 2c$, $y = b - a + 2c$, $z = 2c$. Thus for any vector (a, b, c), its coordinates with respect to the given basis are $a - 2c$, $b - a + 2c$, $2c$.

For example, $(1, 2, 3)$ has $a = 1$, $b = 2$, $c = 3$, and therefore has first coordinate $a - 2c = 1 - 2(3) = -5$ with respect to this basis. Similarly, its second coordinate $b - a + 2c = 2 - 1 + 2(3) = 7$ and third coordinate $2c = 2(3) = 6$.   ●

The method just used to find coordinates is exactly the method used in Sec. 12 to determine if given vectors *span* a given vector space. The fundamental equation is still

Vector = linear combination of given basis vectors

However, the emphasis has changed. In Sec. 12 we only needed to know whether a solution x, y, z *existed;* now we need to know x, y, and z precisely, because x, y, and z are the coordinates we seek.

Example 16-2

Find the coordinates of (a, b, c) with respect to the basis $(0, 0, 1)$, $(1, 0, 0)$, $(0, 1, 0)$ for $\mathcal{R}^3$. We seek the unique numbers x, y, z such that $(a, b, c) = x(0, 0, 1) + y(1, 0, 0) + z(0, 1, 0)$. We immediately obtain $x = c$, $y = a$, $z = b$. Thus the coordinates of (a, b, c) with respect to the "scrambled" usual basis $(0, 0, 1)$, $(1, 0, 0)$, $(0, 1, 0)$ are c, a, b.

This shows that the order in which basis vectors are written down is important; scrambling their order scrambles the coordinates of vectors in the same way. ●

Example 16-3

In $\mathscr{P}^2$, the vector $2x^2 - 5x + 4$ has coordinates $2, -5, 4$ with respect to the basis x^2, x, 1. More generally, $ax^2 + bx + c$ has coordinates a, b, c with respect to the basis x^2, x, 1. This is evident since $ax^2 + bx + c$ equals the linear combination $a(x^2) + b(x) + c(1)$ of x^2, x, 1. Thus with every polynomial in $\mathscr{P}^2$ there corresponds a triple of real numbers, namely, its coordinates. This suggests one use of coordinates: In passing from the vectors in $\mathscr{P}^2$ to their coordinates, we pass from polynomials to the more familiar triples of real numbers. ●

To conclude this section, we note without proof that if vectors X and Y have coordinates $c_1, c_2, \ldots, c_n$ and $d_1, d_2, \ldots, d_n$ respectively, with respect to a given basis, then $X + Y$ has coordinates $c_1 + d_1, c_2 + d_2, \ldots, c_n + d_n$ and rX has coordinates $rc_1, rc_2, \ldots, rc_n$ with respect to the same basis. Thus, in passing to coordinates, we pass to the simplicity and familiarity of adding and scalar multiplying n-tuples of real numbers.

Exercises

16-1 Find the coordinates of the following vectors with respect to the basis $(1, 1)$, $(2, 3)$ of $\mathscr{R}^2$:

(a) (a, b) (b) $(4, -5)$ (c) $(0, 0)$
(d) $(6, 6)$ (e) $(-8, -12)$

16-2 Find the coordinates of the following vectors with respect to the basis $(1, 0, 1)$, $(0, 2, 0)$, $(1, 0, -3)$ for $\mathscr{R}^3$:

(a) (a, b, c) (b) $(2, -1, -2)$
(c) $(0, 2, -4)$ (d) $65(2, -1, -2)$
(e) $(0, 2, -4) + 65(2, -1, -2)$

[Hint: For parts (d) and (e) use the information in the last paragraph of this section.]

16-3   Find the coordinates of $(1, 3, 0)$ with respect to each of these bases for $\mathcal{R}^3$:

(a)   The usual basis
(b) $(2, 3, 1)$, $(0, 2, 0)$, $(1, 1, 1)$
(c) $(1, 0, 0)$, $(0, 3, 0)$, $(0, 0, 4)$

16-4   Find the coordinates of $\begin{bmatrix} 2 & 0 \\ 1 & 3 \end{bmatrix}$ with respect to:

(a)   The usual basis for $\mathcal{M}^{2 \times 2}$

(b)   The basis $\begin{bmatrix} 1 & 0 \\ 0 & 1 \end{bmatrix}$, $\begin{bmatrix} 1 & 0 \\ 1 & 0 \end{bmatrix}$, $\begin{bmatrix} 0 & 0 \\ 1 & 1 \end{bmatrix}$, $\begin{bmatrix} 0 & 1 \\ 1 & 1 \end{bmatrix}$

16-5   Find the coordinates of the following vectors with respect to the basis $x^2 + 2x$, $x + 3$, $x + 4$ of $\mathcal{P}^2$:

(a) $ax^2 + bx + c$ (b) $x^2 - 6$
(c) $2x^2 - x - 1$ (d) $(x^2 - 6) + (2x^2 - x - 1)$
(e) $1000(2x^2 - x - 1)$

16-6   Find the coordinates of $3x^2 - x + 4$ with respect to each of these bases for $\mathcal{P}^2$:

(a) x^2, x, 1
(b) x^2, $2x$, 3
(c) $x^2 - 1$, $2x + 1$, $x^2 + 3$

16-7   A vector space $\mathcal{V}$ has the basis $\mathbf{B}_1$, $\mathbf{B}_2$, $\mathbf{B}_3$, $\mathbf{B}_4$. If $\mathbf{X} = 3\mathbf{B}_2 - 2\mathbf{B}_4$, what are its coordinates with respect to this basis? Explain.

16-8   Suppose that a vector space $\mathcal{V}$ has the basis $\mathbf{X}_1$, $\mathbf{X}_2$, $\mathbf{X}_3$. Let $\mathbf{X}$ have coordinates c_1, c_2, c_3 and $\mathbf{Y}$ have coordinates d_1, d_2, d_3 with respect to $\mathbf{X}_1$, $\mathbf{X}_2$, $\mathbf{X}_3$.

(a)   Prove that $\mathbf{X} + \mathbf{Y}$ has coordinates $c_1 + d_1$, $c_2 + d_2$, $c_3 + d_3$ with respect to $\mathbf{X}_1$, $\mathbf{X}_2$, $\mathbf{X}_3$.
(b)   Prove that $r\mathbf{X}$ has coordinates rc_1, rc_2, rc_3 with respect to $\mathbf{X}_1$, $\mathbf{X}_2$, $\mathbf{X}_3$.

17.   FUNCTIONS

In subsequent sections we will deal with particular types of functions known as isomorphisms and linear transformations. It will be desirable, then, to have a brief treatment of the basic facts about functions.

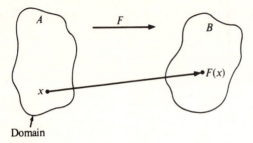

Domain

Fig. 17-1   A function $F : A \rightarrow B$.

Definition

A **function** F is a rule that assigns to each element x of one set a single element $F(x)$ of a second set. The element $F(x)$ is called the *value* of the function at x. The first set is called the *domain* of the function, and is the set of all elements x to which the rule applies.†

Therefore, to define a function F:

(a)   Give the rule for F.

(b)   Specify the domain of F and the set where its values lie.

Occasionally the domain and the set containing the values are clear from the context and are not explicitly mentioned.

For example, we might write "F is defined by $F(x) = x^3 - x$" or "G is defined by the rule $G(x, y) = x^2 + y^2$" or "H has the rule that assigns to each person his social security number." In each case we have a clear-cut rule for computing the values of the function. We assume that the domain of F is $\mathcal{R}$, the domain of G is $\mathcal{R}^2$, and the domain of H is the set of all people having social security numbers. In the first two examples the set of values lies in $\mathcal{R}$, in the third example the set of values lies in the set of all "numbers" of the form xxx-xx-xxxx.

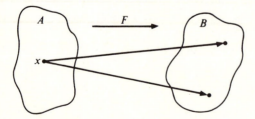

Fig. 17-2 F is not a function since x is assigned two values.

† The domain of a function is much like the domain of a king: the king's domain is the set of things to which *his* rule applies.

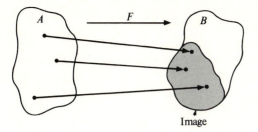

Fig. 17-3   *F* is not onto *B*.

We use the shorthand notation $F : \mathcal{R} \to \mathcal{R}$ to mean "the domain of *F* is $\mathcal{R}$ and its values lie in $\mathcal{R}$." Similarly, $G : \mathcal{R}^2 \to \mathcal{R}$ means "the domain of *G* is $\mathcal{R}^2$ and its values lie in $\mathcal{R}$." Most of the functions studied in calculus are of the form $F : \mathcal{R} \to \mathcal{R}$; in linear algebra most of the functions are of the type $F : \mathcal{V} \to \mathcal{W}$, where $\mathcal{V}$ and $\mathcal{W}$ are vector spaces.

The *image* (or *range*) of a function *F* is the set of all values that result from applying the rule for *F* to every element in its domain. Since every value in the range can be written as $F(x)$ for some *x* in the domain of *F*, in set notation the image of *F* is $\{F(x) : x$ is in the domain of $F\}$. For example, if $F(x) = x^2$, where $F : \mathcal{R} \to \mathcal{R}$, then the image is $[0, \infty)$ since the set of all squares of real numbers gives us all the nonnegative numbers. If $G(x) = \sqrt{4 - x^2}$, where $G : [-2, 2] \to \mathcal{R}$, the image of *G* is $[0, 2]$, since the values of *G* are positive square roots of numbers between 0 and 4.

It often proves useful to illustrate pictorially how a function behaves. One such way is shown in Fig. 17-1. The figure illustrates a function $F : A \to B$. Figure 17-2 illustrates a rule that is *not* a function since *two* values (not *one*) are assigned to *x*.

The image of a function $F : A \to B$ need not be all of *B*. The set of values might be only a part of *B*. When the image happens to be *all* of *B*, we say the function is *onto B*. *Onto* thus means that every element of the second set is used as a value of the function. For example, if $F : \mathcal{R} \to \mathcal{R}$ has the rule $F(x) = x^3$, then *F* is onto the second set specified, $\mathcal{R}$. But if $G : \mathcal{R} \to \mathcal{R}$ with $G(x) = x^2$, then *G* is not onto $\mathcal{R}$, since no negative numbers are values of this function. Figure 17-3 illustrates a function $F : A \to B$ that is *not* onto *B*. Its image is only the indicated part of *B*.

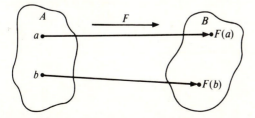

Fig. 17-4   A 1-1 function.

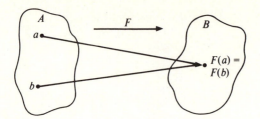

Fig. 17-5   A function that is not 1-1.

If S is a set contained in the domain of F, the *image of S* is the set of all values $F(x)$ that result from applying the rule for F to the elements of the set S. In set notation, the image of S can be written $\{F(x) : x \in S\}$.

A *one-to-one* function (more briefly, 1-1) is a function that always produces different values for different elements of the domain. In symbols, F is 1-1 provided that whenever $a \neq b$, then $F(a) \neq F(b)$.

For example, $F(x) = x^2$ defines a function that is *not* 1-1 since $F(2)$ and $F(-2)$ are both equal to 4. Thus 2 and -2 produce the same value. On the other hand, $F(x) = x^3$ defines a function that *is* 1-1; if $a \neq b$, then their function values a^3 and b^3 also are not equal. The function that assigns to each American his or her social security number is 1-1, since no two people have the same number.

Figure 17-4 illustrates the 1-1 property—the values $F(a)$ and $F(b)$ are distinct whenever a and b are. Figure 17-5 illustrates how a function fails to be 1-1; although $a \neq b$, we have $F(a) = F(b)$.

In later sections we will need to verify that certain functions have the two properties of being 1-1 and onto. We now describe explicitly how to go about verifying these properties. Let $F : \mathcal{U} \to \mathcal{W}$ be a given function.

To show F is 1-1: Begin with two arbitrary unequal elements $\mathbf{X}_1$ and $\mathbf{X}_2$ of the first set $\mathcal{U}$, making sure that $\mathbf{X}_1$ and $\mathbf{X}_2$ are written in the form appropriate for $\mathcal{U}$. Then show that $F(\mathbf{X}_1) \neq F(\mathbf{X}_2)$. [One often proves $F(\mathbf{X}_1) \neq F(\mathbf{X}_2)$ by contradiction: Assume $F(\mathbf{X}_1) = F(\mathbf{X}_2)$ and use this to show that $\mathbf{X}_1 = \mathbf{X}_2$, contradicting the fact that $\mathbf{X}_1 \neq \mathbf{X}_2$.]

To show F is onto $\mathcal{W}$: It is the *second* set, $\mathcal{W}$, that is of interest here. Thus:

(a)   Let $\mathbf{Y}$ be an arbitrary element of the second set, $\mathcal{W}$ (written in the appropriate form for elements of $\mathcal{W}$).

(b)   Write $\mathbf{Y} \overset{?}{=} F(\mathbf{X})$, using the rule for F. [We hope that $\mathbf{Y}$ equals a value $F(\mathbf{X})$ of the function.]

(c)   Solve the supposed equation in part (b) for $\mathbf{X}$.

(d)   Verify that $\mathbf{X}$ is in $\mathcal{U}$, thus available for use in forming $F(\mathbf{X})$.

Here are three examples to illustrate these methods.

Example 17-1

Let $F : \mathcal{R} \to \mathcal{R}$ have the rule $F(x) = 5x^3 + 2$. We will verify that F is 1-1 and onto $\mathcal{R}$.

F is 1-1   Let $a \neq b$ be arbitrary unequal elements of the first set $\mathcal{R}$, that is, unequal real numbers. We prove that $F(a) \neq F(b)$ by contradiction: Suppose $F(a) = F(b)$, that is, $5a^3 + 2 = 5b^3 + 2$ (using the rule for F). Subtracting 2 from both sides and cancelling the 5s yields $a^3 = b^3$. Lastly, $a = b$ since real cube roots are unique. This is the desired contradiction.

F is onto $\mathcal{R}$ Let y be an arbitrary element of the second set $\mathcal{R}$. Does $y = F(x)$ for some x in the domain of F? Solving $y = 5x^3 + 2$ for x yields $x = \sqrt[3]{(y - 2)/5}$. Regardless of what y is (positive, negative, or zero), $(y - 2)/5$ is a real number, and will have a real cube root x. Thus y does equal $F(x)$, for the number $x = \sqrt[3]{(y - 2)/5}$ in the domain $\mathcal{R}$.   ●

Example 17-2

Let $G : \mathcal{R}^2 \to \mathcal{R}^3$ have the rule $G(x, y) = (3x, 5y, x)$.

G is 1-1   Let $(a, b) \neq (c, d)$ be arbitrary unequal elements of the first set $\mathcal{R}^2$. We wish to show that $G(a, b) = (3a, 5b, a) \neq (3c, 5d, c) = G(c, d)$. Since $(a, b) \neq (c, d)$, either $a \neq c$ or $b \neq d$. In the first case $G(a, b)$ and $G(c, d)$ differ in their third entries. In the second case the second entries differ. In either case $(3a, 5b, a) \neq (3c, 5d, c)$, that is, $G(a, b) \neq G(c, d)$.

G is not onto $\mathcal{R}^3$ See Exercise 17-4.   ●

Example 17-3

Let $H : \mathcal{R}^3 \to \mathcal{R}^2$ have the rule $H(x, y, z) = (2y, 3z)$.

H is onto $\mathcal{R}^2$ Let (r, s) be an arbitrary element of the second set $\mathcal{R}^2$. Does $(r, s) = H(x, y, z)$ for some (x, y, z) in the domain $\mathcal{R}^3$? Put $(r, s) = (2y, 3z) = H(x, y, z)$; then $y = r/2$ and $z = s/3$. Then for the triple $(x, y, z) = (x, r/2, s/3)$ we have $(r, s) = H(x, y, z)$. Both $r/2$ and $s/3$ are real numbers, and so the triple $(x, r/2, s/3)$ is in the domain $\mathcal{R}^3$ and thus is available for use.

H is not 1-1   See Exercise 17-2.   ●

Exercises

17-1 For each of these functions, determine its image by inspection. (You may want to compute a few key values of the function.)

(a) $f: \mathcal{R} \to \mathcal{R}$ defined by $f(x) = x^2 + 1$
(b) $g: \mathcal{R} \to \mathcal{R}$ defined by $g(t) = \sin t$
(c) $F: \mathcal{R} \to \mathcal{R}$ defined by $F(x) = 1/(1 + x^2)$
(d) $G: \mathcal{R}^2 \to \mathcal{R}^2$ defined by $G(x, y) = (2x - y, 4x - 2y)$

[Hint: Compute $G(1, 0)$ and $G(0, 1)$ and deduce what must happen for other points (x, y).]

17-2 Each of these rules defines a function that is *not* 1-1. Show this by finding two elements of the domain whose images are equal.

(a) $F(x) = x^3 - x$
(b) $G(x, y) = (3x, 0, x)$
(c) $f(t) = \cos t$
(d) $g(x) = |x - 3|$
(e) The function H of Example 17-3

17-3 Each of these rules defines a function that *is* 1-1. Show this by taking $a \neq b$ and showing carefully that they produce different values.

(a) $f(x) = x^5 - 3$
(b) $g(x) = 3^x$
(c) $H(x, y) = (2x, y)$

17-4 Each of these functions is *not* onto the second set. Show this by finding an element of the second set that is not the image of any element of the first set.

(a) $F: \mathcal{R} \to \mathcal{R}$ defined by $F(t) = t^2 - 3$
(b) $h: \mathcal{R}^2 \to \mathcal{R}$ defined by $h(x, y) = x^2 + 2y^2$
(c) $g: \mathcal{R}^2 \to \mathcal{R}^2$ defined by $g(u, v) = (uv, u^2v^2)$
(d) The function G of Example 17-2

17-5 Each of these functions *is* onto the second set. Show this by taking an arbitrary element of the second set and showing that it *does* equal a value of the function.

(a) $g: \mathcal{R} \to \mathcal{R}$ defined by $g(u) = u^3 - 4$
(b) $h: \mathcal{R} \to (0, 1]$ defined by $h(x) = 1/(1 + x^2)$
(c) $F: \mathcal{R}^2 \to \mathcal{R}^2$ defined by $F(x, y) = (3y, 2x)$

18.   ISOMORPHISM

Recall that in Example 16-3 we used coordinates to associate a triple of numbers with any polynomial in $\mathscr{P}^2$. That is, the polynomial $ax^2 + bx + c$ gave rise to the coordinates a, b, c, which in turn suggest the point (a, b, c) in $\mathscr{R}^3$. We have thereby uncovered a basic similarity between $\mathscr{P}^2$ and $\mathscr{R}^3$. As we will see, the similarity goes further; in fact $\mathscr{P}^2$ and $\mathscr{R}^3$ are essentially the same. In mathematics, two objects such as $\mathscr{P}^2$ and $\mathscr{R}^3$ that are essentially the same (even though they may look different superficially) are said to be *isomorphic*.

By analogy, consider the monetary systems of France and the United States: centimes/francs and cents/dollars. The objects are different in appearance, but the systems are essentially the same, since the relationship of centime to franc is the same as that of cent to dollar. Thus the monetary systems of the two countries are isomorphic.

Similarities between two sets can be uncovered by using a function from one set to the other. For example, a function from one set A to a set B that is both 1-1 and onto B sets up a 1-1 correspondence between the elements of A and B, and shows that the two sets have the same number of elements.

When we wish to show that two vector spaces $\mathcal{V}$ and $\mathcal{W}$ are isomorphic, we will then need a function $F : \mathcal{V} \to \mathcal{W}$ that is 1-1 and onto $\mathcal{W}$. (If $\mathcal{V}$ and $\mathcal{W}$ are essentially the same, they must both have the same number of elements.) The function F will also have to have the properties that guarantee that the additions and scalar multiplications of $\mathcal{V}$ and $\mathcal{W}$ are essentially the same. We therefore give the following definition.

Definition

Let $\mathcal{V}$ and $\mathcal{W}$ be vector spaces. A function $F : \mathcal{V} \to \mathcal{W}$ is an **isomorphism** if it has these four properties:

(a) F is 1-1.

(b) F is onto $\mathcal{W}$.

(c) $F(\mathbf{X} + \mathbf{Y}) = F(\mathbf{X}) + F(\mathbf{Y})$ for   all   vectors $\mathbf{X}$ and $\mathbf{Y}$ in $\mathcal{V}$.   Addition is preserved.

(d) $F(r\mathbf{X}) = rF(\mathbf{X})$ for all vectors $\mathbf{X}$ in $\mathcal{V}$ and all scalars r.   Scalar multiplication is preserved.

When an isomorphism exists between $\mathcal{V}$ and $\mathcal{W}$, we say that $\mathcal{V}$ is *isomorphic* to $\mathcal{W}$ and write $\mathcal{V} \simeq \mathcal{W}$.

The function F in the above definition makes $\mathbf{X}$ correspond to $F(\mathbf{X})$

and $\mathbf{Y}$ correspond to $F(\mathbf{Y})$. Property (c) says that their sum $\mathbf{X} + \mathbf{Y}$ also correspond to the appropriate sum $F(\mathbf{X}) + F(\mathbf{Y})$; that is, adding vectors in $\mathcal{V}$ and adding corresponding vectors in $\mathcal{W}$ produce sums that correspond. Thus the addition in $\mathcal{V}$ and the addition in $\mathcal{W}$ are essentially the same. Property (d) has a similar interpretation.

Example 18-1

$\mathscr{P}^2 \simeq \mathscr{R}^3$. To verify this, we will find a function $F : \mathscr{P}^2 \to \mathscr{R}^3$ that has the four properties required to be an isomorphism. The most obvious choice for a function is one where $ax^2 + bx + c$ corresponds to (a, b, c), and so we *define* F to make this so:

$$F(ax^2 + bx + c) = (a, b, c)$$

To show that F is an isomorphism we must prove that F has the four required properties.

F is 1-1   Let $ax^2 + bx + c \neq dx^2 + ex + f$ be arbitrary unequal elements of $\mathscr{P}^2$. Since they are unequal, either $a \neq d$, $b \neq e$, or $c \neq f$. In any event, $(a, b, c) \neq (d, e, f)$, and so $F(ax^2 + bx + c) \neq F(dx^2 + ex + f)$. Since different elements in $\mathscr{P}^2$ always have different function values, the function is 1-1.

F is onto $\mathscr{R}^3$ Let (s, t, u) be an arbitrary element of $\mathscr{R}^3$. We must find an element $\mathbf{X}$ in $\mathscr{P}^2$ such that $(s, t, u) = F(\mathbf{X})$. Immediately we see that $\mathbf{X} = sx^2 + tx + u$ will do.

F preserves addition   We use this format:

$$F(\mathbf{X} + \mathbf{Y}) \overset{?}{=} F(\mathbf{X}) + F(\mathbf{Y})$$
Property to be verified

$F((ax^2 + bx + c)$ $+ (dx^2 + ex + f))$	$F(ax^2 + bx + c)$ $+ F(dx^2 + ex + f)$	Suitable notation for $\mathbf{X}$ and $\mathbf{Y}$
$F((a + d)x^2$ $+ (b + e)x + (c + f))$		Definition of addition in $\mathscr{P}^2$
$(a + d, b + e, c + f)$	$(a, b, c) + (d, e, f)$	Definition of F
	$(a + d, b + e, c + f)$	Definition of addition in $\mathscr{R}^3$

The left and right sides are now clearly equal, and hence F preserves addition.

F preserves scalar multiplication   We must show that $F(r\mathbf{X}) = rF(\mathbf{X})$. To do this we will work with each side separately:

$$
\begin{aligned}
F(r\mathbf{X}) &= F(r(ax^2 + bx + c)) \\
&= F((ra)x^2 + (rb)x + (rc)) \\
&= (ra, rb, rc)
\end{aligned}
$$

and

$$
\begin{aligned}
rF(\mathbf{X}) &= rF(ax^2 + bx + c) \\
&= r(a, b, c) \\
&= (ra, rb, rc)
\end{aligned}
$$

The student should supply the reason for each step (Exercise 18-1). Since $F(r\mathbf{X}) = (ra, rb, rc) = rF(\mathbf{X})$, F preserves scalar multiplication.

All four properties have now been verified; therefore F is an isomorphism, and we write $\mathscr{P}^2 \simeq \mathscr{R}^3$.   ●

Example 18-1 can be generalized. The reason the isomorphism worked was that vectors in $\mathscr{P}^2$ have three coordinates, just as vectors in $\mathscr{R}^3$ do. In fact, we might suspect that any vector space where vectors can be described by three coordinates is isomorphic to $\mathscr{R}^3$. The following theorem bears this out.

Theorem 18-1   Every n-dimensional vector space (with real numbers for scalars) is isomorphic to $\mathscr{R}^n$.†

Proof   Let $\mathcal{V}$ be an n-dimensional vector space, with a basis $\mathbf{X}_1, \mathbf{X}_2, \ldots, \mathbf{X}_n$. Since every vector in $\mathcal{V}$ can be written as a linear combination of the basis vectors, $a_1\mathbf{X}_1 + a_2\mathbf{X}_2 + \cdots + a_n\mathbf{X}_n$ represents an arbitrary vector in $\mathcal{V}$. A natural candidate for an isomorphism would be to have a vector $a_1\mathbf{X}_1 + a_2\mathbf{X}_2 + \cdots + a_n\mathbf{X}_n$ and the n-tuple $(a_1, a_2, \ldots, a_n)$ correspond. More precisely, we define a function $F : \mathcal{V} \to \mathscr{R}^n$ by the rule

$$
F(a_1\mathbf{X}_1 + a_2\mathbf{X}_2 + \cdots + a_n\mathbf{X}_n) = (a_1, a_2, \ldots, a_n)
$$

We must show that F satisfies the four properties of an isomorphism.

F is 1-1   Let $\mathbf{X} = a_1\mathbf{X}_1 + \cdots + a_n\mathbf{X}_n$ and $\mathbf{Y} = b_1\mathbf{X}_1 + \cdots + b_n\mathbf{X}_n$ be two distinct arbitrary vectors in $\mathcal{V}$. Since they are unequal, at least one $a_i \neq b_i$. But then $F(\mathbf{X}) = (a_1, \ldots, a_n)$ and $F(\mathbf{Y}) = (b_1, \ldots, b_n)$ differ in their i-entries, and so $F(\mathbf{X}) \neq F(\mathbf{Y})$. This proves that F is 1-1.

The balance of the proof is left as an exercise.   ■

† Note that this theorem assumes that the scalars are real numbers. Sometimes other sets of scalars are used, for example, complex numbers. In such cases, analogous theorems would be true.

The reason that Theorem 18-1 is so important is that it allows us to transfer our knowledge about $\mathscr{R}^n$ (which is relatively familiar to us) to *any* n-dimensional vector space. For example, our matrix test for independence of vectors in $\mathscr{R}^n$ (Theorem 13-2) tells us that $n + 1$ or more vectors in $\mathscr{R}^n$ cannot be independent. Theorem 18-1 ensures that a similar result holds for *any* n-dimensional vector space.

One final note on basis and dimension. Our main concern is with finite-dimensional vector spaces, that is, vector spaces having bases with finite numbers of vectors in them. The $\mathscr{R}^n$s, $\mathscr{M}^{m \times n}$s, and $\mathscr{P}^n$s are all finite-dimensional. (Not every vector space is finite-dimensional. For example, the vector space $\mathscr{P}$ of all polynomials is infinite-dimensional (Exercise 14-16). See also Examples 9-5 and 9-6.) We left unanswered in Sec. 14 the question of uniqueness of dimension. Because of the concept of isomorphism, we are now in a position to answer this question for vector spaces with finite bases; the following example is the key.

Example 18-2

If $m \neq n$, then $\mathscr{R}^m$ and $\mathscr{R}^n$ are not isomorphic. We are given that $m \neq n$. To fix ideas, suppose that $n < m$ (the proof is similar if we assume $m < n$). If $\mathscr{R}^m$ and $\mathscr{R}^n$ were isomorphic, then there would be an isomorphism $F : \mathscr{R}^m \to \mathscr{R}^n$. Let $\mathbf{X}_1, \mathbf{X}_2, \ldots, \mathbf{X}_m$ be the usual basis for $\mathscr{R}^m$. Since these vectors are independent, by Exercise 18-7 the m vectors $F(\mathbf{X}_1), \ldots, F(\mathbf{X}_m)$ in $\mathscr{R}^n$ would also be independent. But we cannot have more than n independent vectors in $\mathscr{R}^n$ (Exercise 13-12). Therefore we would have $m \leq n$, which would contradict the fact that $n < m$. Thus $\mathscr{R}^m$ and $\mathscr{R}^n$ are not isomorphic if $m \neq n$.   ●

Theorem 18-2

If a vector space $\mathscr{V}$ has a basis with m vectors and a basis with n vectors, then $m = n$.

Proof

Since $\mathscr{V}$ has a basis with m vectors, $\mathscr{V} \simeq \mathscr{R}^m$ (Theorem 18-1). Similarly $\mathscr{V} \simeq \mathscr{R}^n$. But these two facts together with Exercise 18-9 yield $\mathscr{R}^m \simeq \mathscr{R}^n$. Therefore, by Example 18-2, $m = n$.   ■

Exercises

18-1   Supply the reasons in the proof that $F(r\mathbf{X}) = rF(\mathbf{X})$ in Example 18-1.

18-2   Finish the proof of Theorem 18-1. That is, show that the function F defined in the proof has the remaining three properties of an isomorphism.

18-3 Prove that $\mathcal{M}^{2\times 2} \simeq \mathcal{R}^4$ by defining a suitable function $G : \mathcal{M}^{2\times 2} \to \mathcal{R}^4$ and verifying the four properties of an isomorphism.

18-4 Prove that $\mathcal{R}^2 \simeq \{(a, 0, b) : a, b \in \mathcal{R}\}$ by defining a suitable function $F : \mathcal{R}^2 \to \{(a, 0, b) : a, b \in \mathcal{R}\}$ and verifying the four properties of an isomorphism.

18-5 Prove that $\mathcal{R}^3 \simeq \{(2a, 3b, 0, a - b - c) : a, b, c \in \mathcal{R}\}$ the easy way, by proving that the vector space on the right is three-dimensional and then citing Theorem 18-1.

18-6 Find five different vector spaces all isomorphic to $\mathcal{R}^{12}$.

18-7 Suppose that $\mathcal{V}$ and $\mathcal{W}$ are vector spaces and $F : \mathcal{V} \to \mathcal{W}$ is an isomorphism. Prove the following:

(a)   If $\mathbf{X}_1, \ldots, \mathbf{X}_n$ span $\mathcal{V}$, then $F(\mathbf{X}_1), \ldots, F(\mathbf{X}_n)$ span $\mathcal{W}$.
(b)   If $\mathbf{X}_1, \ldots, \mathbf{X}_n$ are independent vectors in $\mathcal{V}$, then $F(\mathbf{X}_1), \ldots, F(\mathbf{X}_n)$ are independent.
(c)   If $\mathbf{X}_1, \ldots, \mathbf{X}_n$ form a basis for $\mathcal{V}$, then $F(\mathbf{X}_1), \ldots, F(\mathbf{X}_n)$ form a basis for $\mathcal{W}$.
(d) $\mathcal{V}$ and $\mathcal{W}$ have the same dimension.

18-8 Suppose that $\mathcal{V}$ and $\mathcal{W}$ have the same dimension. Prove that $\mathcal{V} \simeq \mathcal{W}$. (Hint: Specify a basis $\mathbf{X}_1, \ldots, \mathbf{X}_n$ for $\mathcal{V}$ and a basis $\mathbf{Y}_1, \ldots, \mathbf{Y}_n$ for $\mathcal{W}$. Then define values $F(a_1\mathbf{X}_1 + \cdots + a_n\mathbf{X}_n)$ appropriately.)

18-9 Suppose $\mathcal{V}$, $\mathcal{W}$, and $\mathcal{Y}$ are vector spaces such that $\mathcal{V} \simeq \mathcal{W}$ and $\mathcal{W} \simeq \mathcal{Y}$. Prove that $\mathcal{V} \simeq \mathcal{Y}$. (Hint: Use Exercises 18-7 and 18-8.)

19.   APPLICATIONS

Example 19-1

Magic Squares

A *magic square* is an $n \times n$ matrix such that there is a constant K (called the magic constant) with the following properties:

(a)   The sum of the entries in any row equals K.

(b)   The sum of the entries in any column equals K.

(c)   The sum of the entries on either diagonal (from upper left to lower right and from upper right to lower left) equals K.

For example,

$$\begin{bmatrix} 4 & 9 & 2 \\ 3 & 5 & 7 \\ 8 & 1 & 6 \end{bmatrix}$$

is a 3×3 magic square whose magic constant is 15. The matrix

$$\begin{bmatrix} -1 & 8 & -1 \\ 2 & 2 & 2 \\ 5 & -4 & 5 \end{bmatrix}$$

is also a magic square where $K = 6$.† We will consider only 3×3 magic squares in this example.

Since we are considering only 3×3 magic squares, each one is a vector in $\mathcal{M}^{3 \times 3}$. If $\mathcal{S}$ is the set of all 3×3 magic squares, then $\mathcal{S}$ is a subspace of $\mathcal{M}^{3 \times 3}$. To see this, let A and B be magic squares with magic constants K and L respectively. The reader can easily verify that $A + B$ is a magic square with magic constant $K + L$ and that rA is a magic square with magic constant rK. Thus $\mathcal{S}$ is closed under addition and scalar multiplication and hence is a subspace of $\mathcal{M}^{3 \times 3}$.

We will now find a basis for $\mathcal{S}$ and thereby determine its dimension. To do this we need two facts regarding magic squares.

(a) If

$$A = \begin{bmatrix} a & b & c \\ d & e & f \\ g & h & i \end{bmatrix}$$

is a magic square, then its magic constant $K = 3e$. This follows by noting that

$$a + e + i = g + h + i$$
$$d + e + f = c + f + i$$
$$g + e + c = c + f + i$$

and therefore

$$a + e = g + h$$
$$d + e = c + i$$
$$g + e = f + i$$

---

† In books of mathematical puzzles it is often required that entries in an $n \times n$ magic square be the consecutive integers $1, 2, \ldots, n^2$. The first magic square above has this property. We do not require this. In fact, it is known that the first magic square above is essentially the only 3×3 magic square with this property.

Adding these last three equations, we have

$$(a + d + g) + 3e = (c + f + i) + (g + h + i)$$

Therefore,

$$K + 3e = K + K$$

and hence $K = 3e$.

(b) Any magic square is completely determined by particular entries in positions a, c, and e. For example, if we begin with

$$\begin{bmatrix} 2 & - & 7 \\ - & 6 & - \\ - & - & - \end{bmatrix}$$

then the entry in the b position (that is, the 1,2 entry) must be 9 since the top row must add up to the magic constant, which by fact (a) must be $3 \cdot 6 = 18$. Similarly the g entry (the 3,1 entry) must be 5. Finding the other entries in similar fashion, we have the magic square

$$\begin{bmatrix} 2 & 9 & 7 \\ 11 & 6 & 1 \\ 5 & 3 & 10 \end{bmatrix}$$

More generally, if we begin with

$$\begin{bmatrix} a & - & c \\ - & e & - \\ - & - & - \end{bmatrix}$$

then the entire magic square, A, must be

$$A = \begin{bmatrix} a & -a - c + 3e & c \\ -a + c + e & e & a - c + e \\ -c + 2e & a + c - e & -a + 2e \end{bmatrix}$$

(For example, since we must have $a +$ "b entry" $+ c = 3e$, then the b entry must equal $-a - c + 3e$.) Therefore every magic square is determined by the three entries in the a, c, and e positions.

If we take this matrix A, we can rewrite it as

$$A = \begin{bmatrix} a & -a & 0 \\ -a & 0 & a \\ 0 & a & -a \end{bmatrix} + \begin{bmatrix} 0 & -c & c \\ c & 0 & -c \\ -c & c & 0 \end{bmatrix} + \begin{bmatrix} 0 & 3e & 0 \\ e & e & e \\ 2e & -e & 2e \end{bmatrix}$$

$$= a \begin{bmatrix} 1 & -1 & 0 \\ -1 & 0 & 1 \\ 0 & 1 & -1 \end{bmatrix} + c \begin{bmatrix} 0 & -1 & 1 \\ 1 & 0 & -1 \\ -1 & 1 & 0 \end{bmatrix} + e \begin{bmatrix} 0 & 3 & 0 \\ 1 & 1 & 1 \\ 2 & -1 & 2 \end{bmatrix}$$

$$= a\mathbf{B}_1 + c\mathbf{B}_2 + e\mathbf{B}_3$$

Since an arbitrary magic square is a linear combination of $\mathbf{B}_1$, $\mathbf{B}_2$, and $\mathbf{B}_3$, $S = \langle \mathbf{B}_1, \mathbf{B}_2, \mathbf{B}_3 \rangle$. These three matrices are easily seen to be independent. Therefore, $\mathbf{B}_1$, $\mathbf{B}_2$, $\mathbf{B}_3$ are a basis for S, and so S is a three-dimensional subspace of $\mathcal{M}^{3 \times 3}$.   ●

Exercises

19-1 Show that $\mathbf{B}_1$, $\mathbf{B}_2$, $\mathbf{B}_3$ are independent. (Hint: Suppose that $x\mathbf{B}_1 + y\mathbf{B}_2 + z\mathbf{B}_3 = 0$; then combine the left side of this equation into one matrix.)

19-2 Show that the d, f, g, h, and i entries of A must be $-a + c + e$, $a - c + e$, $-c + 2e$, $a + c - e$, and $-a + 2e$ respectively.

19-3 Let $\mathcal{W}$ be the set of all 3×3 magic squares whose magic constant K is zero.

(a)   Prove that $\mathcal{W}$ is a subspace of $\mathcal{M}^{3 \times 3}$.
(b)   Find a basis for $\mathcal{W}$ and determine its dimension.

19-4 Describe all 2×2 magic squares; that is, if $\begin{bmatrix} a & b \\ c & d \end{bmatrix}$ is a 2×2 magic square, what simple relationship must hold among the four entries a, b, c, and d?

Review of Chapter 3

1 A vector $\mathbf{X}$ is a linear combination of the vectors $\mathbf{X}_1$, $\mathbf{X}_2$, ..., $\mathbf{X}_k$ if
$\mathbf{X} = $ _____ for some scalars a_1, a_2, ..., a_k.

To show that $X_1, X_2, \ldots, X_n$ span $\mathcal{V}$, we show that any vector $X \in \mathcal{V}$ equals a _____ of $X_1, \ldots, X_n$. This entails showing that the basic equation for spanning

$$X = a_1 X_1 + a_2 X_2 + \cdots + a_n X_n$$

can be solved for $a_1, a_2, \ldots, a_n$.

To show that $X_1, X_2, \ldots, X_n$ are independent, we must show that the basic equation for independence

_____

has the unique solution $a_1 = $ ____, $a_2 = $ ____, $\ldots$, $a_n = $ ____.

2   Vectors $X_1, X_2, \ldots, X_n$ are a basis for $\mathcal{V}$ if they _____, and are _____. The _____ of $\mathcal{V}$ is the number of vectors in a basis for $\mathcal{V}$.

3   Given n vectors in $\mathcal{R}^n$, we have a shortcut matrix method for checking whether they form a basis. Write the vectors as _____ of a matrix A. If the rank of A equals _____, they form a basis for $\mathcal{R}^n$; if the rank of A is _____, they do not form a basis for $\mathcal{R}^n$. This test was developed from a similar test for independence of vectors in $\mathcal{R}^n$: given k vectors in $\mathcal{R}^n$, write them as _____ of a matrix A. If the rank of A equals ____, they are _____; if the rank of A is _____, they are dependent.

4   Complete the following table:

	$\mathcal{R}^n$	$\mathcal{M}^{m \times n}$	$\mathcal{P}^n$
Usual basis			
Dimension			

5   Subspaces of $\mathcal{R}^3$ can be of dimensions 0, 1, 2, or 3. They are, respectively, $\{(0, 0, 0)\}$, _____ through the origin, _____ through the origin, and $\mathcal{R}^3$ itself.

6   The set of solutions of a homogeneous system of equations with n unknowns is a subspace of $\mathcal{R}^n$. Its dimension equals _____, where r is the _____ of the unaugmented matrix of coefficients.

If the rank of the unaugmented matrix of the system is n, then the solution is _____; if the rank is less than n, there are _____ solutions.

The set of solutions of a nonhomogeneous system is never a subspace. However, if the system has a solution, the solutions are all of the form

_____   +   _____

a particular solution plus a subspace part of dimension _____.
The subspace part consists of all solutions of the corresponding _____ system with matrix ____ instead of $[A \mid B]$.

7 Let $X_1, X_2, \ldots, X_n$ be a basis for $\mathcal{V}$. Then any vector X in $\mathcal{V}$ can be written in exactly one way as a _____ _____ $a_1 X_1 + \cdots + a_n X_n$ of basis vectors. The unique numbers $a_1, a_2, \ldots, a_n$ are called the _____ of X with respect to the basis $X_1, X_2, \ldots, X_n$.

8 An isomorphism is a function $F : \mathcal{V} \to \mathcal{W}$ with the four properties

(a)_____

(b)_____

(c)_____

(d)_____

To show that F is onto $\mathcal{W}$, we begin with an arbitrary vector Y in ____ and show that _____. To show that F is 1-1, we begin with arbitrary unequal vectors X_1 and X_2 in ____ and show that _____.

9 Two vector spaces with real numbers for scalars are isomorphic if they have the same _____. In particular, $\mathcal{M}^{2 \times 3}$ is isomorphic to $\mathcal{R}^{\text{___}}$ and to $\mathcal{P}^{\text{___}}$.

Review Exercises

1 Which of the following are bases for $\mathcal{R}^3$? Explain, or show work.

(a) $(1, 2, 0), (0, 1, 1), (2, 5, 1)$ (b) $(6, 1, 3), (2, 0, 4)$
(c) $(1, 2, 3), (4, 5, 6), (7, 8, 9)$ (d) $(1, 1, 0), (0, 1, 1), (2, 1, 3), (0, 2, 1)$

2 Which of the following are bases for the subspace of $\mathcal{R}^4$ given by $\{(2a, b, 0, a + b): a, b \in \mathcal{R}\}$? Explain, or show work.

(a) (2, 1, 0, 2)   (b) (1, 0, 0, 0), (0, 1, 0, 0), (0, 0, 0, 1)
(c) (2, 0, 0, 1), (0, 1, 0, 1)   (d) (2, 1, 0, 2), (0, 2, 0, 2)

3 Which of the following are bases for the subspace

$$\left\{ \begin{bmatrix} a & a+b \\ a-b & 0 \end{bmatrix} : a, b \in \mathcal{R} \right\} \text{ of } \mathcal{M}^{2 \times 2}?$$

Explain, or show work.

(a) $\begin{bmatrix} 1 & 0 \\ 0 & 0 \end{bmatrix}, \begin{bmatrix} 0 & 1 \\ 0 & 0 \end{bmatrix}, \begin{bmatrix} 0 & 0 \\ 1 & 0 \end{bmatrix}$ (b) $\begin{bmatrix} 1 & 1 \\ 1 & 0 \end{bmatrix}, \begin{bmatrix} 0 & 1 \\ -1 & 0 \end{bmatrix}$

(c) $\begin{bmatrix} 1 & 2 \\ 0 & 0 \end{bmatrix}, \begin{bmatrix} 1 & 0 \\ 2 & 0 \end{bmatrix}$ (d) $\begin{bmatrix} 1 & 1 \\ 1 & 1 \end{bmatrix}, \begin{bmatrix} 1 & 2 \\ -2 & 1 \end{bmatrix}$

4 Which of the following are bases for $\mathscr{P}^2$? Explain, or show work.

(a) $2x^2, -4x, 7$ (b) $x + 3, 3x^2 - x, 3x^2 + 3$
(c) $x^2 + 1, x - 1, x + 1$ (d) $x + \frac{1}{2}, x^2 + x, 3x^2, \frac{1}{4}$

5 (a) Are (1, 2, 3), (4, 5, 6), (9, 12, 15) independent? What is the dimension of $\langle (1, 2, 3), (4, 5, 6), (9, 12, 15) \rangle$? (Hint: If the vectors are dependent, discard one that depends on the others and retest for independence.)
(b) Answer the same two questions for (4, 0, 3, 2), (3, −1, −1, −1), (2, −2, 2, −3), (1, −3, 5, 5).
(c) Answer the same two questions for (1, 1, 0, 1, 2), (2, 0, 3, 0, −4), (0, $\frac{1}{2}$, −$\frac{3}{4}$, −$\frac{1}{2}$, 2).
(d) Why is it impossible for seven vectors in $\mathcal{R}^6$ to be independent?

6 Determine the dimensions of the following vector spaces (by exhibiting specific vectors and proving that they form a basis):

(a) The subspace of $\mathcal{R}^4$ consisting of all 4-tuples that satisfy $2x_1 + 3x_2 - x_3 - 2x_4 = 0$
(b) The subspace of solutions of the system

$$\begin{aligned} 2x + y &= 0 \\ x + 3y &= 0 \end{aligned}$$

(c) The subspace of solutions of the system

$$\begin{aligned} x + 2y + 3z &= 0 \\ 2x - y - z &= 0 \\ -x + 3y + 2z &= 0 \end{aligned}$$

(d) $\{(a, b, b + c, 2b + 3c, 5a - 2b + 3c) : a, b, c \in \mathcal{R}\}$
(e) The subspace of solutions of

$$
\begin{aligned}
x_1 - 3x_2 + \quad 2x_3 \quad\quad - 4x_5 &= 0 \\
-2x_1 \quad\quad + 3x_3 - 4x_4 \quad\quad &= 0
\end{aligned}
$$

(f) $\{(0, 4d, 0, 2d - 3e, e, 2e) : d, e \in \mathcal{R}\}$
(g) All polynomials of the form $ax^4 + bx^2 + c$

7 (a) Prove that $\mathcal{M}^{2 \times 3}$ is six-dimensional.
(b) Prove that $\{(a, 3a + c, 5a - 2c, 0) : a, c \in \mathcal{R}\}$ is two-dimensional.

8 Find a basis for $\mathcal{R}^3$ containing $(1, 0, 0)$, $(2, 1, 1)$.

9 Determine a basis for each of the following:

(a) The plane through the origin, $(1, 1, 1)$, $(1, 0, 0)$
(b) The plane with equation $x + 4y - z = 0$
(c) The subspace that is the intersection of $x - 2y + z = 0$ and $2x + y - 5z = 0$
(d) The subspace that is the intersection of $x + y + z = 0$, $2x - y + 2z = 0$, and $x - 2y + z = 0$
(e) The plane passing through the origin that is parallel to the plane $x - y + 4z = 1$
(f) The line through the origin that is parallel to the line $3x - y = 6$
(g) The line through the origin that is parallel to the intersection of the planes $x - 2y + z = 2$ and $2x + y - 5z = 4$

10 Find the coordinates of the vector $(-1, 2, 0, 4)$ with respect to:

(a) The usual basis for $\mathcal{R}^4$
(b) The basis $(0, 0, 1, 0)$, $(0, 1, 0, 0)$, $(0, 0, 0, 1)$, $(1, 0, 0, 0)$
(c) The basis $(1, 1, 0, 0)$, $(0, 1, -1, 0)$, $(0, 0, 1, 0)$, $(0, 0, 2, -3)$

11 Find the coordinates of the vector $(-1, 0, -3, 0)$ with respect to the basis $(1, 1, -2, 1)$, $(3, 1, 4, 1)$ for the subspace $\{(a + 3b, a + b, -2a + 4b, a + b) : a, b \in \mathcal{R}\}$.

12 (a) Prove that $\mathcal{M}^{2 \times 1} \simeq \mathcal{R}^2$ by defining a suitable function and verifying that it has the four required properties.
(b) Set up an isomorphism between $\mathcal{R}^3$ and the subspace $\mathcal{W} = \{(x, y, x + y, z) : x, y, z \in \mathcal{R}\}$ of $\mathcal{R}^4$ and verify that the function has the four required properties of an isomorphism.
(c) Prove that $\{a + bx^2 : a, b \in \mathcal{R}\}$ (a subspace of $\mathcal{P}^2$) is isomorphic to $\mathcal{R}^2$.

CHAPTER 4

LINEAR TRANSFORMATIONS

It is important to realize that the *theory* of vector spaces (theorems about dimension, linear independence, existence and uniqueness of solutions, etc.) is also of great *practical* importance. There are various reasons for this: The theory (a) provides shortcuts for computation, (b) helps us to know in advance what results to expect, and (c) unifies many areas of natural or social science where the mathematical model turns out to be a vector space.

We are about to consider a topic—linear transformations—that, although theoretical and abstract, also has its practical side. In addition to the three items mentioned in the above paragraph, linear transformations play a central role in a very widely used problem-solving technique—the *transform-solve-invert* method.† The basic idea is to *transform* a given problem into a simpler one, *solve* the simpler problem, and then *invert* the results to apply them to the original problem. The technique pervades mathematics. Here are some examples:

(a) We wish to multiply the roman numerals XXIII and XLV. We *transform* to arabic numerals: XXIII → 23 and XLV → 45. We *solve* by multiplying: $23 \times 45 = 1035$. We *invert* back to roman numerals: 1035 → MXXXV.

---

† This terminology is from M. S. Klamkin and D. J. Newman. For a further treatment of this concept, see M. S. Klamkin and D. J. Newman, "The Philosophy and Applications of Transform Theory," *SIAM Review*, Vol. 3, No. 1, January 1961, pp. 10–36.

(b) We wish to solve a given system of equations. We *transform* the problem by taking the matrix for the system. We *solve* by reducing the matrix. We *invert* by writing the system of equations corresponding to the reduced matrix and reading off the solutions.

(c) We wish to multiply $\sqrt[3]{25}$ and $1/\sqrt{12}$. We *transform* by taking logarithms: $\sqrt[3]{25} \to \frac{1}{3}\log 25$ and $1/\sqrt{12} \to -\frac{1}{2}\log 12$. We *solve* by finding the logarithms and adding: $\frac{1}{3}\log 25 + -(\frac{1}{2})\log 12 = \frac{1}{3}3.2189 - \frac{1}{2} \times 2.4849 = -0.1695$. We *invert* by taking the antilogarithm of -0.1695 to obtain 0.8441.

(d) We wish to show that the function f defined by the rule $f(x) = x^3 + x$ is an increasing function. We *transform* the function to its derivative: $f(x) \to f'(x) = 3x^2 + 1$. We *solve* by noting by inspection that $f'(x) > 0$ for all real numbers x. We *invert* from the derivative back to the original function by recalling the theorem that states: If $f'(x) > 0$ for all x, then $f(x)$ is an increasing function.

(e) We wish to find the area of an ellipse with semiaxes of lengths a and b. We *transform* the ellipse to a circle of radius 1 by means of a function F. (Specific details on how to do this will be given later in

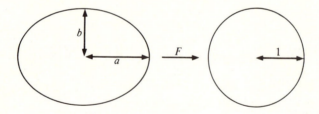

this chapter.) We *solve* the problem by noting that the area of this circle is $\pi r^2 = \pi$. We *invert* by using the fact that the function F changes all areas by a factor of $1/ab$ (details later), and hence the area of the ellipse is πab.

20. LINEAR TRANSFORMATIONS

As indicated above, a transformation is often the mathematical device for simplifying a problem and for reducing the study of a complicated object to the study of the transformed (hopefully simpler) object. Each area of mathematics has its own transformations of special interest; if the mathematical model dealt with happens to be a vector space, then the transformations used will most likely be *linear* transformations.

Any transformation is actually a function, and linear transformations are functions that have the geometric effect of carrying lines to lines (hence the name *linear*). That is, if £ is a line, the image of £ (the set of function values that come from £) will lie on a straight line.†

A precise definition of linear transformation will be given below. First a slightly different notation for values of functions will be introduced: If T is a function, a typical value of the function will be written as xT or $(x)T$ instead of the more familiar $T(x)$. Reasons for this notation will emerge as we go along.

Definition

A **linear transformation** is a function $T : \mathcal{V} \to \mathcal{W}$, where $\mathcal{V}$ and $\mathcal{W}$ are vector spaces, such that

(a)	$(\mathbf{X} + \mathbf{Y})T = \mathbf{X}T + \mathbf{Y}T$	T preserves addition
(b)	$(r\mathbf{X})T = r(\mathbf{X}T)$	T preserves scalar multiplication

for all vectors $\mathbf{X}$ and $\mathbf{Y}$ and for all scalars r.

The reader may recognize the above conditions as the last two properties of an isomorphism (Sec. 18).

We now note some examples of linear transformations, including an efficient method for verifying that they satisfy the two properties of the definition.

Example 20-1

Let $T : \mathcal{R}^2 \to \mathcal{R}^2$ have the rule $(a, b)T = (3b, 0)$. To show that T is a linear transformation we must show that T preserves addition and that T preserves scalar multiplication.

T preserves addition   We will use the format of setting up the desired equation $(\mathbf{X} + \mathbf{Y})T \overset{?}{=} \mathbf{X}T + \mathbf{Y}T$ and working on both sides, using appropriate notation, definitions, and known laws until equality is demonstrated.

---

† The image of £ may degenerate to a single point, but in that case the image still lies on a line.

REASONS

$$(\mathbf{X} + \mathbf{Y})T \overset{?}{=} \mathbf{X}T + \mathbf{Y}T$$

		Desired property
$((a, b) + (c, d))T$	$(a, b)T + (c, d)T$	Rewriting, using appropriate notation for $\mathcal{R}^2$
$(a + c, b + d)T$		Definition of addition in $\mathcal{R}^2$
$(3(b + d), 0)$	$(3b, 0) + (3d, 0)$	Definition of T
	$(3b + 3d, 0 + 0)$	Definition of addition in $\mathcal{R}^2$
	$(3b + 3d, 0)$	Property of real number 0
$(3b + 3d, 0)$		Distributive property for real numbers

The two sides are now clearly equal.

T preserves scalar multiplication

REASONS

$$(r\mathbf{X})T \overset{?}{=} r(\mathbf{X}T)$$

		Desired property
$(r(a, b))T$	$r((a, b)T)$	Rewriting, using appropriate notation for $\mathcal{R}^2$
$(ra, rb)T$		Definition of scalar multiplication for $\mathcal{R}^2$
$(3(rb), 0)$	$r(3b, 0)$	Definition of T
	$(r(3b), 0)$	Definition of scalar multiplication for $\mathcal{R}^2$
	$(3(rb), 0)$	Associative and commutative laws for multiplication of real numbers

Thus T is a linear transformation.   ●

Example 20-2

Let $T : \mathcal{R}^2 \to \mathcal{R}^2$ have the rule: $\mathbf{X}T$ is the vector obtained from $\mathbf{X}$ by perpendicular projection onto the y axis. Figure 20-1 shows an arbitrary vector $\mathbf{X}$ and its projection $\mathbf{X}T$. To obtain $\mathbf{X}T$ we drop a perpendicular from $\mathbf{X}$ to the y axis, and thus obtain the numerical rule $(a, b)T = (0, b)$. Both the geometric and the numerical rules for T have their uses—in fact, the interplay between geometry and algebra is one of the central features of linear algebra. We will use the numerical rule $(a, b)T = (0, b)$ to show

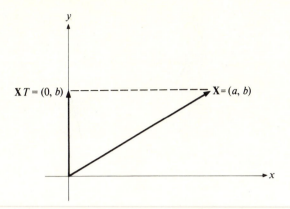

Fig. 20-1 Perpendicular projection onto the *y* axis.

that T is a linear transformation. To do this we must verify the two equations of the definition, as in Example 20-1.

T preserves addition   We must show that $(\mathbf{X} + \mathbf{Y})T = \mathbf{X}T + \mathbf{Y}T$. We will work on each side separately and show that the two are equal. The reader should supply the reasons.

$$
\begin{aligned}
(\mathbf{X} + \mathbf{Y})T &= ((a, b) + (c, d))T \\
&= (a + c, b + d)T \\
&= (0, b + d) \\
\mathbf{X}T + \mathbf{Y}T &= (a, b)T + (c, d)T \\
&= (0, b) + (0, d) \\
&= (0 + 0, b + d) \\
&= (0, b + d)
\end{aligned}
$$

The proof that T preserves scalar multiplication is left to the reader.

●

Projections, as in Example 20-2, are important in geometry, algebra, and physics. In this text we will consider *perpendicular* projections: in $\mathcal{R}^2$, we consider perpendicular projections onto lines through the origin; in $\mathcal{R}^3$, perpendicular projections onto planes through the origin. In both cases, if T is the projection, $\mathbf{X}T$ is found by dropping a perpendicular from $\mathbf{X}$ to the line or plane specified. See Fig. 20-2*a*, which depicts the perpendicular projection of a vector $\mathbf{X}$ onto a plane $\mathcal{P}$ in $\mathcal{R}^3$, and Fig. 20-2*b*, which depicts the perpendicular projection of a vector $\mathbf{X}$ onto a line $\mathcal{L}$ in $\mathcal{R}^2$.

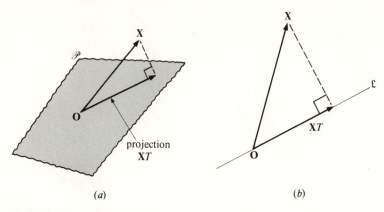

Fig. 20-2 Perpendicular projections: (*a*) onto a plane $\mathscr{P}$, (*b*) onto a line $\mathscr{L}$.

Example 20-3

Let $U : \mathscr{R}^2 \rightarrow \mathscr{R}^2$ be defined by the rule: $\mathbf{X}U$ is the reflection of $\mathbf{X}$ in the y axis (Fig. 20-3). When we speak of reflection in the y axis we treat the y axis as a mirror, and thus the image of a vector becomes its mirror image. The image $\mathbf{X}U$ of any vector $\mathbf{X}$ can be found by dropping a perpendicular from $\mathbf{X}$ to the y axis and marking off an equal distance beyond the y axis. For performing computations, and for verifying that U is a linear transformation, it helps to have the numerical rule for U. Clearly $(a, b)U = (-a, b)$, since the y entry remains the same (b in this case) and the x entry is reversed in sign (a becomes $-a$). The verification that U is a linear transformation is left as an exercise.   ●

Reflections are another important class of geometric transformations. For our purposes, any reflection in $\mathscr{R}^2$ has a line as its "mirror"; in $\mathscr{R}^3$, a plane as its "mirror." If T is a reflection, $\mathbf{X}T$ is found by dropping a

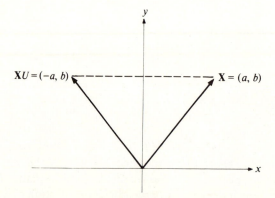

Fig. 20-3 Reflection in the y axis.

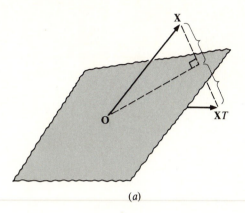

(a)

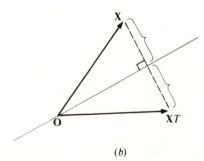

(b)

Fig. 20-4   Reflections: (a) in a plane, (b) in a line.

perpendicular from **X** through the mirror, and marking off an equal
distance beyond the mirror. See Fig. 20-4a, which shows the reflection of
X in a plane, and Fig. 20-4b, which shows the reflection of **X** in a line. The
"mirror image" **X**T obtained by applying a reflection lies on the opposite
side of the mirror, as far from the mirror as the original vector **X**.

Example 20-4

Let $T : \mathcal{R}^2 \rightarrow \mathcal{R}^2$ be the counterclockwise rotation about the origin by θ
radians. At this stage it is not very easy to state a numerical rule for T;
however, it can be shown geometrically that T is a linear transformation.
We will show that T preserves scalar multiplication.

The vectors **X** and r**X** lie on the same straight line. Rotating by θ
radians leaves the images **X**T and $(r$**X**$)T$ collinear (Fig. 20-5). (The figure
and the following argument assume $r > 0$. A slight change would cover
the case $r \leq 0$.) T also leaves the lengths of the images unchanged, and so
$(r$**X**$)T$ is still r times as long as **X**T and collinear with **X**T. Therefore,
$(r$**X**$)T = r($**X**$T)$.

The other law for linear transformations, $($**X** $+$ **Y**$)T =$ **X**$T +$ **Y**T, can
also be shown geometrically.   ●

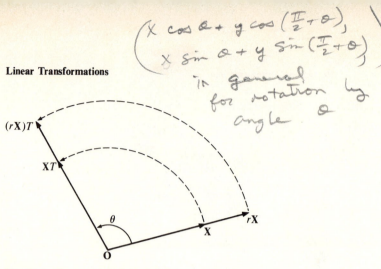

$$\left(\begin{array}{c} x\cos\alpha + y\cos\left(\frac{\pi}{2}+\alpha\right), \\ x\sin\alpha + y\sin\left(\frac{\pi}{2}+\alpha\right) \end{array} \right)$$

in general
for rotation by
angle . α

Fig. 20-5   Counterclockwise rotation by θ radians.

Example 20-5

Let $S : \mathcal{R}^3 \to \mathcal{R}^3$ have the rule $(x, y, z)S = (-y, x, z)$. Again, S can be shown to be a linear transformation by verifying that $(X + Y)S = XS + YS$ and $(rX)S = r(XS)$. What is not so easily seen is that S has a simple geometric effect. To analyze this, we examine what S does to the usual basis:

$$(1, 0, 0)S = (0, 1, 0)$$
$$(0, 1, 0)S = (-1, 0, 0)$$
$$(0, 0, 1)S = (0, 0, 1)$$

Note that $(1, 0, 0)$ is rotated $\pi/2$ radians onto the y axis; $(0, 1, 0)$ is rotated $\pi/2$ radians onto the negative x axis; and $(0, 0, 1)$ is left unchanged. Thus it appears that S is a rotation by $\pi/2$ radians about the z axis. (See Fig. 20-6.)   •

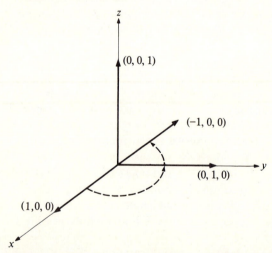

Fig. 20-6   The effect on the usual basis of a rotation by $\pi/2$ radians.

Example 20-6

Any isomorphism $T : \mathcal{V} \to \mathcal{W}$ is necessarily a linear transformation, since the definition of isomorphism includes the two properties of a linear transformation. (See Sec. 18.) In fact, an isomorphism $T : \mathcal{V} \to \mathcal{W}$ is simply a linear transformation that is also 1-1 and onto $\mathcal{W}$.   ●

Example 20-7

When we have a linear transformation $T : \mathcal{V} \to \mathcal{W}$, the two vector spaces need not be the same. For example, the function $T : \mathcal{R}^3 \to \mathcal{R}^2$ defined by the rule $(x, y, z)T = (x - z, y)$ is a linear transformation. The reader can verify this using the method of Example 20-1.   ●

Part of the process of analyzing and understanding a linear transformation consists of determining its effect on various sets in the domain. What shapes are changed? How drastically? What shapes are left unaltered? We noted earlier that lines are carried to lines (or single points). Also, planes are carried to planes (or subsets of planes). The next examples illustrate these statements.

Example 20-8

Suppose that T is the linear transformation with the rule $(x, y, z)T = (x + y, x + z, 0)$. We wish to find the image of the line $S = \langle (1, 0, 0) \rangle$. Since every element of S is a multiple $a(1, 0, 0)$, we compute $(a(1, 0, 0))T$. But

$$[a(1, 0, 0)]T = a[(1, 0, 0)T] \qquad \text{Since } T \text{ is a linear transformation}$$
$$= a(1, 1, 0) \qquad \text{Using the rule for } T$$

Hence the image of S consists of all multiples $a(1, 1, 0)$. That is, the image is the line through the origin and the point $(1, 1, 0)$.   ●

Example 20-9

Let T be as in the previous example. We wish to find the image of the plane $\mathcal{U} = \langle (1, 2, 0), (0, 1, 2) \rangle$. Since $\mathcal{U}$ consists of all linear combinations $a(1, 2, 0) + b(0, 1, 2)$, the image of $\mathcal{U}$ will consist of all of the values $(a(1, 2, 0) + b(0, 1, 2))T$. But

$$[a(1, 2, 0) + b(0, 1, 2)]T = a[(1, 2, 0)T] + b[(0, 1, 2)T]$$
$$= a(3, 1, 0) + b(1, 2, 0)$$

Thus the image of $\mathcal{U}$ is the plane spanned by $(3, 1, 0)$ and $(1, 2, 0)$, that is, $\langle(3, 1, 0), (1, 2, 0)\rangle$.   ●

Example 20-10

Let T be the linear transformation of Example 20-8, and consider the line $\mathcal{W} = \langle(1, -1, -1)\rangle$. Since $(1, -1, -1)T = (0, 0, 0)$, every point of $\mathcal{W}$ is sent to $(0, 0, 0)$ [since every point of $\mathcal{W}$ is a multiple of $(1, -1, -1)$]. Thus $\mathcal{W}T$ is the single point $(0, 0, 0)$. Although the image of a line always lies *on* a line, in certain cases (such as this) it can degenerate to a single point. Similarly, the image of a plane can degenerate to a single point or a line, but will always be contained in a plane.   ●

Examples 20-8 to 20-10 have a deeper significance than is at first apparent. In each example we were able to compute the effect of the linear transformation T on a subspace simply by computing the effect of T on a *basis* for the subspace.† And by examining what T did to a basis, we were able to know that in those examples T carried lines to lines and planes to planes. We now generalize these remarks in the following two theorems.

Theorem 20-1

Let $T : \mathcal{V} \to \mathcal{W}$ be a linear transformation.

1 The image of the subspace $\langle \mathbf{X}_1, \mathbf{X}_2, ..., \mathbf{X}_k \rangle$ of $\mathcal{V}$ is the subspace $\langle \mathbf{X}_1 T, \mathbf{X}_2 T, ..., \mathbf{X}_k T \rangle$ of $\mathcal{W}$. Therefore, the image of a subspace of $\mathcal{V}$ of dimension k is a subspace of $\mathcal{W}$ of dimension $\leq k$.

2 For a specific vector $\mathbf{X}_0$, the image of the subset $\mathbf{X}_0 + \langle \mathbf{X}_1, \mathbf{X}_2, ..., \mathbf{X}_k \rangle$ is the subset $\mathbf{X}_0 T + \langle \mathbf{X}_1 T, \mathbf{X}_2 T, ..., \mathbf{X}_k T \rangle$. In particular, lines and planes are carried to lines and planes (or subsets of them).

Proof

To show part (2) note that the subset $\mathbf{X}_0 + \langle \mathbf{X}_1, \mathbf{X}_2, ..., \mathbf{X}_k \rangle$ consists of all vectors of the form

$$\mathbf{X}_0 + a_1\mathbf{X}_1 + a_2\mathbf{X}_2 + \cdots + a_k\mathbf{X}_k$$

† This is far different from the behavior of functions that are *not* linear transformations; it may be exceedingly difficult to find the image of a set under a function that is not linear.

Its image, therefore, consists of all vectors of the form

$$(\mathbf{X}_0 + a_1\mathbf{X}_1 + a_2\mathbf{X}_2 + \cdots + a_k\mathbf{X}_k)T$$

and since T satisfies the two properties of a linear transformation, the latter equals

$$\mathbf{X}_0\,T + a_1(\mathbf{X}_1\,T) + a_2(\mathbf{X}_2\,T) + \cdots + a_k(\mathbf{X}_k\,T)$$

This vector is an arbitrary vector in the subset $\mathbf{X}_0\,T + \langle \mathbf{X}_1\,T, \mathbf{X}_2\,T, \ldots, \mathbf{X}_k\,T\rangle$, as desired.

Conversely, every vector in the subset $\mathbf{X}_0\,T + \langle \mathbf{X}_1T, \mathbf{X}_2\,T, \ldots, \mathbf{X}_k\,T\rangle$ is the image of a vector in the set $\mathbf{X}_0 + \langle \mathbf{X}_1, \mathbf{X}_2, \ldots, \mathbf{X}_k\rangle$. The student can verify this, which proves part (2).

When $\mathbf{X}_0 = \mathbf{0}$, we have part (1), since $\mathbf{0}T = \mathbf{0}$ (Exercise 20-11).

The fact that $\langle \mathbf{X}_1, \mathbf{X}_2, \ldots, \mathbf{X}_k\rangle$ and $\langle \mathbf{X}_1\,T, \mathbf{X}_2\,T, \ldots, \mathbf{X}_k\,T\rangle$ are subspaces was proved in Sec. 12. The dimension of $\langle \mathbf{X}_1\,T, \mathbf{X}_2\,T, \ldots, \mathbf{X}_k\,T\rangle$ could be less than k since the vectors $\mathbf{X}_1, \mathbf{X}_2, \ldots, \mathbf{X}_k$ could be dependent. A set

$$\mathbf{X}_0 + \langle \mathbf{X}_1\rangle = \{(c, d, e) + a(f, g, h) : a \in \mathcal{R}\}$$

is a line (Sec. 4). Its image

$$(c, d, e)T + a((f, g, h)T)$$

is also a line unless $(f, g, h)T = (0, 0, 0)$, in which case the image degenerates to a single point. A set $\mathbf{X}_0 + \langle \mathbf{X}_1, \mathbf{X}_2\rangle$ is a plane if $\mathbf{X}_1$ and $\mathbf{X}_2$ are independent (Sec. 4), and its image $\mathbf{X}_0\,T + \langle \mathbf{X}_1\,T, \mathbf{X}_2\,T\rangle$ is likewise a plane unless $\mathbf{X}_1\,T$ and $\mathbf{X}_2\,T$ are dependent. ∎

Theorem 20-2

The rule for a linear transformation $T : \mathcal{V} \to \mathcal{W}$ is completely determined by its effect on a basis for $\mathcal{V}$.

Proof

Suppose that $\mathbf{X}_1, \mathbf{X}_2, \ldots, \mathbf{X}_n$ is a basis for $\mathcal{V}$ and that $\mathbf{X}_1\,T, \mathbf{X}_2\,T, \ldots, \mathbf{X}_n\,T$ are known. (That is, the effect of T on the basis is known.) We wish to show that for an arbitrary vector $\mathbf{X}$ in $\mathcal{V}$, $\mathbf{X}T$ is then determined. Since the vectors $\mathbf{X}_1, \mathbf{X}_2, \ldots, \mathbf{X}_n$ form a basis for $\mathcal{V}$, we can write $\mathbf{X}$ in exactly one way as

$$\mathbf{X} = a_1\mathbf{X}_1 + a_2\mathbf{X}_2 + \cdots + a_n\mathbf{X}_n$$

Thus

$$\mathbf{X}T = (a_1\mathbf{X}_1 + a_2\mathbf{X}_2 + \cdots + a_n\mathbf{X}_n)T$$

But

$$(a_1 X_1 + a_2 X_2 + \cdots + a_n X_n)T = (a_1 X_1)T + (a_2 X_2)T + \cdots + (a_n X_n)T$$
$$= a_1(X_1 T) + a_2(X_2 T) + \cdots + a_n(X_n T)$$

by the two defining properties of a linear transformation. Since $X_1 T$, $X_2 T, \ldots, X_n T$ are assumed known, then XT, which equals the linear combination $a_1(X_1 T) + a_2(X_2 T) + \cdots + a_2(X_n T)$, is uniquely determined. ∎

Example 20-11

Suppose we know that $(1, 0)T = (1, 2)$ and $(0, 1)T = (-1, 0)$. Then for an arbitrary vector (a, b) we must have

$$(a, b)T = (a(1, 0) + b(0, 1))T$$
$$= a((1, 0)T) + b((0, 1)T)$$
$$= a(1, 2) + b(-1, 0)$$
$$= (a - b, 2a)$$

Thus the rule for T is $(a, b)T = (a - b, 2a)$, and is uniquely determined by its effect on the basis vectors $(1, 0)$, $(0, 1)$. ●

Exercises

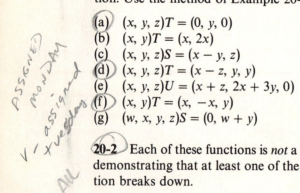

20-1 Prove that each of the following functions is a linear transformation. Use the method of Example 20-1.

(a) $(x, y, z)T = (0, y, 0)$
(b) $(x, y)T = (x, 2x)$
(c) $(x, y, z)S = (x - y, z)$
(d) $(x, y, z)T = (x - z, y, y)$
(e) $(x, y, z)U = (x + z, 2x + 3y, 0)$
(f) $(x, y)T = (x, -x, y)$
(g) $(w, x, y, z)S = (0, w + y)$

20-2 Each of these functions is *not* a linear transformation. Show this by demonstrating that at least one of the conditions for a linear transformation breaks down.

(a) $(x, y)T = (x^2, y)$
(b) $(x, y)S = (x + 1, y)$
(c) $(x, y, z)U = (xy, x, 0)$
(d) $(x, y, z)T = (2, y, z)$

20-3 Suppose that T is the perpendicular projection of $\mathcal{R}^2$ onto the x axis. Find a numerical rule for T and then show (as in Example 20-1) that T is a linear transformation. [Hint: First draw an arbitrary vector (a, b) and drop a perpendicular to the x axis.]

20-4 Let $T : \mathcal{R}^2 \to \mathcal{R}^2$ be the reflection in the x axis. Find the numerical rule for T and verify that T is a linear transformation.

20-5 Let $S : \mathcal{R}^2 \to \mathcal{R}^2$ be a 90° clockwise rotation about the origin. Find the numerical rule for S and verify that S is a linear transformation.

20-6 Suppose that $T : \mathcal{R}^2 \to \mathcal{R}^2$ is a linear transformation such that $(1, 0)T = (5, 0)$ and $(0, 1)T = (0, -3)$. Find a numerical rule for T. (Hint: Use the method of Example 20-11.)

20-7 Suppose that $T : \mathcal{R}^2 \to \mathcal{R}^2$ is a linear transformation where $(1, 0)T = (1, 1)$ and $(0, 1)T = (-2, 1)$. Find a numerical rule for T.

20-8 Suppose that $T : \mathcal{R}^2 \to \mathcal{R}^3$ is a linear transformation where $(1, 0)T = (1, 0, 1)$ and $(0, 1)T = (0, 2, 2)$. Find a numerical rule for T.

20-9 Let $\mathcal{S} = \langle (1, 0, 1) \rangle$. Determine its image $(\mathcal{S})T$ using each of the linear transformations in parts (a), (c), (d), and (e) of Exercise 20-1.

20-10 Let $\mathcal{X} = \langle (1, 0, 0), (3, -2, 3) \rangle$. Determine the image of $\mathcal{X}$ for each of the linear transformations in parts (a), (c), (d), and (e) of Exercise 20-1. Also, find the dimension of each of the images.

20-11 Suppose that $T : \mathcal{V} \to \mathcal{W}$ is a linear transformation. Prove that $(\mathbf{0})T = \mathbf{0}$.

20-12 Let $T : \mathcal{V} \to \mathcal{W}$ be a linear transformation. Let $\mathcal{K} = \{\mathbf{X} : \mathbf{X}T = \mathbf{0}\}$; that is, $\mathcal{K}$ is the set of all vectors in $\mathcal{V}$ that T sends to the zero vector. $\mathcal{K}$ is called the *kernel (or null space)* of T. Prove that $\mathcal{K}$ is a subspace of $\mathcal{V}$.

20-13 Find the kernel of each of the following linear transformations:

(a) $T : \mathcal{R}^2 \to \mathcal{R}^2$, where $(x, y)T = (0, y)$
(b) $T : \mathcal{R}^3 \to \mathcal{R}^2$, where $(x, y, z)T = (x - y, z)$
(c) $T : \mathcal{R}^3 \to \mathcal{R}^3$, where $(x, y, z)T = (x + z, z, y)$

Hint: Finding the kernel amounts to solving the system of linear equations $\mathbf{X}T = \mathbf{0}$. For instance, in part (b) the system is

$$\begin{aligned} x - y \quad &= 0 \\ z &= 0 \end{aligned}$$

20-14 Let $T : \mathcal{V} \to \mathcal{W}$ be a linear transformation. Prove the following:

(a) If T is 1-1, then $\mathcal{K} = \{\mathbf{0}\}$.

(b) If $\mathcal{K} = \{\mathbf{0}\}$, then T is 1-1. [Hint: Suppose that T is not 1-1. Then there are vectors $\mathbf{X} \neq \mathbf{Y}$ such that $\mathbf{X}T = \mathbf{Y}T$. Consider $(\mathbf{X} - \mathbf{Y})T$.]

20-15 Let $T : \mathcal{R}^3 \to \mathcal{R}^3$ be a linear transformation. Prove that if T is 1-1, the images of lines and planes *are* lines and planes, respectively; that is, they do not degenerate.

21. THE MATRIX OF A LINEAR TRANSFORMATION

We have used matrices in a variety of problems (solving systems of equations, testing vectors for independence and spanning, etc.). Matrices generally lead to quite efficient computational methods. The present situation is no exception. We will see how to obtain a matrix for any given linear transformation $T : \mathcal{R}^m \to \mathcal{R}^n$, and then show how to use the matrix effectively.

An example will show the approach to be used; as indicated in Theorem 20-2, the key ingredient is the effect of T on a *basis*.

Suppose that $T : \mathcal{R}^2 \to \mathcal{R}^2$ is the linear transformation with the rule $(x, y)T = (2x + y, 3y)$. Using the usual basis for $\mathcal{R}^2$, we have

$$(1, 0)T = (2, 0)$$
$$(0, 1)T = (1, 3)$$

The *usual matrix* for T is defined to be the 2×2 matrix

$$\begin{bmatrix} 2 & 0 \\ 1 & 3 \end{bmatrix}$$

whose first row contains the entries of the image of $(1, 0)$ and whose second row contains the entries of the image of $(0, 1)$.

In general, if $T : \mathcal{R}^m \to \mathcal{R}^n$ is a linear transformation, we can form a matrix for T whose rows contain the entries of $(1, 0, \ldots, 0)T$, $(0, 1, \ldots, 0)T$, etc. Thus we have the following definition.

Definition

If $T : \mathcal{R}^m \to \mathcal{R}^n$ is a linear transformation, the **usual matrix** for T is the $m \times n$ matrix

$$\begin{bmatrix} a_{11} & a_{12} & \cdots & a_{1n} \\ a_{21} & a_{22} & \cdots & a_{2n} \\ \cdots\cdots\cdots\cdots\cdots\cdots \\ a_{m1} & a_{m2} & \cdots & a_{mn} \end{bmatrix}$$

where

$$(1, 0, \ldots, 0)T = (a_{11}, a_{12}, \ldots, a_{1n})$$
$$(0, 1, \ldots, 0)T = (a_{21}, a_{22}, \ldots, a_{2n})$$
$$\vdots$$
$$(0, 0, \ldots, 1)T = (a_{m1}, a_{m2}, \ldots, a_{mn})$$

Example 21-1

Let $T : \mathcal{R}^3 \to \mathcal{R}^3$ have the rule $(x, y, z)T = (x + y - 2z, 3y, 5x - z)$. To find the usual matrix for T, we compute the effect of T on the usual basis

$$(1, 0, 0)T = (1, 0, 5)$$
$$(0, 1, 0)T = (1, 3, 0)$$
$$(0, 0, 1)T = (-2, 0, -1)$$

Using these entries to form the rows, we obtain the usual matrix

$$A = \begin{bmatrix} 1 & 0 & 5 \\ 1 & 3 & 0 \\ -2 & 0 & -1 \end{bmatrix}$$

●

Example 21-2

Let $U : \mathcal{R}^3 \to \mathcal{R}^2$ have the rule $(x, y, z)U = (2x - z, 7y + z)$. The effect of U on the usual basis for $\mathcal{R}^3$ is

$$(1, 0, 0)U = (2, 0)$$
$$(0, 1, 0)U = (0, 7)$$
$$(0, 0, 1)U = (-1, 1)$$

and hence the usual matrix for U is the 3×2 matrix

$$A = \begin{bmatrix} 2 & 0 \\ 0 & 7 \\ -1 & 1 \end{bmatrix}$$

●

Example 21-3

Let $T : \mathcal{R}^2 \to \mathcal{R}^2$ be a counterclockwise rotation about the origin by $\pi/4$ radians. In order to obtain the usual matrix for T, we must first find $(1, 0)T$ and $(0, 1)T$. To compute $(1, 0)T$ and $(0, 1)T$, some diagrams are helpful.

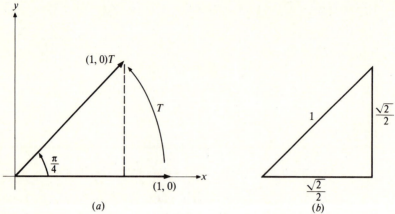

Fig. 21-1

Figure 21-1a shows the effect of T on the vector $(1, 0)$. Using the triangle in Fig. 21-1b, we obtain

$$(1, 0)T = \left(\frac{\sqrt{2}}{2}, \frac{\sqrt{2}}{2}\right)$$

Similarly, Fig. 21-2a shows the effect of T on the vector $(0, 1)$, and from the triangle in Fig. 21-2b we obtain

$$(0, 1)T = \left(-\frac{\sqrt{2}}{2}, \frac{\sqrt{2}}{2}\right)$$

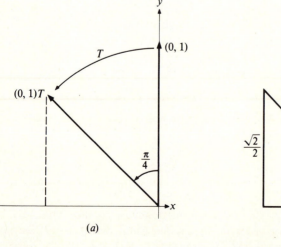

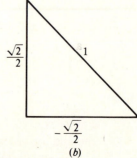

Fig. 21-2

Thus the usual matrix for T is the matrix

$$A = \begin{bmatrix} \dfrac{\sqrt{2}}{2} & \dfrac{\sqrt{2}}{2} \\[2ex] -\dfrac{\sqrt{2}}{2} & \dfrac{\sqrt{2}}{2} \end{bmatrix}$$

●

Example 21-4

Let $I : \mathcal{R}^n \to \mathcal{R}^n$ be the linear transformation with the rule $\mathbf{X}I = \mathbf{X}$ for every vector $\mathbf{X}$. I is called the *identity linear transformation*. Since $(1, 0, \ldots, 0)I = (1, 0, \ldots, 0)$, $(0, 1, \ldots, 0)I = (0, 1, \ldots, 0)$, etc., the usual matrix for I is

$$\begin{bmatrix} 1 & 0 & \cdots & 0 \\ 0 & 1 & \cdots & 0 \\ \cdots\cdots\cdots\cdots\cdots \\ 0 & 0 & \cdots & 1 \end{bmatrix}$$

This matrix is called the *identity matrix* and is also denoted I.   ●

Example 21-5

Let $Z : \mathcal{R}^n \to \mathcal{R}^n$ be the linear transformation with the rule $\mathbf{X}Z = \mathbf{0}$ for all vectors $\mathbf{X}$ in $\mathcal{R}^n$. Z is called the *zero linear transformation*. Since the image of each usual basis vector is $\mathbf{0}$, the usual matrix for Z is

$$\begin{bmatrix} 0 & 0 & \cdots & 0 \\ 0 & 0 & \cdots & 0 \\ \cdots\cdots\cdots\cdots\cdots \\ 0 & 0 & \cdots & 0 \end{bmatrix}$$

the *zero matrix*.   ●

One advantage of using the usual matrices for linear transformations is that we can find the rules for various combinations of linear transformations (sums, scalar multiples, products) simply by carrying out the appropriate operation on their matrices. For instance, if S and T are linear transformations from $\mathcal{R}^m$ to $\mathcal{R}^n$, their *sum* $S + T$ is defined to be the function whose rule is given by

$$\mathbf{X}(S + T) = \mathbf{X}S + \mathbf{X}T$$

It turns out that $S + T$ is a linear transformation and the usual matrix for

$S + T$ is $A + B$, where A and B are the usual matrices for S and T, respectively. In like fashion, defining the scalar multiple rS by the rule

$$\mathbf{X}(rS) = r(\mathbf{X}S)$$

yields a linear transformation whose usual matrix is rA.†

Later we will define a product of linear transformations and a corresponding product of matrices that will make the matrix product AB the usual matrix for the product transformation ST.

We now will see how the usual matrix can be used to compute values $\mathbf{X}T$ of a linear transformation. Suppose that A is the usual matrix for a linear transformation $T : \mathcal{R}^m \to \mathcal{R}^n$. Computation of values $\mathbf{X}T$ is accomplished by considering the vector $\mathbf{X} = (x_1, x_2, \ldots, x_m)$ to be a $1 \times m$ matrix and defining matrix multiplication in such a way that the matrix product $(x_1, x_2, \ldots, x_m)A$ gives the value $(x_1, x_2, \ldots, x_m)T$ of the transformation.

Before defining the product $\mathbf{X}A$ we first define the dot product of vectors in $\mathcal{R}^m$.

Definition

Let $\mathbf{X} = (x_1, x_2, \ldots, x_m)$ and $\mathbf{Y} = (y_1, y_2, \ldots, y_m)$ be vectors in $\mathcal{R}^m$. The **dot product** of $\mathbf{X}$ and $\mathbf{Y}$, denoted $\mathbf{X} \cdot \mathbf{Y}$, is defined by the rule

$$\mathbf{X} \cdot \mathbf{Y} = x_1 y_1 + x_2 y_2 + \cdots + x_m y_m$$

For example,

$$(2, 1, -3) \cdot (6, 0, 5) = 2(6) + 1(0) + (-3)(5) = -3$$

and

$$(1, 7, 4, -6) \cdot (0, \tfrac{2}{3}, -2, -1) = 1(0) + 7(\tfrac{2}{3}) + 4(-2) + (-6)(-1) = \tfrac{8}{3}$$

Note that the dot product of two vectors is a *number*.

Now suppose that $\mathbf{X} = (x_1, x_2, \ldots, x_m) \in \mathcal{R}^m$ and

$$A = \begin{bmatrix} a_{11} & a_{12} & \cdots & a_{1n} \\ a_{21} & a_{22} & \cdots & a_{2n} \\ \cdots\cdots\cdots\cdots\cdots\cdots \\ a_{m1} & a_{m2} & \cdots & a_{mn} \end{bmatrix}$$

† In fact, the correspondence between linear transformations and matrices is an *isomorphism* (Sec. 18), and it has the typical advantage of isomorphisms: any information we have about linear transformations can be applied to their matrices, and vice versa.

To define $\mathbf{X}A$ it is advantageous to regard each *column* of A as a vector in $\mathcal{R}^m$.

Definition

The product $\mathbf{X}A$ is the vector in $\mathcal{R}^n$ whose entries are the dot products of $\mathbf{X}$ with the columns of A, taken in order. Thus the first entry of $\mathbf{X}A$ is $x_1 a_{11} + x_2 a_{21} + \cdots + x_m a_{m1}$,   the   second   entry   of $\mathbf{X}A$ is $x_1 a_{12} + x_2 a_{22} + \cdots + x_m a_{m2}$, etc.

Example 21-6

To compute

$$(2,\ 3)\begin{bmatrix} 4 & 7 \\ -1 & 0 \end{bmatrix}$$

we first form the dot product of $(2, 3)$ with the first column of the matrix, obtaining

$$2(4) + 3(-1) = 5$$

Next we form the dot product of $(2, 3)$ with the second column of the matrix, obtaining

$$2(7) + 3(0) = 14$$

The product has the two entries just computed, and hence

$$(2,\ 3)\begin{bmatrix} 4 & 7 \\ -1 & 0 \end{bmatrix} = (5,\ 14) \qquad \bullet$$

Example 21-7

To compute

$$(3,\ -1)\begin{bmatrix} 15 & 4 & 0 \\ -1 & 7 & 9 \end{bmatrix}$$

we form the dot product of $(3, -1)$ with each of the three columns of the matrix, obtaining

$$3(15) + (-1)(-1) = 46$$
$$3(4) + (-1)(7) = 5$$
$$3(0) + (-1)(9) = -9$$

The product is $(46, 5, -9)$.   ●

Example 21-8

The product

$$(2, -1, 4)\begin{bmatrix} 1 & -2 & 3 \\ 4 & 7 & -8 \end{bmatrix}$$

is not defined since the vector $(2, -1, 4)$ does not "fit" the columns of the matrix.   ●

We now prove that the product $\mathbf{X}A$ does equal $\mathbf{X}T$, as claimed.

Theorem 21-1   Let $T : \mathcal{R}^m \to \mathcal{R}^n$ be a linear transformation and let A be the usual matrix for T. Then $\mathbf{X}T = \mathbf{X}A$ for all vectors $\mathbf{X}$ in $\mathcal{R}^m$.

Proof   We will verify only the 2×2 case, that is, where $T : \mathcal{R}^2 \to \mathcal{R}^2$. (The proof for any positive integers m and n is similar.)
Let

$$A = \begin{bmatrix} a & b \\ c & d \end{bmatrix}$$

be the usual matrix for T. From the rows of A we know what T does to the usual basis:

$$(1, 0)T = (a, b)$$
$$(0, 1)T = (c, d)$$

Since we wish to prove that

$$(x, y)T = (x, y)A$$

we will compute $(x, y)T$ and $(x, y)A$ separately, and then see that the results are equal. On the one hand,

$$
\begin{aligned}
(x, y)T &= [x(1, 0) + y(0, 1)]T \\
&= x((1, 0)T) + y((0, 1)T) \\
&= x(a, b) + y(c, d) \\
&= (xa + yc, xb + yd)
\end{aligned}
$$

On the other hand,

$$
(x, y)A = (x, y)\begin{bmatrix} a & b \\ c & d \end{bmatrix}
$$

$$
= (xa + yc, xb + yd)
$$

Therefore, $(x, y)T = (x, y)A$, as desired.   ∎

We now summarize the results of the preceding few pages. If we are given a linear transformation $T : \mathcal{R}^m \to \mathcal{R}^n$, then T has an $m \times n$ matrix A associated with it, called the usual matrix for T. The rows of A are found by determining what T does to the usual basis for $\mathcal{R}^m$. For any vector $\mathbf{X}$ the value $\mathbf{X}T$ can be calculated by finding the matrix product $\mathbf{X}A$. The following example illustrates this process.

Example 21-9

We wish to find $(3, 1)T$, where T is the counterclockwise rotation about the origin by $\pi/2$ radians. We first find the usual matrix A for T, by determining $(1, 0)T$ and $(0, 1)T$. From Fig. 21-3 we see that $(1, 0)T = (0, 1)$ and $(0, 1)T = (-1, 0)$. Hence the matrix A for T is

$$
A = \begin{bmatrix} 0 & 1 \\ -1 & 0 \end{bmatrix}
$$

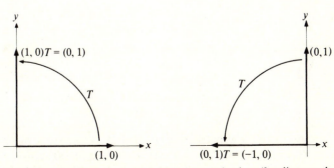

Fig. 21-3   The effect of a counterclockwise rotation by $\pi/2$ radians on the usual basis.

Finally,

$$(3, 1)T = (3, 1)A = (3, 1)\begin{bmatrix} 0 & 1 \\ -1 & 0 \end{bmatrix} = (-1, 3)$$

We can easily find a numerical rule for T by computing $\mathbf{X}A$ where $\mathbf{X} = (x, y)$ is an arbitrary vector. Doing this, we obtain

$$\mathbf{X}T = \mathbf{X}A = (x, y)\begin{bmatrix} 0 & 1 \\ -1 & 0 \end{bmatrix} = (-y, x)$$

Thus $(x, y)T = (-y, x)$.   ●

Exercises

In each of Exercises 21-1 to 21-9:

(a)   Determine what the transformation does to the usual basis of the domain, either numerically or else by drawing pictures.
(b)   Use this information to find the usual matrix A for the transformation.

21-1 $T : \mathcal{R}^3 \to \mathcal{R}^3$, where $(x, y, z)T = (0, 2y, 3z + x)$.

21-2 $T : \mathcal{R}^3 \to \mathcal{R}^2$, where $(x, y, z)T = (4x - y, 2x - y + 5z)$.

21-3 $U : \mathcal{R}^2 \to \mathcal{R}^4$, where $(x, y)U = (x + y, 0, 3y, -x + y)$.

21-4 $T : \mathcal{R}^2 \to \mathcal{R}^2$, where T is the perpendicular projection onto the y axis, as in Example 20-2.

21-5 $U : \mathcal{R}^2 \to \mathcal{R}^2$, the reflection in the y axis, as in Example 20-3.

21-6 $S : \mathcal{R}^2 \to \mathcal{R}^2$, where S is a clockwise rotation about the origin by $\pi/2$ radians. (Hint: Draw two pictures showing the effect of S on the usual basis.)

21-7 $W : \mathcal{R}^2 \to \mathcal{R}^2$, where W is the counterclockwise rotation by $\pi/6$ radians (30°) about the origin. (Hint: Draw two pictures showing the effect of W on the usual basis. Recall that a 30°-60°-90° triangle has the proportions given in Fig. 21-4.)

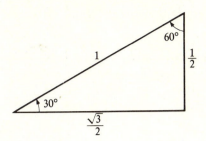

Fig. 21-4

21-8 $T: R^2 \to R^2$, where T is the counterclockwise rotation about the origin by θ radians.

21-9 $T: R^3 \to R^3$, where T is the counterclockwise rotation by $\pi/3$ radians about the z axis (viewed from the positive side of the z axis).

21-10 Let $T: R^3 \to R^3$ be the perpendicular projection onto the xy plane. Find the usual matrix A, then compute $(1, 2, 3)T$ and $(-3, 0, -5)T$ by using the matrix A, and finally give the numerical rule for T by finding $(x, y, z)T$.

21-11 Let $U: R^2 \to R^2$ be the reflection in the line $y = x$. Find the usual matrix and then compute $(1, -2)U$ and $(x, y)U$.

21-12 Find the usual matrix for the transformation T where $(x, y, z)T = (x + y - 2z, y, x - z)$. Then compute $(1, 2, -3)T$ in two ways:

(a) Using the usual matrix
(b) Using the given numerical rule for T

21-13 Calculate $(2, 4)W$, where W is the linear transformation of Exercise 21-7.

21-14 Suppose that $S: R^3 \to R^3$ has the usual matrix

$$\begin{bmatrix} 0 & -1 & 0 \\ 1 & 0 & 0 \\ 0 & 0 & 1 \end{bmatrix}$$

First describe the effect of S on the usual basis (easily determined from the rows of the matrix). Then describe the geometric effect of S (as deduced from what S does to the basis).

21-15 Suppose that T has the usual matrix

$$\begin{bmatrix} 1 & 0 & 0 \\ 0 & 0 & 0 \\ 0 & 0 & 1 \end{bmatrix}$$

First describe the effect of T on the usual basis, and then deduce the geometric effect of T.

21-16 Suppose that $T : \mathcal{R}^m \to \mathcal{R}^n$ is a linear transformation and A is its usual matrix. Prove that the rows of A span the image of T. Hint: Note that $(x_1, \ldots, x_m)T = [x_1(1, 0, \ldots, 0) + \cdots + x_m(0, 0, \ldots, 1)]T.$

21-17 Let $\mathcal{C}$ be the set of all linear transformations from $\mathcal{R}^m$ to $\mathcal{R}^n$. Prove the following:

(a) If S and T are linear transformations, then $S + T$ is a linear transformation. This is law A1 (closure under addition) for a vector space. [Recall that $S + T$ is the function defined by the rule $\mathbf{X}(S + T) = \mathbf{X}S + \mathbf{X}T$.]
(b) If S is a linear transformation and r is a scalar, then rS is a linear transformation. This is law S1 (closure under scalar multiplication) for a vector space. [Recall that rS is defined by the rule $\mathbf{X}(rS) = r(\mathbf{X}S)$.]
(c) The additive identity law (A4) for $\mathcal{C}$.
(d) The additive inverse law (A5) for $\mathcal{C}$.

(Note: The four parts of this exercise prove four of the ten laws for a vector space. The other six are also true. Hence $\mathcal{C}$ is a vector space.)

21-18 (a) Prove that $\mathcal{C}$ (see Exercise 21-17) is isomorphic to $\mathcal{M}^{m \times n}$, using the function $F : \mathcal{C} \to \mathcal{M}^{m \times n}$ where $F(T) =$ the usual matrix for T.
(b) What is the dimension of $\mathcal{C}$?

***21-19** Write a computer program that computes dot products $\mathbf{X} \cdot \mathbf{Y}$.

22. MATRIX MULTIPLICATION AND PRODUCTS OF LINEAR TRANSFORMATIONS

In the last section we dealt with products such as

$$(1, 2, 3)\begin{bmatrix} 4 & 1 \\ 5 & -2 \\ 6 & 0 \end{bmatrix} = (32, -3)$$

It is a simple matter to generalize this and extend it to a matrix product such as

$$\begin{bmatrix} 1 & 2 & 3 \\ 2 & -1 & 5 \end{bmatrix}\begin{bmatrix} 4 & 1 \\ 5 & -2 \\ 6 & 0 \end{bmatrix}$$

We deal with each row of the first matrix *separately*. Forming the dot product of (1 2 3) with each of the two columns of the second matrix yields the first row of the product:

$$1(4) + 2(5) + 3(6) = 32 \qquad \text{and} \qquad 1(1) + 2(-2) + 3(0) = -3$$

Forming the dot product of the second row (2 -1 5) with each of the columns of the second matrix yields the second row of the product:

$$2(4) + (-1)(5) + 5(6) = 33 \qquad \text{and} \qquad 2(1) + (-1)(-2) + 5(0) = 4$$

Thus we have

$$\begin{bmatrix} 1 & 2 & 3 \\ 2 & -1 & 5 \end{bmatrix}\begin{bmatrix} 4 & 1 \\ 5 & -2 \\ 6 & 0 \end{bmatrix} = \begin{bmatrix} 32 & -3 \\ 33 & 4 \end{bmatrix}$$

Recall that the i, j entry of a matrix is the element that is located in row i and column j of that matrix. The essence of the above example is that the i, j entry of a matrix product AB is found by taking the dot product of row i of A with column j of B. For example, to find the 2, 1 entry of the product in the example above, we form the dot product of row 2 of the first matrix with column 1 of the second matrix and obtain $2(4) + (-1)(5) + 5(6) = 33$.

In order to be able to form the product AB, each row of A must have the same number of entries as each column of B. Thus the number of columns of A must equal the number of rows of B. That is, if A is an $m \times n$ matrix and B is a $p \times q$ matrix, we must have $n = p$ for the product AB to be defined.

We may summarize the previous paragraphs in a formal definition.

Definition

Let A be an $m \times n$ matrix and B be an $n \times p$ matrix where

$$A = \begin{bmatrix} a_{11} & a_{12} & \cdots & a_{1n} \\ a_{21} & a_{22} & \cdots & a_{2n} \\ \cdots\cdots\cdots\cdots\cdots\cdots \\ a_{m1} & a_{m2} & \cdots & a_{mn} \end{bmatrix} \quad \text{and} \quad B = \begin{bmatrix} b_{11} & b_{12} & \cdots & b_{1p} \\ b_{21} & b_{22} & \cdots & b_{2p} \\ \cdots\cdots\cdots\cdots\cdots\cdots \\ b_{n1} & b_{n2} & \cdots & b_{np} \end{bmatrix}$$

The **product** AB is the $m \times p$ matrix whose i, j entry equals $a_{i1} b_{1j} + a_{i2} b_{2j} + \cdots + a_{in} b_{nj}$; that is, the i, j entry of AB is the dot product of row i of A with column j of B.

Example 22-1

Let

$$A = \begin{bmatrix} 1 & 0 & -3 \\ 2 & \frac{1}{2} & 4 \\ 0 & 1 & 2 \end{bmatrix}, \quad B = \begin{bmatrix} 3 & 1 & -1 \\ 0 & 4 & 2 \\ 1 & -1 & 5 \end{bmatrix}$$

$$C = \begin{bmatrix} 2 & 0 \\ 1 & 3 \\ -1 & 2 \end{bmatrix}, \quad I = \begin{bmatrix} 1 & 0 & 0 \\ 0 & 1 & 0 \\ 0 & 0 & 1 \end{bmatrix}$$

Then

$$AB = \begin{bmatrix} 1 & 0 & -3 \\ 2 & \frac{1}{2} & 4 \\ 0 & 1 & 2 \end{bmatrix} \begin{bmatrix} 3 & 1 & -1 \\ 0 & 4 & 2 \\ 1 & -1 & 5 \end{bmatrix} = \begin{bmatrix} 0 & 4 & -16 \\ 10 & 0 & 19 \\ 2 & 2 & 12 \end{bmatrix}$$

where, for example, the 2, 3 entry of AB is obtained by forming the dot product of the second row of A with the third column of B: $(2 \quad \frac{1}{2} \quad 4) \cdot (-1 \quad 2 \quad 5) = 19$.

Similarly,

$$BA = \begin{bmatrix} 3 & 1 & -1 \\ 0 & 4 & 2 \\ 1 & -1 & 5 \end{bmatrix} \begin{bmatrix} 1 & 0 & -3 \\ 2 & \frac{1}{2} & 4 \\ 0 & 1 & 2 \end{bmatrix} = \begin{bmatrix} 5 & -\frac{1}{2} & -7 \\ 8 & 4 & 20 \\ -1 & 4\frac{1}{2} & 3 \end{bmatrix}$$

Note that $AB \neq BA$. This says that matrix multiplication is not commutative.

We can form the matrix product BC:

$$BC = \begin{bmatrix} 3 & 1 & -1 \\ 0 & 4 & 2 \\ 1 & -1 & 5 \end{bmatrix} \begin{bmatrix} 2 & 0 \\ 1 & 3 \\ -1 & 2 \end{bmatrix} = \begin{bmatrix} 8 & 1 \\ 2 & 16 \\ -4 & 7 \end{bmatrix}$$

However, we cannot form the product CB, since the rows of C do not "fit" the columns of B.

The reader can verify that $AI = IA = A$ and similarly $BI = IB = B$. The matrix I is the multiplicative identity for 3×3 matrices. ●

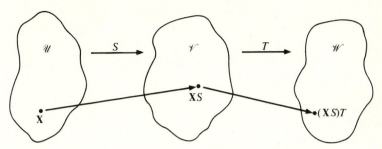

Fig. 22-1 The product ST; first perform S, then perform T.

Some laws, true for multiplication of real numbers, are also true for multiplication of $n \times n$ matrices; for example, the associative law $A(BC) = (AB)C$, the left distributive law $A(B + C) = AB + AC$, and the right distributive law $(A + B)C = AC + BC$. Other laws, however, fail. (See Exercise 22-2.)

We know that every linear transformation has a matrix. Thus it is natural to ask the following: If S and T are linear transformations with usual matrices A and B, respectively, how is the matrix AB related to S and T? The answer lies in the concept of *product*, or *composition*, of linear transformations.

Suppose that $S : \mathcal{U} \to \mathcal{V}$ and $T : \mathcal{V} \to \mathcal{W}$ are two linear transformations. The product of S and T is the function $ST : \mathcal{U} \to \mathcal{W}$, whose rule is: First perform the rule for S, and then perform the rule for T. (See Fig. 22-1.) For example, if S is a rotation and T is a projection, we would compute $\mathbf{X}(ST)$ by first rotating $\mathbf{X}$, then projecting the result.

More precisely, we have the following definition.

Definition

Let $S : \mathcal{U} \to \mathcal{V}$ and $T : \mathcal{V} \to \mathcal{W}$ be linear transformations, where $\mathcal{U}$, $\mathcal{V}$, and $\mathcal{W}$ are vector spaces. The **product** of S and T is the function $ST : \mathcal{U} \to \mathcal{W}$ defined by the rule $\mathbf{X}(ST) = (\mathbf{X}S)T$.†

Note that in order for ST to be defined we must have the image (or range) of S contained in the domain of T.

Example 22-2

Suppose that $S : \mathcal{R}^2 \to \mathcal{R}^2$ has the rule $(x, \ y)S = (x, \ x + y)$ and $T : \mathcal{R}^2 \to \mathcal{R}^2$ has the rule $(x, \ y)T = (0, \ 2y)$. We wish to find the rule for ST. By direct computation $(x, \ y)ST = ((x, \ y)S)T = (x, \ x + y)T =$

---

† In the terminology of calculus, ST is the *composite* function obtained by performing the two functions S and T in succession. Calculus notation for this function would be $T \circ S$; its values would be written $(T \circ S)(x)$ or $T(S(x))$. Writing functions on the right has the benefit that ST, read from left to right, conveys the correct order: first S, then T.

$(0, 2(x + y))$. Thus the rule for ST is $(x, y)ST = (0, 2x + 2y)$. [Note: T has the verbal rule "replace the first entry by zero and double the second entry." Thus when we apply T to the vector $(x, x + y)$, the first entry, x, becomes zero, and the second entry, $x + y$, is doubled to become $2(x + y)$.]

●

Example 22-3

Suppose that S is a counterclockwise rotation of $\mathcal{R}^2$ by $\pi/6$ radians and T is a counterclockwise rotation of $\mathcal{R}^2$ by $\pi/3$ radians. When ST is applied to a vector $\mathbf{X}$, we first rotate $\mathbf{X}$ by $\pi/6$ radians, and then rotate further by $\pi/3$ radians. Hence we conclude that ST is a counterclockwise rotation of $\mathcal{R}^2$ by $\pi/6 + \pi/3$ radians, that is, a counterclockwise rotation by $\pi/2$ radians.

●

Matrix multiplication gives an efficient way of finding the rule for a product of linear transformations: If A is the usual matrix for S and if B is the usual matrix for T, then AB is the usual matrix for ST. (In fact, this is one of the reasons for defining matrix multiplication as it was defined in this text.)

Theorem 22-1

If $S: \mathcal{R}^m \to \mathcal{R}^n$ and $T: \mathcal{R}^n \to \mathcal{R}^p$ are linear transformations with usual matrices A and B, then ST is a linear transformation whose usual matrix is AB.

Proof

It is left to the reader to verify that the function ST is a linear transformation.

The proof that AB is the matrix for ST will be given only for the case where $m = n = p = 2$. (A similar proof may be given for any positive integers m, n, and p.) Suppose that

$$A = \begin{bmatrix} a & b \\ c & d \end{bmatrix} \quad \text{and} \quad B = \begin{bmatrix} e & f \\ g & h \end{bmatrix}$$

Then

$$AB = \begin{bmatrix} ae + bg & af + bh \\ ce + dg & cf + dh \end{bmatrix}$$

We need to show that

(a) $(1, 0)ST = (ae + bg, af + bh)$ The first row of AB
(b) $(0, 1)ST = (ce + dg, cf + dh)$ The second row of AB

Statement (a) is true since

$$(1, 0)ST = ((1, 0)S)T$$
$$= (a, b)T$$
$$= (a, b)B$$
$$= (ae + bg, af + bh)$$

We were able to replace $(1, 0)S$ by (a, b), since the first row of A gives the effect of S on $(1, 0)$.

Statement (b) is true since

$$(0, 1)ST = ((0, 1)S)T$$
$$= (c, d)T$$
$$= (c, d)B$$
$$= (ce + dg, cf + dh)$$

Therefore, AB is the usual matrix for ST.   ∎

Example 22-4

Suppose that $S : \mathcal{R}^3 \to \mathcal{R}^3$ has the rule $(x, y, z)S = (x + y - 2z, y, x - z)$ and $T : \mathcal{R}^3 \to \mathcal{R}^2$ has the rule $(x, y, z)T = (2x - z, 7y + z)$. S has the usual matrix

$$A = \begin{bmatrix} 1 & 0 & 1 \\ 1 & 1 & 0 \\ -2 & 0 & -1 \end{bmatrix}$$

and T has the usual matrix

$$B = \begin{bmatrix} 2 & 0 \\ 0 & 7 \\ -1 & 1 \end{bmatrix}$$

To find the usual matrix for ST, we use Theorem 22-1 and form the matrix product

$$AB = \begin{bmatrix} 1 & 1 \\ 2 & 7 \\ -3 & -1 \end{bmatrix}$$

To find the usual matrix for S^3 (by definition, $S^3 = SSS$), we form the matrix product A^3 (which, by definition, is AAA):

$$A^3 = \begin{bmatrix} -1 & 0 & -1 \\ 1 & 1 & 1 \\ 2 & 0 & 1 \end{bmatrix}$$

●

Example 22-5

Suppose that S is the counterclockwise rotation of $\mathcal{R}^2$ about the origin by $\pi/2$ radians and T is the reflection of $\mathcal{R}^2$ in the y axis. Then S and T have A and B as their usual matrices, where

$$A = \begin{bmatrix} 0 & 1 \\ -1 & 0 \end{bmatrix} \quad \text{and} \quad B = \begin{bmatrix} -1 & 0 \\ 0 & 1 \end{bmatrix}$$

Consequently ST has the usual matrix

$$AB = \begin{bmatrix} 0 & 1 \\ 1 & 0 \end{bmatrix}$$

and TS has the usual matrix

$$BA = \begin{bmatrix} 0 & -1 \\ -1 & 0 \end{bmatrix}$$

The reader can verify that ST is a reflection in the line $y = x$ and TS is a reflection in the line $y = -x$.   ●

Exercises

22-1 Let

$$A = \begin{bmatrix} 2 & 4 & 0 \\ 1 & -1 & 2 \end{bmatrix}, \quad B = \begin{bmatrix} 2 & 1 & 0 \\ 0 & 0 & -1 \\ 5 & 3 & -2 \end{bmatrix}, \quad C = \begin{bmatrix} 3 & 1 \\ 2 & 5 \\ 1 & -1 \end{bmatrix}$$

$$D = \begin{bmatrix} 3 & 2 & -1 \\ -5 & -4 & 2 \\ 0 & -1 & 0 \end{bmatrix}, \quad I = \begin{bmatrix} 1 & 0 & 0 \\ 0 & 1 & 0 \\ 0 & 0 & 1 \end{bmatrix}$$

(a)   Find the products AB, BC, BD, DB, IB, B^2 $(= BB)$.
(b)   Which of the following products do not exist: CA, BA, AC, C^2, IA, CI?

22-2 We saw in Sec. 7 that $\mathcal{M}^{m \times n}$ is a vector space. Hence matrices obey the same laws for addition that the real numbers obey. Under multiplication, however, the following laws, true for real numbers, are *false* for matrices. Give examples of 2×2 matrices showing that these laws are false.

(a)   The commutative law: $AB = BA$.
(b)   The inverse law: For any nonzero A there is an A^{-1} such that $AA^{-1} = A^{-1}A = I$. $A = \begin{pmatrix} 1 & 1 \\ 1 & 1 \end{pmatrix}$
(c)   The cancellation laws: If $AB = AC$, then $B = C$; if $BA = CA$, then $B = C$. $A = \vec{0}$

22-3 Let $S: \mathcal{R}^3 \to \mathcal{R}^2$ have the rule $(x, y, z)S = (2x, y)$, and let $T: \mathcal{R}^2 \to \mathcal{R}^2$ have the rule $(x, y)T = (3x + y, x)$.

(a)   Find the rule for ST by applying S and T in succession to an arbitrary vector (x, y, z).
(b)   Explain why TS is not defined.

22-4 Let S be the perpendicular projection of $\mathcal{R}^2$ onto the y axis of $\mathcal{R}^2$, and let T be the perpendicular projection of $\mathcal{R}^2$ onto the x axis of $\mathcal{R}^2$. Find the rule for ST by means of a picture: apply S and T in succession to an arbitrary vector (x, y).

22-5 Let S be the perpendicular projection of $\mathcal{R}^2$ onto the y axis of $\mathcal{R}^2$, and let T be the counterclockwise rotation of $\mathcal{R}^2$ about the origin by $\pi/2$ radians.

(a)   Find the matrices A and B for S and T, respectively.
(b)   Find the rule for ST by multiplying the matrices A and B and applying the result to an arbitrary vector (x, y).
(c)   Find the rule for TS. (Note that $ST \neq TS$. In general, linear transformations do not commute and hence it is crucial to know which transformation is to be performed first.)

22-6 Suppose that $S: \mathcal{U} \to \mathcal{V}$ and $T: \mathcal{V} \to \mathcal{W}$ are linear transformations. Verify that $ST: \mathcal{U} \to \mathcal{W}$ is a linear transformation. (Hint: Use the format of Example 20-1.)

22-7(a)   Find the usual matrix C for the linear transformation ST of Example 22-2 by first computing the effect of ST on the usual basis.
(b)   Find the usual matrices A and B for the linear transformations S and T of Example 22-2.
(c)   Verify that $AB = C$.

22-8 Let $S:\mathcal{R}^2 \to \mathcal{R}^2$ have the rule $(x, y)S = (x + 2y, x)$ and let $T:\mathcal{R}^2 \to \mathcal{R}^2$ have the rule $(x, y)T = (x + y, y)$.

(a) Find the usual matrix A for S and the usual matrix B for T.
(b) Find the usual matrix C for ST by matrix multiplication.
(c) Find the usual matrix D for TS.

22-9 Let S be the $45°$ counterclockwise rotation (as viewed from the positive z axis) about the z axis in $\mathcal{R}^3$. Let T be the perpendicular projection onto the xy plane in $\mathcal{R}^3$.

(a) Find the usual matrix for ST, and compute $(x, y, z)ST$.
(b) Find the usual matrix for TS, and compute $(x, y, z)TS$.

22-10 Suppose that S has the usual matrix

$$\begin{bmatrix} 1 & -1 & 0 \\ 2 & 3 & 0 \\ 0 & 5 & -2 \end{bmatrix}$$

and T has the usual matrix

$$\begin{bmatrix} \frac{3}{5} & \frac{1}{5} & 0 \\ -\frac{2}{5} & \frac{1}{5} & 0 \\ -1 & \frac{1}{2} & -\frac{1}{2} \end{bmatrix}$$

Describe what ST and TS do geometrically.

22-11 Suppose that S has the usual matrix

$$\begin{bmatrix} 0 & -1 \\ -1 & 0 \end{bmatrix}$$

and T has the usual matrix

$$\begin{bmatrix} -1 & 0 \\ 0 & 1 \end{bmatrix}$$

Describe ST and TS geometrically.

22-12 Suppose that S has the usual matrix

$$\begin{bmatrix} \dfrac{\sqrt{3}}{2} & \dfrac{1}{2} \\ -\dfrac{1}{2} & \dfrac{\sqrt{3}}{2} \end{bmatrix}$$

and T has the usual matrix

$$\begin{bmatrix} \dfrac{1}{2} & \dfrac{\sqrt{3}}{2} \\ -\dfrac{\sqrt{3}}{2} & \dfrac{1}{2} \end{bmatrix}$$

Describe ST and TS geometrically.

*22-13(a)   Write a computer program that multiplies matrices.
(b)   Use this program to compute the image of $(3, 7)$ under each of the linear transformations of Exercises 21-3 to 21-8.
(c)   Use this program to compute the matrices for the products of the linear transformations in Exercises 22-5 and 22-8 to 22-12.

23.   INVERSES

As was mentioned in the introduction to this chapter, the final step in the solution of many problems is *inverting* the solution of the transformed problem so that it applies to the original problem.

For example, in using logarithms, once we have determined that $\log x = -0.1695$, we use the inverse of the logarithm function—exponentiation—to conclude that $x = e^{-0.1695}$. We can consider this as using the inverse function on both sides of the transformed equation $\log x = -0.1695$ to strip away the unwanted log and leave in its place just the original unknown x.

As a second example, suppose we wish to find an unknown acute angle x, already knowing that $\tan x = \sqrt{3}$. If we apply the inverse function—arctangent—to both sides of the equation, we obtain $\arctan (\tan x) = \arctan \sqrt{3}$, or $x = \arctan \sqrt{3}$. Thus the original unknown is a $60°$ angle.

In many problems involving concepts from linear algebra, the *transform* part of the process involves the use of a linear transformation or a matrix. Thus, to carry out the *invert* part of the process, we often need to have an inverse for the linear transformation or the matrix. In this section we will study the problem of determining when a linear transformation has an inverse, and a method for finding inverses.

Definition

If A is an $n \times n$ matrix, its **inverse** A^{-1} is an $n \times n$ matrix that satisfies $AA^{-1} = I$ and $A^{-1}A = I$, where

$$I = \begin{bmatrix} 1 & 0 & \cdots & 0 \\ 0 & 1 & \cdots & 0 \\ \multicolumn{4}{c}{\dotfill} \\ 0 & 0 & \cdots & 1 \end{bmatrix}$$

(the $n \times n$ identity matrix).

The **inverse** T^{-1} of a linear transformation $T : \mathcal{V} \to \mathcal{V}$ is a function satisfying $TT^{-1} = I$ and $T^{-1}T = I$, where I is the identity linear transformation (that is, $\mathbf{X}I = \mathbf{X}$ for all vectors $\mathbf{X} \in \mathcal{V}$).

A^{-1} and T^{-1} are unique. (See Exercise 23-2.)
The inverse of

$$\begin{bmatrix} 1 & 2 \\ 5 & 8 \end{bmatrix}$$

is

$$\begin{bmatrix} -4 & 1 \\ \frac{5}{2} & -\frac{1}{2} \end{bmatrix}$$

since

$$\begin{bmatrix} 1 & 2 \\ 5 & 8 \end{bmatrix} \begin{bmatrix} -4 & 1 \\ \frac{5}{2} & -\frac{1}{2} \end{bmatrix} = \begin{bmatrix} 1 & 0 \\ 0 & 1 \end{bmatrix}$$

and

$$\begin{bmatrix} -4 & 1 \\ \frac{5}{2} & -\frac{1}{2} \end{bmatrix} \begin{bmatrix} 1 & 2 \\ 5 & 8 \end{bmatrix} = \begin{bmatrix} 1 & 0 \\ 0 & 1 \end{bmatrix}$$

If $T : \mathcal{R}^3 \to \mathcal{R}^3$ has the rule $(x, y, z)T = (x - y, 2z, y)$, then T^{-1} has the rule $(x, y, z)T^{-1} = (x + z, z, \frac{1}{2}y)$, since

$$\begin{aligned} (x, y, z)TT^{-1} &= (x - y, 2z, y)T^{-1} \\ &= ((x - y) + y, y, \tfrac{1}{2}(2z)) \\ &= (x, y, z) \end{aligned}$$

and similarly

$$(x, y, z)T^{-1}T = (x, y, z)$$

Not all $n \times n$ matrices or linear transformations have inverses. If a matrix (or a linear transformation) has an inverse, it is *invertible*; otherwise, it is *noninvertible*.†
We will now show how to find matrix inverses. An example will clarify the theory behind the method.
We wish to find the inverse of

$$A = \begin{bmatrix} 1 & 0 & 2 \\ 2 & 1 & 2 \\ 0 & 0 & 1 \end{bmatrix}$$

† Some texts refer to a noninvertible matrix as *singular* and to an invertible matrix as *nonsingular*.

We let

$$A^{-1} = \begin{bmatrix} x_1 & y_1 & z_1 \\ x_2 & y_2 & z_2 \\ x_3 & y_3 & z_3 \end{bmatrix}$$

The x's, y's, and z's are unknown, and A^{-1} must satisfy $AA^{-1} = I$, or

$$\begin{bmatrix} 1 & 0 & 2 \\ 2 & 1 & 2 \\ 0 & 0 & 1 \end{bmatrix} \begin{bmatrix} x_1 & y_1 & z_1 \\ x_2 & y_2 & z_2 \\ x_3 & y_3 & z_3 \end{bmatrix} = \begin{bmatrix} 1 & 0 & 0 \\ 0 & 1 & 0 \\ 0 & 0 & 1 \end{bmatrix}$$

If we consider only the *first column*

$$\begin{bmatrix} x_1 \\ x_2 \\ x_3 \end{bmatrix}$$

of A^{-1}, we see that we must have

$$\begin{bmatrix} 1 & 0 & 2 \\ 2 & 1 & 2 \\ 0 & 0 & 1 \end{bmatrix} \begin{bmatrix} x_1 \\ x_2 \\ x_3 \end{bmatrix} = \begin{bmatrix} 1 \\ 0 \\ 0 \end{bmatrix}$$

This is so since we form the dot product of the rows of A with the *first column* of A^{-1} to obtain the entries in the first column of I. Written out, this becomes

$$\begin{aligned} x_1 \qquad\quad + 2x_3 &= 1 \\ 2x_1 + x_2 + 2x_3 &= 0 \\ x_3 &= 0 \end{aligned}$$

a system of equations with matrix

$$\left[\begin{array}{ccc:c} 1 & 0 & 2 & 1 \\ 2 & 1 & 2 & 0 \\ 0 & 0 & 1 & 0 \end{array} \right]$$

or

$$\left[\begin{array}{c:c} A & \begin{matrix} 1 \\ 0 \\ 0 \end{matrix} \end{array} \right]$$

for short. The unaugmented matrix A is the same matrix we began with. Reducing the displayed matrix would yield solutions for x_1, x_2, x_3.

Similarly, the second column entries y_1, y_2, y_3 of A^{-1} can be found by reducing

$$\left[\begin{array}{c:c} A & \begin{matrix} 0 \\ 1 \\ 0 \end{matrix} \end{array}\right]$$

and the third column entries z_1, z_2, z_3 of A^{-1} can be found by reducing

$$\left[\begin{array}{c:c} A & \begin{matrix} 0 \\ 0 \\ 1 \end{matrix} \end{array}\right]$$

Notice that solving each of the three systems entails reducing the same matrix A. Rather than reduce A three separate times, we form the "superaugmented" matrix

$$\left[\begin{array}{c:c:c:c} A & \begin{matrix} 1 \\ 0 \\ 0 \end{matrix} & \begin{matrix} 0 \\ 1 \\ 0 \end{matrix} & \begin{matrix} 0 \\ 0 \\ 1 \end{matrix} \end{array}\right]$$

and reduce it to the form

$$\begin{bmatrix} 1 & 0 & 0 & a_1 & b_1 & c_1 \\ 0 & 1 & 0 & a_2 & b_2 & c_2 \\ 0 & 0 & 1 & a_3 & b_3 & c_3 \end{bmatrix}$$

thus obtaining the entries of A^{-1} all at once. For instance, the first row permits reading not only $x_1 = a_1$, but also $y_1 = b_1$, and $z_1 = c_1$.

We now carry out this process of reducing the superaugmented matrix

$$\begin{bmatrix} 1 & 0 & 2 & 1 & 0 & 0 \\ 2 & 1 & 2 & 0 & 1 & 0 \\ 0 & 0 & 1 & 0 & 0 & 1 \end{bmatrix} \longrightarrow \begin{bmatrix} 1 & 0 & 2 & 1 & 0 & 0 \\ 0 & 1 & -2 & -2 & 1 & 0 \\ 0 & 0 & 1 & 0 & 0 & 1 \end{bmatrix}$$

$$\longrightarrow \begin{bmatrix} 1 & 0 & 0 & 1 & 0 & -2 \\ 0 & 1 & 0 & -2 & 1 & 2 \\ 0 & 0 & 1 & 0 & 0 & 1 \end{bmatrix}$$

(Explanation: In the first step, twice the first row was subtracted from the second row; in the second step, twice the third row was subtracted from the first row and twice the third row was added to the second row.) The entries to the right of the dotted line are the entries of A^{-1}; that is,

$$A^{-1} = \begin{bmatrix} 1 & 0 & -2 \\ -2 & 1 & 2 \\ 0 & 0 & 1 \end{bmatrix}$$

This can be verified by computing AA^{-1}; it does equal I.

The above method works for any $n \times n$ matrix and can be summarized briefly as follows.

To find A^{-1}:

(a)   Form the matrix $[A \vdots I]$.

(b)   Reduce this matrix until A is reduced to I. (If this is impossible, then A has no inverse.)

(c)   The reduced form $[I \vdots A^{-1}]$ gives the matrix A^{-1}.

The above method finds a *right* inverse, that is, a matrix A^{-1} (if it exists) such that $AA^{-1} = I$. We did not bother to verify that $A^{-1}A = I$, which is also required of an inverse. That was not necessary because of the following theorem (whose proof is omitted).

Theorem 23-1   Let A be an $n \times n$ matrix, and B a matrix such that $AB = I$. Then $BA = I$ also, so that B is the inverse of A.

Example 23-1

We wish to find the inverse of

$$\begin{bmatrix} 1 & 2 \\ 3 & 7 \end{bmatrix}$$

We first form the matrix

$$\begin{bmatrix} 1 & 2 & \vdots & 1 & 0 \\ 3 & 7 & \vdots & 0 & 1 \end{bmatrix}$$

and reduce it:

$$\begin{bmatrix} 1 & 2 & \vdots & 1 & 0 \\ 3 & 7 & \vdots & 0 & 1 \end{bmatrix} \longrightarrow \begin{bmatrix} 1 & 2 & \vdots & 1 & 0 \\ 0 & 1 & \vdots & -3 & 1 \end{bmatrix}$$

$$\longrightarrow \begin{bmatrix} 1 & 0 & \vdots & 7 & -2 \\ 0 & 1 & \vdots & -3 & 1 \end{bmatrix}$$

Hence the inverse is

$$\begin{bmatrix} 7 & -2 \\ -3 & 1 \end{bmatrix}$$

The answer can always be checked by multiplying the matrix and its inverse to obtain I:

$$\begin{bmatrix} 1 & 2 \\ 3 & 7 \end{bmatrix} \begin{bmatrix} 7 & -2 \\ -3 & 1 \end{bmatrix} = \begin{bmatrix} 1 & 0 \\ 0 & 1 \end{bmatrix}$$

Theorem 23-1 saves us the trouble of verifying that

$$\begin{bmatrix} 7 & -2 \\ -3 & 1 \end{bmatrix}$$

works on the left.   ●

Example 23-2

We wish to find the inverse of

$$\begin{bmatrix} 1 & 2 & 4 \\ 0 & 1 & 0 \\ 2 & 3 & 5 \end{bmatrix}$$

We reduce

$$\left[\begin{array}{ccc:ccc} 1 & 2 & 4 & 1 & 0 & 0 \\ 0 & 1 & 0 & 0 & 1 & 0 \\ 2 & 3 & 5 & 0 & 0 & 1 \end{array} \right]$$

to obtain

$$\left[\begin{array}{ccc:ccc} 1 & 0 & 0 & -\frac{5}{3} & -\frac{2}{3} & \frac{4}{3} \\ 0 & 1 & 0 & 0 & 1 & 0 \\ 0 & 0 & 1 & \frac{2}{3} & -\frac{1}{3} & -\frac{1}{3} \end{array} \right]$$

and hence the inverse of the given matrix is

$$\begin{bmatrix} -\frac{5}{3} & -\frac{2}{3} & \frac{4}{3} \\ 0 & 1 & 0 \\ \frac{2}{3} & -\frac{1}{3} & -\frac{1}{3} \end{bmatrix}$$

Again, the answer can be checked.   ●

Example 23-3

We wish to find the inverse of

$$A = \begin{bmatrix} 1 & 2 & 5 \\ 3 & 1 & -1 \\ 4 & 3 & 4 \end{bmatrix}$$

We form the matrix

$$\begin{bmatrix} 1 & 2 & 5 & \vdots & 1 & 0 & 0 \\ 3 & 1 & -1 & \vdots & 0 & 1 & 0 \\ 4 & 3 & 4 & \vdots & 0 & 0 & 1 \end{bmatrix}$$

Reducing yields

$$\begin{bmatrix} 1 & 2 & 5 & \vdots & 1 & 0 & 0 \\ 3 & 1 & -1 & \vdots & 0 & 1 & 0 \\ 4 & 3 & 4 & \vdots & 0 & 0 & 1 \end{bmatrix}$$

$$\longrightarrow \begin{bmatrix} 1 & 2 & 5 & \vdots & 1 & 0 & 0 \\ 0 & -5 & -16 & \vdots & -3 & 1 & 0 \\ 0 & -5 & -16 & \vdots & -4 & 0 & 1 \end{bmatrix}$$

$$\longrightarrow \begin{bmatrix} 1 & 2 & 5 & \vdots & 1 & 0 & 0 \\ 0 & -5 & -16 & \vdots & -3 & 1 & 0 \\ 0 & 0 & 0 & \vdots & -1 & -1 & 1 \end{bmatrix}$$

At this point it is clear that A cannot be reduced to I, and hence A^{-1} does not exist. ●

The inverse of a matrix A provides us with the numerical rule for the inverse of the linear transformation T that corresponds to A. We now prove this.

Theorem 23-2   If A is the usual matrix for the linear transformation $T : \mathcal{R}^n \to \mathcal{R}^n$, then A^{-1} is the usual matrix for T^{-1}. Hence, to compute values $\mathbf{X}T^{-1}$, we need only compute $\mathbf{X}A^{-1}$.

Proof T^{-1} is a linear transformation (Exercise 23-5). Hence T^{-1} has a usual matrix B. By definition, $TT^{-1} = I = T^{-1}T$, and for the corresponding usual matrices we have $AB = I = BA$ by Theorem 22-1. Thus B is the inverse of A, as claimed. ∎

Example 23-4

Suppose that a linear transformation T has the rule $(x, y)T = (2y, x - 4y)$. To find the rule for T^{-1} we first find A^{-1}, where A is the usual matrix for T. The reader can verify that

$$A = \begin{bmatrix} 0 & 1 \\ 2 & -4 \end{bmatrix}$$

We reduce $[A \mathrel{\vdots} I]$:

$$\begin{bmatrix} 0 & 1 & \vdots & 1 & 0 \\ 2 & -4 & \vdots & 0 & 1 \end{bmatrix} \longrightarrow \begin{bmatrix} 2 & -4 & \vdots & 0 & 1 \\ 0 & 1 & \vdots & 1 & 0 \end{bmatrix}$$

$$\longrightarrow \begin{bmatrix} 1 & -2 & \vdots & 0 & \frac{1}{2} \\ 0 & 1 & \vdots & 1 & 0 \end{bmatrix} \longrightarrow \begin{bmatrix} 1 & 0 & \vdots & 2 & \frac{1}{2} \\ 0 & 1 & \vdots & 1 & 0 \end{bmatrix}$$

Hence

$$A^{-1} = \begin{bmatrix} 2 & \frac{1}{2} \\ 1 & 0 \end{bmatrix}$$

Since A^{-1} corresponds to T^{-1}, the rule for T^{-1} is

$$(x, y)T^{-1} = (x, y)A^{-1} = (x, y)\begin{bmatrix} 2 & \frac{1}{2} \\ 1 & 0 \end{bmatrix} = (2x + y, \tfrac{1}{2}x) \qquad \bullet$$

In practical situations it is often desirable to *test* for invertibility by means of some other property, rather than to try to find the inverse directly (and possibly wasting a great deal of work if no inverse exists).

For example, in calculus, we might ask: Does $f(x) = x^3 + 2x - 4$ have an inverse? Before trying to compute f^{-1} we can determine that f^{-1} does indeed exist. We note that $f'(x) = 3x^2 + 2$, which is always positive, and hence f is a strictly increasing function. Since strictly increasing functions have inverses, we conclude that f^{-1} exists. We used a positive derivative as the test for invertibility of this function f. Since we know that f^{-1} does exist, we can confidently look for it.

The following theorem lists various properties that are useful in determining whether a linear transformation T has an inverse.

Theorem 23-3   Let $T : \mathcal{R}^n \to \mathcal{R}^n$ be a linear transformation. The following are equivalent:

1   T is invertible.

2   T is 1-1.

3   The only solution to $\mathbf{X}T = \mathbf{0}$ is $\mathbf{X} = \mathbf{0}$.

4 T carries the usual basis for $\mathcal{R}^n$ to a basis for $\mathcal{R}^n$.

5 T is onto $\mathcal{R}^n$ (that is, the image of T is all of $\mathcal{R}^n$).

Proof

[Note: When we say that these statements are *equivalent*, we mean that if any one of the statements is true, then all are true; if any one is false, then all are false. Thus any one of the conditions is a test for all the rest. It suffices to prove that condition $(1) \Rightarrow (2) \Rightarrow (3) \Rightarrow (4) \Rightarrow (5) \Rightarrow (1)$; that is, (1) implies (2), (2) implies (3), etc.]

(1) $\Rightarrow$ *(2)*   We are assuming that T is invertible. If T were *not* 1-1, then we would have two distinct vectors $\mathbf{X}$ and $\mathbf{Y}$ such that $\mathbf{X}T = \mathbf{Y}T$. Then we would have $(\mathbf{X}T)T^{-1} = (\mathbf{Y}T)T^{-1}$ and hence $\mathbf{X} = \mathbf{Y}$, a contradiction of the assumption that $\mathbf{X}$ and $\mathbf{Y}$ were distinct. Therefore, T is 1-1.

(2) $\Rightarrow$ *(3)*   From Exercise 20-11 we know that $\mathbf{0}T = \mathbf{0}$. Since we are given that T is 1-1, no nonzero vector $\mathbf{Y}$ can satisfy $\mathbf{Y}T = \mathbf{0}$, or else we would have two vectors $\mathbf{0}$ and $\mathbf{Y}$ both assigned the value $\mathbf{0}$ by T.

(3) $\Rightarrow$ *(4)*   We must show that $(1, 0, \ldots, 0)T, (0, 1, \ldots, 0)T, \ldots, (0, 0, \ldots, 1)T$ is a basis for $\mathcal{R}^n$. Let us show that these n vectors are independent. The fundamental equation for independence of these vectors is

$$c_1[(1, 0, \ldots, 0)T] + \cdots + c_n[(0, 0, \ldots, 1)T] = \mathbf{0}$$

Reworking the left side of this equation, using the two defining properties of a linear transformation, we obtain

$$[c_1(1, 0, \ldots, 0) + \cdots + c_n(0, 0, \ldots, 1)]T = \mathbf{0}$$

or

$$(c_1, \ldots, c_n)T = \mathbf{0}$$

By (3) the only vector $\mathbf{X}$ satisfying $\mathbf{X}T = \mathbf{0}$ is $\mathbf{X} = (0, 0, \ldots, 0)$. Thus all the numbers c_i are zero, as required in the definition of independence. Hence $(1, 0, \ldots, 0)T, (0, 1, \ldots, 0)T, \ldots, (0, 0, \ldots, 1)T$ are independent. By Corollary 14-2 these n independent vectors must form a basis.

(4) $\Rightarrow$ *(5)*   Theorem 20-1 shows that the image of a vector space $\langle \mathbf{X}_1, \mathbf{X}_2, \ldots, \mathbf{X}_n \rangle$ is $\langle \mathbf{X}_1 T, \mathbf{X}_2 T, \ldots, \mathbf{X}_n T \rangle$. Taking the $\mathbf{X}_i$'s as the usual basis vectors, (4) says that the $\mathbf{X}_i T$'s are a basis for $\mathcal{R}^n$. Hence the image is all of $\mathcal{R}^n$.

(5) $\Rightarrow$ *(1)*   This is left as an exercise.   ∎

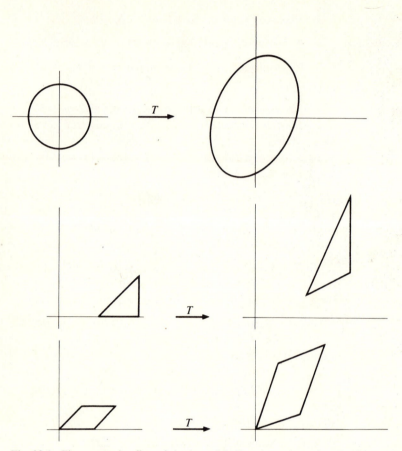

Fig. 23-1 The geometric effect of the invertible linear transformation T whose rule is $(x, y)T = (2x, 3y + x)$.

If $T : \mathcal{R}^2 \to \mathcal{R}^2$ or $T : \mathcal{R}^3 \to \mathcal{R}^3$ is invertible, it cannot change shapes "too much." For example, if $T : \mathcal{R}^2 \to \mathcal{R}^2$ is the invertible linear transformation whose rule is given by $(x, y)T = (2x, 3y + x)$, then Fig. 23-1 shows how various shapes are changed by T. The images of circles are ellipses, the images of triangles are triangles, the images of parallelograms are parallelograms. This is always true for invertible linear transformations.

On the other hand, if T is *not* invertible, shapes can be changed considerably. If $T : \mathcal{R}^3 \to \mathcal{R}^3$ is the perpendicular projection onto the xy plane, a sphere becomes a disk and a vertical line becomes a single point (Fig. 23-2). However, lack of invertibility is not necessarily a drawback. In fact, maps, blueprints, and photographs are all useful two-dimensional versions of three-dimensional objects (in which the images are obtained by projection onto a plane).

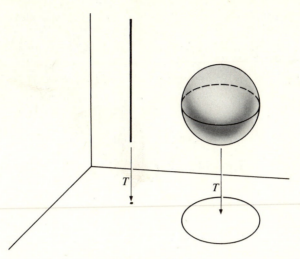

Fig. 23-2 Some geometric effects of the perpendicular projection onto the xy plane.

Exercises

23-1 Find the inverse (if it exists) of each of the following matrices:

(a) $\begin{bmatrix} 1 & 2 \\ 0 & 5 \end{bmatrix}$

(b) $\begin{bmatrix} 2 & 3 & 4 \\ -2 & 0 & 2 \\ 1 & 3 & 0 \end{bmatrix}$

(c) $\begin{bmatrix} 1 & 3 & -1 \\ 0 & 2 & 1 \\ 1 & 5 & 0 \end{bmatrix}$

(d) $\begin{bmatrix} 2 & 0 & 0 & 1 \\ 1 & 2 & 0 & -1 \\ 0 & 1 & 3 & 5 \\ 0 & 0 & 2 & 4 \end{bmatrix}$

(e) $\begin{bmatrix} 2 & 2 & 4 \\ -1 & 3 & -2 \\ -1 & 7 & -2 \end{bmatrix}$

23-2 Suppose that A is invertible. Prove that A^{-1} is unique. (Hint: Suppose there were another matrix B such that $AB = BA = I$. Show that $B = A^{-1}$.)

$AB(A^{-1}) = (BA)A^{-1}$
$(AB)A^{-1} \leftarrow A(BA^{-1}) = B$

23-3 Let A and B be invertible matrices. Show that $(AB)^{-1} = B^{-1}A^{-1}$; that is, show that the inverse of the matrix AB is $B^{-1}A^{-1}$. (Hint: Multiply AB and $B^{-1}A^{-1}$ together.)

23-4 Suppose that $S: \mathcal{R}^n \to \mathcal{R}^n$ and $T: \mathcal{R}^n \to \mathcal{R}^n$ are invertible. Show that $(ST)^{-1} = T^{-1}S^{-1}$.

[handwritten: $(ax)T^{-1} = (a y)T)T^{-1}\{(ay)T)T = ay = a(x)T^{-1}$]

[handwritten margin: LT — scalar mult. & vector addition]

23-5 Suppose that $T : \mathcal{V} \to \mathcal{V}$ is a linear transformation. Prove that T^{-1}, if it exists, is a linear transformation. (Hint: If T^{-1} exists, T must be *onto* $\mathcal{V}$, and so an arbitrary vector $\mathbf{X}$ in $\mathcal{V}$ can be written as $\mathbf{W}T$ for some $\mathbf{W} \in \mathcal{V}$.)

[handwritten: $(x+y)T^{-1} = ((w)T + (v)T)T^{-1} = (u+v)T)T^{-1} = u+v = xT^{-1} + yT^{-1}$]

23-6 Let

$$A = \begin{bmatrix} 2 & 3 & 4 \\ -2 & 0 & 0 \\ 1 & 3 & 0 \end{bmatrix}$$

Solve these systems of equations:

(a) $(x, y, z)A = (2, 0, 4)$
(b) $(x, y, z)A = (0, 0, 0)$
(c) $(x, y, z)A = (-5, 1, \frac{1}{2})$
(d) $(x, y, z)A = (\frac{4}{9}, -\frac{1}{8}, \frac{7}{3})$

(Hint: Each of the four systems has the form $\mathbf{X}A = \mathbf{Y}$. By multiplying both members of this equation on the right by A^{-1} we have $\mathbf{X} = \mathbf{Y}A^{-1}$, and hence the solutions can be obtained by multiplying the appropriate $\mathbf{Y}$ by A^{-1}.)

23-7 Explain in words why any rotation about the origin in $\mathcal{R}^3$ is invertible.

23-8 Explain why any reflection in $\mathcal{R}^2$ or $\mathcal{R}^3$ is invertible.

23-9 Suppose that T is the perpendicular projection onto the line $y = x$ in $\mathcal{R}^2$. Find T^{-1}, if it exists.

23-10 Suppose that $(x, y, z)T = (2x, -y - z, 4x + 2y - 5z)$. Find T^{-1}, if it exists.

23-11 Determine which of these linear transformations have inverses without actually computing the inverses. [Hint: Use condition (4) or (2) of Theorem 23-3.]

(a) $(x, y)T = (0, x)$
(b) $(x, y, z)T = (y, y - 2z, 6z)$
(c) $(x, y, z)T = (x + y, z, 3y)$

23-12 Prove the implication $(5) \Rightarrow (1)$ of Theorem 23-3. [Hint: Let A be the usual matrix for T. By Exercise 21-16, the rows of A span the image of T. Use condition (5) to explain why A has an inverse, and use this fact to explain why T has an inverse.]

23-13 Prove directly the implication $(4) \Rightarrow (3)$ of Theorem 23-3.

23-14   The *rank* of a linear transformation T is defined to be the dimension of the image of T. Explain why the rank of $T: \mathcal{R}^m \to \mathcal{R}^n$ equals the rank of the usual matrix A for T. (Hint: See Exercise 21-16.)

23-15   Recall that if $T: \mathcal{V} \to \mathcal{W}$ is a linear transformation, then its kernel is the set of all vectors $\mathbf{X}$ such that $\mathbf{X}T = \mathbf{0}$. Complete the table below. For example, the kernel of the first linear transformation may be described as " the set of solutions to the system $x = 0$"; thus $\mathcal{K} = \{(0, a, b): a, b \in \mathcal{R}\}$, which is two-dimensional.

Linear transformation	Description of its kernel $\mathcal{K}$	Dimension of kernel	Usual matrix	Rank of T
$(x, y, z)T = (x, 0, 0)$				
$(x, y, z)T = (x, 0, z)$				
$(x, y, z)T = (z, 2x, y)$				
$(x, y, z)T = (0, 0, 0)$				
$(x, y, z)T = (x - y, y + z, x + z)$				

23-16   In the table of Exercise 23-15, what is the numerical relationship between the dimension of the kernel and the rank of T?

23-17   Suppose that $T: \mathcal{V} \to \mathcal{W}$ is a linear transformation. Using the result of Exercise 23-16, formulate a general theorem relating the dimension of the kernel and the rank of T.

***23-18**   Write a computer program for finding the inverse of a matrix by reducing $[A \mid I]$ to $[I \mid A^{-1}]$.

24.   DETERMINANTS

As noted in the previous section, even invertible linear transformations distort shapes somewhat. The topic of this section—the determinant—assigns to each $n \times n$ matrix a number that reveals (among other things) how much the corresponding linear transformation magnifies areas or volumes. For example, noninvertible linear transformations (such as projections) crush areas (in $\mathcal{R}^2$) or volumes (in $\mathcal{R}^3$) down to zero, and, as we will see, each of their matrices has a determinant equal to zero.

We now define the determinant.

Definition

|det. A| = factor by which A multiplies area. (area in 2×2)

The **determinant** of a 2 × 2 matrix $\begin{bmatrix} a & b \\ c & d \end{bmatrix}$ is the number $ad - bc$. We write $\det \begin{bmatrix} a & b \\ c & d \end{bmatrix} = ad - bc$.

For example, the determinant of $\begin{bmatrix} 5 & 8 \\ 3 & 2 \end{bmatrix}$ is $5(2) - 8(3) = -14$.

Definition

The **determinant** of a 3 × 3 matrix

$$\begin{bmatrix} a_1 & a_2 & a_3 \\ b_1 & b_2 & b_3 \\ c_1 & c_2 & c_3 \end{bmatrix}$$

is the number

$$a_1 \det \begin{bmatrix} b_2 & b_3 \\ c_2 & c_3 \end{bmatrix} - a_2 \det \begin{bmatrix} b_1 & b_3 \\ c_1 & c_3 \end{bmatrix} + a_3 \det \begin{bmatrix} b_1 & b_2 \\ c_1 & c_2 \end{bmatrix}$$

where the 2 × 2 matrices are obtained by striking out the row and column of the a_i term.

For example,

$$\det \begin{bmatrix} 1 & 2 & 3 \\ 4 & 5 & 6 \\ 0 & 0 & 2 \end{bmatrix} = 1 \det \begin{bmatrix} 5 & 6 \\ 0 & 2 \end{bmatrix} - 2 \det \begin{bmatrix} 4 & 6 \\ 0 & 2 \end{bmatrix} + 3 \det \begin{bmatrix} 4 & 5 \\ 0 & 0 \end{bmatrix}$$

$$= 1(10) - 2(8) + 3(0)$$
$$= -6$$

Definition

The **determinant** of an $n \times n$ matrix

$$A = \begin{bmatrix} a_{11} & a_{12} & \cdots & a_{1n} \\ a_{21} & a_{22} & \cdots & a_{2n} \\ \multicolumn{4}{c}{\dotfill} \\ a_{n1} & a_{n2} & \cdots & a_{nn} \end{bmatrix}$$

is the number

$$\det A = a_{11} \det \begin{bmatrix} a_{22} & \cdots & a_{2n} \\ \cdots\cdots\cdots\cdots\cdots \\ a_{n2} & \cdots & a_{nn} \end{bmatrix} - a_{12} \det \begin{bmatrix} a_{21} & a_{23} & \cdots & a_{2n} \\ \cdots\cdots\cdots\cdots\cdots\cdots \\ a_{n1} & a_{n3} & \cdots & a_{nn} \end{bmatrix}$$

$$+ \cdots \pm a_{1n} \det \begin{bmatrix} a_{21} & \cdots & a_{2\,n-1} \\ \cdots\cdots\cdots\cdots\cdots \\ a_{n1} & \cdots & a_{n\,n-1} \end{bmatrix}$$

where the $(n-1) \times (n-1)$ matrices are obtained by striking out the row and column in which the term a_{1i} appears. The signs alternate $+$, $-$, $+$, $-$, ...

This definition uses the first row in a special way, but actually *any row or column* can be used to compute det A. Once we choose a row (or column), det A is found as follows: Go across the row (or down the column), forming the products

$$\pm a_{ij} \det M_{ij}$$

where a_{ij} is a typical entry of the row (or column), M_{ij} is the $(n-1) \times (n-1)$ matrix obtained by striking out a_{ij}'s row and column, and the $\pm$ signs are determined by mentally superimposing on A this checkerboard pattern:

$$\begin{array}{ccccc} + & - & + & - & \cdots \\ - & + & - & + & \cdots \\ + & - & + & - & \cdots \end{array}$$

For example, the signs used for the second row would be $- + - + \cdots$.

Example 24-1

Find

$$\det \begin{bmatrix} 1 & 2 & 0 \\ 5 & 6 & 7 \\ 2 & 3 & 0 \end{bmatrix}$$

Using the second row, we have

$$-5 \det \begin{bmatrix} 2 & 0 \\ 3 & 0 \end{bmatrix} + 6 \det \begin{bmatrix} 1 & 0 \\ 2 & 0 \end{bmatrix} - 7 \det \begin{bmatrix} 1 & 2 \\ 2 & 3 \end{bmatrix} = 7$$

If we use the third column, we have

$$+0 \det \begin{bmatrix} 5 & 6 \\ 2 & 3 \end{bmatrix} - 7 \det \begin{bmatrix} 1 & 2 \\ 2 & 3 \end{bmatrix} + 0 \det \begin{bmatrix} 1 & 2 \\ 5 & 6 \end{bmatrix} = 7$$

It is usually most efficient to choose a row or a column with the most zeros. This simplifies computation.　●

Example 24-2

Find

$$\det \begin{bmatrix} 1 & 1 & 2 & 1 \\ 3 & 0 & 0 & 3 \\ 0 & 5 & 0 & 2 \\ 1 & 2 & 3 & 0 \end{bmatrix}$$

Using the third row, since it has two zeros in it, we have

$$0 - 5 \det \begin{bmatrix} 1 & 2 & 1 \\ 3 & 0 & 3 \\ 1 & 3 & 0 \end{bmatrix} + 0 - 2 \det \begin{bmatrix} 1 & 1 & 2 \\ 3 & 0 & 0 \\ 1 & 2 & 3 \end{bmatrix}$$

We must now find each of these 3×3 determinants. Using the second row in each case, we have

$$-5\left(-3 \det \begin{bmatrix} 2 & 1 \\ 3 & 0 \end{bmatrix} + 0 - 3 \det \begin{bmatrix} 1 & 2 \\ 1 & 3 \end{bmatrix}\right) - 2\left(-3 \det \begin{bmatrix} 1 & 2 \\ 2 & 3 \end{bmatrix} + 0 - 0\right)$$

$$= -5[-3(-3) + 0 - 3(3 - 2)] - 2[-3(3 - 4) + 0 - 0]$$

$$= -36$$　●

Example 24-3

Find the determinant of

$$A = \begin{bmatrix} 3 & 1 & 16 & -7 & 4 \\ 0 & 2 & 4 & 5 & 9 \\ 0 & 0 & 2 & -12 & 9 \\ 0 & 0 & 0 & 4 & 7 \\ 0 & 0 & 0 & 0 & 2 \end{bmatrix}$$

Using the first column, since it has four zeros, we have

$$\det A = +3 \det \begin{bmatrix} 2 & 4 & 5 & 9 \\ 0 & 2 & -12 & 9 \\ 0 & 0 & 4 & 7 \\ 0 & 0 & 0 & 2 \end{bmatrix} - 0 + 0 - 0 + 0$$

$$= 3\left(2 \det \begin{bmatrix} 2 & -12 & 9 \\ 0 & 4 & 7 \\ 0 & 0 & 2 \end{bmatrix} - 0 + 0 - 0 \right)$$

$$= 3\left(2\left(2 \det \begin{bmatrix} 4 & 7 \\ 0 & 2 \end{bmatrix} - 0 + 0 \right) \right)$$

$$= 3(2)(2)(4)(2)$$

$$= 96$$ •

The large number of zero entries made this example relatively easy. Had we begun with a 5 × 5 matrix with no zero entries, the definition of determinant would have forced us to compute five 4 × 4 determinants. Each of these 4 × 4 matrices would have given rise to four 3 × 3 determinants, and each of the 3 × 3 matrices would have given three 2 × 2 determinants. The potential length of this process leads us to look for a shorter method for computing the determinant.

Example 24-3 involves a matrix of the best type for computing determinants. This matrix has a column in which four of the five entries are zero; in the second step we had a column in which three of the four entries were zero; in the third step we had a column in which two of three entries were zero, etc. Because of these features in this example, the determinant, 96, was the product of all the entries on the main diagonal†: 3(2)(2)(4)(2). This is always true for an "upper triangular" matrix.

Definition

An $n \times n$ matrix is **upper triangular** if it is of the form

$$\begin{bmatrix} a_{11} & a_{12} & a_{13} & \cdots & a_{1n} \\ 0 & a_{22} & a_{23} & \cdots & a_{2n} \\ 0 & 0 & a_{33} & \cdots & a_{3n} \\ & & & \cdots & \\ 0 & 0 & 0 & \cdots & a_{nn} \end{bmatrix}$$

that is, it has zeros in all positions below the main diagonal.

† The *main diagonal* of a matrix is the one from upper left to lower right.

Theorem 24-1   If A is an $n \times n$ matrix in upper triangular form, then $\det A = a_{11} a_{22} \cdots a_{nn}$.

We leave it to the reader to formulate a definition and an analogous theorem for *lower* triangular matrices.

Since upper triangular matrices have determinants that can be computed so painlessly, it pays to first put a matrix into upper triangular form and then to find its determinant. The process we use to put a matrix into upper triangular form is the same process we used in Chap. 1 to reduce a matrix. The three row operations used in reducing a matrix have the following effect on determinants:

OPERATION	EFFECT ON DETERMINANT
(a) Interchanging two rows	Multiplies determinant by -1
(b) Multiplying a row by a constant $k \neq 0$	Multiplies determinant by k
(c) Adding a multiple of one row to another	No change in determinant

The verification of these facts is left as an exercise.

Example 24-4

Find
$$\det \begin{bmatrix} 2 & 3 & -2 \\ 2 & 4 & 1 \\ 6 & 0 & 3 \end{bmatrix}$$

$$\det \begin{bmatrix} 2 & 3 & -2 \\ 2 & 4 & 1 \\ 6 & 0 & 3 \end{bmatrix} = \det \begin{bmatrix} 2 & 3 & -2 \\ 0 & 1 & 3 \\ 0 & -9 & 9 \end{bmatrix} = \det \begin{bmatrix} 2 & 3 & -2 \\ 0 & 1 & 3 \\ 0 & 0 & 36 \end{bmatrix} = 72$$

[Explanation: In the first step we used operation (c) twice to clear the first column below the main diagonal. In the second step we used operation (c) to clear the second column below the main diagonal, thus putting the matrix into upper triangular form. At each step the determinant remained the same, since operation (c) does not change the determinant.]

●

Example 24-5

Find
$$\det \begin{bmatrix} 0 & 3 & 1 & 7 \\ 0 & 3 & 4 & 6 \\ 2 & 5 & 9 & -7 \\ 0 & 0 & 2 & 5 \end{bmatrix}$$

$$
\det \begin{bmatrix} 0 & 3 & 1 & 7 \\ 0 & 3 & 4 & 6 \\ 2 & 5 & 9 & -7 \\ 0 & 0 & 2 & 5 \end{bmatrix} = -\det \begin{bmatrix} 2 & 5 & 9 & -7 \\ 0 & 3 & 4 & 6 \\ 0 & 3 & 1 & 7 \\ 0 & 0 & 2 & 5 \end{bmatrix} =
$$

$$
-\det \begin{bmatrix} 2 & 5 & 9 & 7 \\ 0 & 3 & 4 & 6 \\ 0 & 0 & -3 & 1 \\ 0 & 0 & 2 & 5 \end{bmatrix} = -\det \begin{bmatrix} 2 & 5 & 9 & -7 \\ 0 & 3 & 4 & 6 \\ 0 & 0 & -3 & 1 \\ 0 & 0 & 0 & \frac{17}{3} \end{bmatrix} = 102
$$

(Explanation: The first step interchanged the first and third rows; hence the -1. The remaining two steps cleared the second and third columns below the main diagonal.)   ●

The method just used for finding determinants by reducing matrices to upper triangular form also gives a very useful test for invertibility: A matrix A is invertible exactly when det $A \neq 0$. The following theorem lists this and other tests for invertibility of matrices.

Theorem 24-2

Let A be an $n \times n$ matrix. The following are equivalent:

1 A is invertible.

2   det $A \neq 0$.

3   The rows of A are independent.

4 A has rank n.

Proof

This can be proved by proving the chain of implications $(1) \Rightarrow (2) \Rightarrow (3) \Rightarrow (4) \Rightarrow (1)$.

$(1) \Rightarrow (2)$ Suppose that A is invertible. Then our method for finding A^{-1} (by reducing $[A \mid I]$ to $[I \mid A^{-1}]$) shows that A reduces to I. Each step of this reduction process at most changes the sign of det A or multiplies det A by a nonzero constant. Since det $I \neq 0$, we must also have det $A \neq 0$.
The balance of the proof is deferred to the exercises.   ■

This theorem also yields one more test for the invertibility of a linear transformation: A linear transformation is invertible if and only if its usual matrix A has a nonzero determinant.

Determinants also have a geometric interpretation. Suppose that $T : \mathcal{R}^2 \to \mathcal{R}^2$ is an invertible linear transformation where $(1, 0)T = (a, b)$ and $(0, 1)T = (c, d)$. Then the square determined by the vectors $(1, 0)$ and $(0, 1)$ is changed to the parallelogram determined by the vectors (a, b)

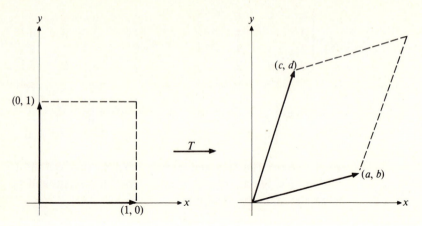

Fig. 24-1

and (c, d). See Fig. 24-1, which illustrates the case where a, b, c, and d are all positive. Points inside the square will be carried to points inside the parallelogram.

The area of the square is 1. The parallelogram has an area equal to $|ad - bc|$, the absolute value of the determinant of the usual matrix

$$\begin{bmatrix} a & b \\ c & d \end{bmatrix}$$

for T. Figure 24-2 indicates this. The area of the parallelogram is equal to the area of the rectangle minus the areas of the four triangles labeled 1, 2, 3, and 4. Hence the area of the parallelogram is

$$(a + c)(b + d) - \tfrac{1}{2}(a + c)b - \tfrac{1}{2}(b + d)c - \tfrac{1}{2}(a + c)b - \tfrac{1}{2}(b + d)c$$

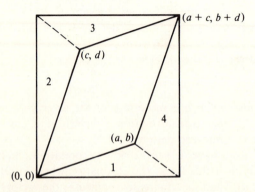

Fig. 24-2

which equals $ad - bc$.† Thus the square has its area multiplied by the factor $|ad - bc|$. When the parallelogram lies in any other quadrant, similar figures lead to the same conclusion.

It can be shown that all shapes have their areas changed by this same factor. The proof would proceed first to all squares with sides parallel to $(1, 0)$ and $(0, 1)$, and then to more irregular shapes, all of which can be filled out approximately with squares of various sizes.

If T is not invertible, then by Theorem 23-3, $(1, 0)T$ and $(0, 1)T$ are not a basis and thus are dependent. That is, (a, b) and (c, d) are multiples of each other, and hence $ad - bc = 0$. On the other hand, all shapes have their areas reduced to zero. Thus, even in this case, $|ad - bc|$ is the factor by which areas are changed by T. Thus the absolute value of the determinant of T's matrix A tells the factor by which T multiplies areas. For transformations $T : \mathcal{R}^3 \to \mathcal{R}^3$ the absolute value of the determinant of A gives the factor by which T multiplies volumes.

Example 24-6

We can now easily derive the formula for the area of an ellipse. Suppose that we wish to find the area of an ellipse, such as the one in Fig. 24-3b, with equation $(x/a)^2 + (y/b)^2 = 1$. We begin with the unit circle, as in Fig. 24-3a and transform it to the ellipse by using a linear transformation T where $(1, 0)T = (a, 0)$ and $(0, 1)T = (0, b)$. Thus the usual matrix A for T is

$$A = \begin{bmatrix} a & 0 \\ 0 & b \end{bmatrix}$$

The preceding discussion shows that T multiplies areas by the factor $|\det A| = ab$. Since the unit circle has area $\pi 1^2 = \pi$, the ellipse has area πab.

The reader should compare the simplicity of this derivation with that usually given in calculus, which entails evaluating a rather tedious integral.   ●

As mentioned earlier in this chapter, problems often arise that are best analyzed by considering a product of linear transformations. If S and T are two linear transformations, it is natural to ask how ST changes areas. It should impress the reader as intuitively clear that, for example, if S doubles areas and T triples areas, then ST changes areas by a factor of 6 ($= 2 \cdot 3$). If A is the usual matrix for S and B is the usual matrix for T

---

† If the positions of the vectors (a, b) and (c, d) were reversed, we would obtain the number $bc - ad$. In either case the area is $|ad - bc|$.

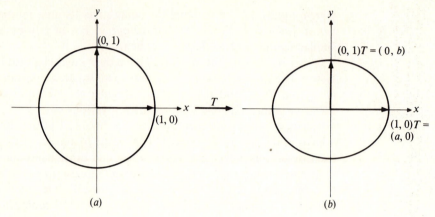

Fig. 24-3

(and hence AB is the usual matrix for ST), then from our previous discussion, det $A = \pm 2$, det $B = \pm 3$, and det $(AB) = \pm 6$. This suggests an important general formula: det $(AB) = (\det A)(\det B)$, which in fact is correct.

Theorem 24-3
If A and B are $n \times n$ matrices, then det $(AB) = (\det A)(\det B)$.
The proof is omitted.

Exercises

24-1 For each of the following matrices:
(1) Find its determinant, using only the definition of determinant.
(2) Find its determinant by reducing the matrix to upper triangular form.

(a) $\begin{bmatrix} 2 & 4 & 6 \\ -2 & 3 & 5 \\ -6 & -5 & 3 \end{bmatrix}$
(b) $\begin{bmatrix} 1 & 0 & 0 \\ 0 & 1 & 0 \\ 0 & 0 & 1 \end{bmatrix}$

(c) $\begin{bmatrix} 0 & 0 & 1 \\ 0 & 1 & 0 \\ 1 & 0 & 0 \end{bmatrix}$
(d) $\begin{bmatrix} 0 & 0 & 3 & -2 \\ -5 & 2 & 1 & 3 \\ 2 & 2 & 2 & -5 \\ 1 & 2 & 4 & 0 \end{bmatrix}$

(e) $\begin{bmatrix} 4 & 7 & 1 & 5 \\ 2 & 3 & 1 & 9 \\ 0 & 0 & 0 & 0 \\ 2 & 7 & -8 & 9 \end{bmatrix}$

24-2 Let

$$A = \begin{bmatrix} a & b & c \\ d & e & f \\ g & h & i \end{bmatrix}$$

(a) Suppose that B is the matrix A with the first two rows interchanged. Verify that det $A = -$ det B.

(b) Suppose that C is the matrix A where the first row has been multiplied by k. Verify that det $C = k$ det A.

(c) Suppose that D is the matrix A where k times the first row is added to the second row. Verify that det $A =$ det D.

24-3 Let

$$A = \begin{bmatrix} 2 & 4 & 25 \\ 0 & -3 & 6 \\ 0 & 0 & 1 \end{bmatrix} \quad \text{and} \quad B = \begin{bmatrix} 3 & 0 & -2 \\ 0 & -2 & 29 \\ 3 & 0 & 4 \end{bmatrix}$$

(a) Find AB, and evaluate its determinant.

(b) Find det A and det B.

(c) Use Theorem 24-3 to find det (AB).

24-4 Let

$$A = \begin{bmatrix} 3 & 4 & 0 & 1 \\ 1 & -1 & 0 & 2 \\ 2 & 2 & 3 & 4 \\ 7 & 7 & 0 & 5 \end{bmatrix} \quad \text{and} \quad B = \begin{bmatrix} 2 & 0 & 0 & 0 \\ 1 & 3 & 4 & 1 \\ -1 & -1 & 2 & 7 \\ 1 & 0 & 3 & 8 \end{bmatrix}$$

Use Theorem 24-3 to find det (AB) and det (BA).

24-5 Prove that det $(AB) =$ det (BA) for any $n \times n$ matrices A and B. (Hint: Use Theorem 24-3.)

24-6 Prove the implications $(2) \Rightarrow (3)$, $(3) \Rightarrow (4)$, $(4) \Rightarrow (1)$ in Theorem 24-2.

24-7(a) Let A be an $n \times n$ invertible matrix. How is det (A^{-1}) related to det A? (Hint: Start with $AA^{-1} = I$.)

(b) Use the result of (a) to give an alternate proof of the implication $(1) \Rightarrow (2)$ in Theorem 24-2.

24-8 Suppose that $S : \mathcal{R}^2 \to \mathcal{R}^2$ doubles all areas and $T : \mathcal{R}^2 \to \mathcal{R}^2$ is noninvertible. Find the area multiplication factor for ST.

24-9 Suppose that $T : \mathcal{R}^3 \to \mathcal{R}^3$ multiplies volumes by the factor $\frac{1}{2}$. What effect does T^2 have on volumes? How about T^{10}?

24-10 Suppose that $T : \mathcal{R}^2 \to \mathcal{R}^2$ is the counterclockwise rotation by θ radians about the origin.

(a) Compute the determinant of the usual matrix for T (see Exercise 21-8).
(b) Does this agree with what you would expect on geometric grounds?

24-11 Suppose that $T : \mathcal{R}^2 \to \mathcal{R}^2$ is the reflection in the line $y = x$.

(a) Compute the determinant of the usual matrix for T (see Exercise 21-11).
(b) Does this agree with what you would expect on geometric grounds?

24-12 Suppose that $T : \mathcal{R}^2 \to \mathcal{R}^2$ has as its usual matrix

$$A = \begin{bmatrix} 1 & 0 \\ 2 & 1 \end{bmatrix}$$

(a) What effect does T have on areas?
(b) To what parallelogram would T carry the square of Fig. 24-1?

*24-13 Write a computer program for computing 3×3 determinants using the first column.

*24-14 Write a computer program for computing $n \times n$ determinants by reducing to upper triangular form and multiplying the main diagonal entries.

25. APPLICATIONS

Example 25-1

Human Consumption of Pesticides

As is well known, the use of pesticides has permitted significant increase in crop output. However, people pay a price for this increase. Some of the pesticides are absorbed by the plants; the plants are consumed by various species of herbivorous animals; these herbivores are in turn eaten by carnivorous animals; finally, the carnivores are eaten by human beings. We will use matrix multiplication to determine the quantities of pesticides that pass through this particular food chain.

Assume we are dealing with m types of pesticides that have been sprayed on n species of plants. Let

$$A = \begin{bmatrix} a_{11} & \cdots & a_{1n} \\ \hdotsfor{3} \\ a_{m1} & \cdots & a_{mn} \end{bmatrix}$$

where a_{ij} is the amount of pesticide i in the average plant of species j. Let us consider p species of herbivores. Let

$$B = \begin{bmatrix} b_{11} & \cdots & b_{1p} \\ \hdotsfor{3} \\ b_{n1} & \cdots & b_{np} \end{bmatrix}$$

where b_{ij} is the number of plants of species i that a herbivore of species j eats per year. If we multiply A and B, then the i, j entry of AB, $a_{i1}b_{1j} + a_{i2}b_{2j} + \cdots + a_{in}b_{nj}$, gives the total amount of pesticide of type i that a herbivore of species j consumes per year (assuming that the herbivores eat only the plants of species $1, 2, \ldots, n$).

To analyze the next link in the food chain, we consider q species of carnivores. Let

$$C = \begin{bmatrix} c_{11} & \cdots & c_{1q} \\ \hdotsfor{3} \\ c_{p1} & \cdots & c_{pq} \end{bmatrix}$$

where c_{ij} is the number of herbivores of species i that are eaten per year by a carnivore of species j. If the carnivores eat only the herbivores of species $1, 2, \ldots, p$, then the i, j entry of ABC gives the total amount of pesticide i that a carnivore of species j consumes per year.

To study the final link in this chain, let

$$D = \begin{bmatrix} d_1 \\ \vdots \\ d_q \end{bmatrix}$$

where d_i is the number of carnivores of species i that an average human consumes per year. If we form the product

$$ABCD = \begin{bmatrix} e_1 \\ \vdots \\ e_m \end{bmatrix}$$

then the entry e_i equals the amount of pesticide i that an average person consumes per year by virtue of eating carnivorous animals. (The astute reader will note that not all of each pesticide is completely retained in each type of animal. Therefore, the actual amounts consumed would be less than those predicted by the above computation.)   ●

Example 25-2

Counting the Number of Routes on a Map

In this example we see how matrix methods may be used to count the number of different routes between two points on a map.

To visualize the method we will refer to the map in Fig. 25-1. More precisely, such a diagram is called a *graph*. The six numbered dots are called *vertices* and the connecting lines are called *edges*. Two vertices are *adjacent* if there is an edge directly connecting them. For example, vertex 2 and vertex 4 are adjacent, whereas vertices 3 and 5 are not. A *route* in a graph is a sequence of edges such that the terminal point of one is the initial point of the next edge in the sequence. For example, $6 \rightarrow 3 \rightarrow 2 \rightarrow 4$ denotes the route starting at vertex 6 and proceeding to vertex 3, then continuing to vertex 2, and finally ending at vertex 4. The same edge can appear more than once in a route; for example, the route $3 \rightarrow 4 \rightarrow 2 \rightarrow 4 \rightarrow 5 \rightarrow 1$ uses the edge joining 2 and 4 twice.

A graph can be numerically described by its *adjacency matrix*. The size of the adjacency matrix is determined by the number of vertices in the graph. In our example the adjacency matrix is a 6×6 matrix where the i, j entry a_{ij} equals the number of edges joining vertex i to vertex j. For example, $a_{16} = 1$ since vertices 1 and 6 have one edge joining them, while $a_{52} = 0$ since there is no edge joining vertex 5 to vertex 2. The adjacency matrix A for the above graph is

$$A = \begin{bmatrix} 0 & 1 & 0 & 0 & 1 & 1 \\ 1 & 0 & 1 & 1 & 0 & 0 \\ 0 & 1 & 0 & 1 & 0 & 1 \\ 0 & 1 & 1 & 0 & 1 & 1 \\ 1 & 0 & 0 & 1 & 0 & 1 \\ 1 & 0 & 1 & 1 & 1 & 0 \end{bmatrix}$$

The *length* of a route is the number of edges in the sequence that forms the route. If the same edge is used more than once, it is counted more than once in determining the length of the route. For example, the route $4 \rightarrow 2 \rightarrow 3 \rightarrow 4 \rightarrow 2 \rightarrow 1$ has length 5. The i, j entry in the adjacency matrix can be interpreted as the number of routes of length 1 from vertex i to vertex j, since there is a route of length 1 from i to j exactly when there is an edge joining i and j.

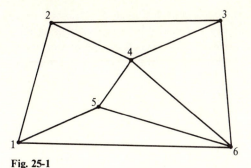

Fig. 25-1

If we form the product

$$A^2 = AA = \begin{bmatrix} 3 & 0 & 2 & 3 & 1 & 1 \\ 0 & 3 & 1 & 1 & 2 & 3 \\ 2 & 1 & 3 & 2 & 2 & 1 \\ 3 & 1 & 2 & 4 & 1 & 2 \\ 1 & 2 & 2 & 1 & 3 & 2 \\ 1 & 3 & 1 & 2 & 2 & 4 \end{bmatrix}$$

we observe that the i, j entry of this matrix gives the number of routes from i to j of length 2. This follows since

$$i, j \text{ entry of } A^2 = a_{i1} a_{1j} + a_{i2} a_{2j} + \cdots + a_{i6} a_{6j}$$

The first product, $a_{i1} a_{1j}$, counts the number of routes that run from i to 1 to j; the second product, $a_{i2} a_{2j}$, counts the number of routes that run from i to 2 to j; similarly for the other four products that appear in this sum. Thus this sum counts all the routes from i to j of length 2. For example, the entry 3 in the 1, 4 position of A^2 counts the three routes $1 \rightarrow 2 \rightarrow 4$, $1 \rightarrow 5 \rightarrow 4$, $1 \rightarrow 6 \rightarrow 4$.

In general, if k is any positive integer, the i, j entry of A^k is the number of routes of length k from vertex i to vertex j. Using $k = 3$, we have

$$A^3 = \begin{bmatrix} 2 & 8 & 4 & 4 & 7 & 9 \\ 8 & 2 & 7 & 9 & 4 & 4 \\ 4 & 7 & 4 & 7 & 5 & 9 \\ 4 & 9 & 7 & 6 & 9 & 10 \\ 7 & 4 & 5 & 9 & 4 & 7 \\ 9 & 4 & 9 & 10 & 7 & 6 \end{bmatrix}$$

The entry 9 in the 4, 2 position counts the nine routes of length 3 from vertex 4 to vertex 2: $4 \rightarrow 2 \rightarrow 1 \rightarrow 2$, $4 \rightarrow 2 \rightarrow 3 \rightarrow 2$, $4 \rightarrow 2 \rightarrow 4 \rightarrow 2$, $4 \rightarrow 3 \rightarrow 4 \rightarrow 2$, $4 \rightarrow 5 \rightarrow 1 \rightarrow 2$, $4 \rightarrow 5 \rightarrow 4 \rightarrow 2$, $4 \rightarrow 6 \rightarrow 1 \rightarrow 2$, $4 \rightarrow 6 \rightarrow 3 \rightarrow 2$, $4 \rightarrow 6 \rightarrow 4 \rightarrow 2$.

In this example each edge has been a "two-way street," since an edge from vertex i to vertex j has also been regarded as an edge from vertex j to vertex i. In the exercises at the end of this section we will also consider *directed graphs* where the edges are "one-way streets."   ●

Example 25-3

Round-Robin Tournaments

In this example we see how a matrix may be used to find the "strongest" player in a round-robin tournament.

A round-robin tournament is a series of matches played by n persons in which each person meets every other person exactly once. The diagram in Fig. 25-2 represents the final results of a hypothetical round-robin tournament played by six people. An arrow pointing along an edge from player i to player j means that player i beat j in their match. (We are assuming that there are no ties.) From this diagram we can observe the following: players 1, 2, 3, and 6 are tied for first place with a 3-2 win-loss record; player 4 is in fifth place with a 2-3 win-loss record; and finally, player 5 is in last place with a 1-4 win-loss record.

A matrix can be used to describe the results of the tournament. Let A

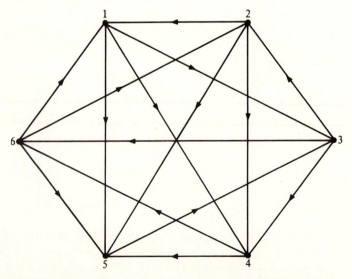

Fig. 25-2 Graph of the results of a six-person round-robin tournament.

be the 6×6 matrix where $a_{ij} = 1$ if player i beat player j in their match; otherwise, $a_{ij} = 0$. Hence

$$A = \begin{bmatrix} 0 & 0 & 1 & 1 & 1 & 0 \\ 1 & 0 & 0 & 1 & 1 & 0 \\ 0 & 1 & 0 & 1 & 0 & 1 \\ 0 & 0 & 0 & 0 & 1 & 1 \\ 0 & 0 & 1 & 0 & 0 & 0 \\ 1 & 1 & 0 & 0 & 1 & 0 \end{bmatrix}$$

If we add the entries across in any row, we obtain the number of wins for that player. We have

Player #	# wins
1	3
2	3
3	3
4	2
5	1
6	3

and hence the ranking of the players is

Place	Player #
1st	1, 2, 3, 6 (tied)
5th	4
6th	5

Using the information we have (the results of each match), we will observe one possible way of breaking the four-way tie for first place. We do this by studying the strength of the competition that each of these four players beat.

We first form the product of A with itself to obtain

$$A^2 = \begin{bmatrix} 0 & 1 & 1 & 1 & 1 & 2 \\ 0 & 0 & 2 & 1 & 2 & 1 \\ 2 & 1 & 0 & 1 & 3 & 1 \\ 1 & 1 & 1 & 0 & 1 & 0 \\ 0 & 1 & 0 & 1 & 0 & 1 \\ 1 & 0 & 2 & 2 & 2 & 0 \end{bmatrix}$$

As in Example 25-2, the i, j entry of A^2 gives the number of routes of length 2 from i to j. However, in this example a route may be given a second interpretation: if $i \to k \to j$ is a route, then this indicates that player i beat player k, who in turn beat player j. We call this a *two-step win* for i over j. Thus the i, j entry of A^2 gives the total number of two-step wins for i over j.

For example, the entry 3 in the 3, 5 position of A^2 counts the three two-step wins: $3 \to 2 \to 5$, $3 \to 4 \to 5$, $3 \to 6 \to 5$. Even though player 5 beat player 3 in their match, player 3 has exhibited considerable strength relative to player 5, since player 3 beat three players who in turn beat player 5.

Let $B = A + A^2$. Then

$$B = \begin{bmatrix} 0 & 1 & 2 & 2 & 2 & 2 \\ 1 & 0 & 2 & 2 & 3 & 1 \\ 2 & 2 & 0 & 2 & 3 & 2 \\ 1 & 1 & 1 & 0 & 2 & 1 \\ 0 & 1 & 1 & 1 & 0 & 1 \\ 2 & 1 & 2 & 2 & 3 & 0 \end{bmatrix}$$

The i, j entry of B equals the number of direct wins plus the number of two-step wins that player i has over player j. If we add the entries across in any row, we obtain the total number of direct and two-step wins for that player over all the competition. In this case we have

Player #	# of direct wins + # of 2-step wins
1	9
2	9
3	11
4	6
5	4
6	10

Thus each player gets a "score," determined by adding his number of direct wins and his number of two-step wins. If we take each number as representing the "strength" of that player, then the ranking becomes

Place	Player #
1st	3
2nd	6
3rd	1, 2 (tied)
5th	4
6th	5

Assuming that this method measures a player's strength, we see that player 3 is the strongest player.

The reader should realize that this is only one possible mathematical model that can be used to find the relative strengths of the players. Another possible model would be the use of the matrix $2A + A^2$, which gives twice as much weight to a direct win as to a two-step win. The choice of the model is a decision the person doing the modelling must make.   ●

Example 25-4   Differential Equations

Consider the differential equation $y'' + 3y' - 10y = 0$. We let $\mathcal{D}$ be the set of all functions defined on the real numbers whose second derivative exists for all real numbers. It can easily be verified that the set $\mathcal{D}$ with the usual definitions of addition and scalar multiplication is a vector space.

Define a function T whose domain is $\mathcal{D}$ by the rule $(f)T = f'' + 3f' - 10f$. The function T is a linear transformation. This follows since

$$
\begin{aligned}
(f + g)T &= (f + g)'' + 3(f + g)' - 10(f + g) \\
&= f'' + 3f' - 10f + g'' + 3g' - 10g \\
&= (f)T + (g)T
\end{aligned}
$$

and since

$$
\begin{aligned}
(rf)T &= (rf)'' + 3(rf)' - 10(rf) \\
&= r(f'' + 3f' - 10f) \\
&= r(fT)
\end{aligned}
$$

If f is a solution to the given differential equation, then $f'' + 3f' - 10f = 0$; that is, $fT = 0$. Conversely, if $fT = 0$, then f is a solution to the differential equation. Thus, if $\mathcal{S}$ is the set of solutions, then $\mathcal{S} = \{f : fT = 0\}$. The set $\mathcal{S}$ is the *kernel* of the transformation T, in the terminology of Exercise 20-12. Hence, by Exercise 20-12, the set $\mathcal{S}$ is a subspace of $\mathcal{D}$.

Although we cannot prove it here, the subspace $\mathcal{S}$ is two-dimensional. The reader can verify that e^{-5x} and e^{2x} are both solutions to the differential equation and that they are linearly independent. Hence e^{-5x} and e^{2x} form a basis for $\mathcal{S}$. It therefore follows that every solution to $y'' + 3y' - 10y = 0$ is of the form

$$
ae^{-5x} + be^{2x}
$$

where a and b are scalars.   ●

Example 25-5   Secret Codes

An invertible matrix provides a powerful method for transmitting messages in secret code. For example, if we wish to send the message "return at noon," we first translate the letters into numbers according to some substitution, for example, $a \to 1$, $b \to 2$, ..., $z \to 26$. In this case the message becomes 18, 5, 20, 21, 18, 14, 1, 20, 14, 15, 15, 14. We then take an invertible matrix, for example,

$$A = \begin{bmatrix} 1 & 2 & -1 \\ 2 & 5 & 2 \\ -1 & -2 & 2 \end{bmatrix}$$

and break the message into vectors each of which has three entries:

$$\mathbf{W} = (18, 5, 20) \qquad \mathbf{X} = (21, 18, 14)$$
$$\mathbf{Y} = (1, 20, 14) \qquad \mathbf{Z} = (15, 15, 14)$$

We then form the four products $\mathbf{W}A$, $\mathbf{X}A$, $\mathbf{Y}A$, and $\mathbf{Z}A$. The 12 numbers in these four products would then be transmitted. In this case the message sent would be 8, 21, 32, 43, 104, 43, 27, 74, 67, 31, 77, 43. The receiver would then break this string of numbers back down into vectors with three entries each and multiply each vector on the right by A^{-1}. Since $(\mathbf{W}A)A^{-1} = \mathbf{W}$, etc., the 12 numbers sent, 8, 21, ..., 43, would be translated back to the original string of numbers, 18, 5, ..., 14, and these would then be translated back to their letter equivalents, r, e, t, u, ...

●

Exercises

25-1(a)   How many routes from vertex 1 to vertex 3 of length 4 are there in the graph of Fig. 25-3?
(b)   How many routes from vertex 4 to vertex 2 of length 6 are there?

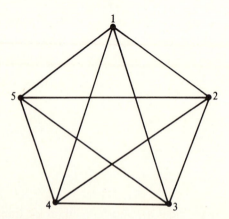

Figure 25-3

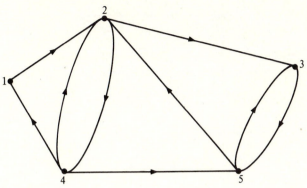

Fig. 25-4

25-2 Consider the one-way street map in Fig. 25-4. This picture is called a *directed graph* since each edge has a direction associated with it (indicated by the arrow placed on each edge). A route in a directed graph is defined in the same way as it is in a graph, with the additional condition that the route must follow the directions on the edges. The adjacency matrix is defined with the same modification: the i, j entry a_{ij} equals the number of edges *from i to j*. For example, $a_{45} = 1$ and $a_{54} = 0$.

(a)   Find the adjacency matrix for this directed graph.
(b)   Find the number of routes of length 3 from vertex 2 to vertex 3.
(c)   Find the number of routes of length 3 from vertex 3 to vertex 2.
(d)   Find the number of routes of length 4 from vertex 5 back to vertex 5.

25-3 The directed graph in Fig. 25-5 shows the results of a five-man round-robin tournament. Players 1, 3, and 4 are in a three-way tie for second place with 2-2 win-loss records. Using the method of Example 25-3, determine which of these three players is the strongest.

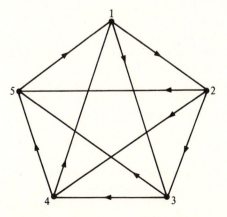

Fig. 25-5

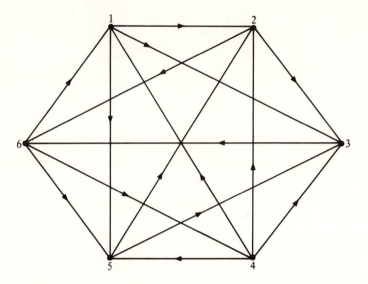

Fig. 25-6

25-4 The directed graph in Fig. 25-6 shows the results of a six-man round-robin tournament.

(a)   What are the results based on win-loss records only?

(b)   What are the results obtained by using the matrix A^2, as in the notation of Example 25-3? [Note that the winners determined by (a) and (b) are different. This happens because the winner determined in (a) has his wins against weak competition.]

25-5 The method used in Example 25-3 can be used to determine the strengths of the players in more general situations, for example, in the following cases: (1) Each player does not meet every other player; (2) Ties occur; (3) Two players meet more than once. In any of these cases we can set up a matrix whose i, j entry is equal to the number of wins player i has over player j. Thus, for example, if players i and j met once and tied, then the i, j entry and the j, i entry are zero. If i and j met three times and i beat j twice and j beat i once, then the i, j entry is 2 and the j, i entry is 1. If i and j did not meet at all, then the i, j entry and the j, i entry are both zero.

(a)   Determine the strengths of the players in the tournament of Fig. 25-7.

(b)   Suppose that we consider a tie between players i and j as counting 1 for each player; that is, if player i and player j met once and tied, then the i, j entry and the j, i entry are both 1 rather than zero. How does this affect the strengths of the players?

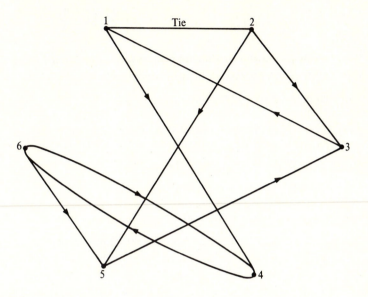

Fig. 25-7

25-6(a)   The following message has been encoded using the substitution and the matrix of Example 25-5: 29, 67, 19, 31, 74, 22, 40, 102, 53, 57, 135, 32. Decode it.
(b)   The following reply was received: 11, 23, 4, 18, 50, 53, 41, 96, 17, 35, 89, 65. Decode it. (Note: If the message does not break up evenly into vectors of the appropriate length, extra letters can be added at the end of the message. This has been done here.)
(c)   Explain why an *invertible* matrix was used in Example 25-5.

*25-7   Write a computer program for computing powers of a single matrix A (for counting routes, two-step wins, etc.).

*25-8   Write a computer program for coding and decoding messages, using a given $n \times n$ matrix.

Review of Chapter 4

1   A linear transformation is a function $T : \mathcal{V} \to \mathcal{W}$, where $\mathcal{V}$ and $\mathcal{W}$ are

_____ that has the following two properties:

(a)   _____   (T preserves addition)

(b) _____ (*T* preserves scalar
multiplication)

2 A linear transformation may be given either by a numerical rule or
else by a geometric rule. A linear transformation $T : \mathcal{V} \rightarrow \mathcal{W}$ is completely
determined by its effect on a _____ for $\mathcal{V}$.

3 If $T : \mathcal{R}^m \rightarrow \mathcal{R}^n$ is a linear transformation, then there is an $m \times n$
matrix associated with T. The rows of the usual matrix A are found by
determining the effect of T on the _____ _____ for $\mathcal{R}^m$. If
A is the usual matrix for T, then $\mathbf{X}T =$ _____ for every vector $\mathbf{X} \in \mathcal{R}^m$.

4 In order for the product AB of two matrices A and B to be defined, the
number of _____ of A must equal the number of _____ of B.
To find the i, j entry of AB we form the dot product of
_____ with _____.
    If $S : \mathcal{U} \rightarrow \mathcal{V}$ and $T : \mathcal{V} \rightarrow \mathcal{W}$ are linear transformations, then ST is the
linear transformation defined by the rule $\mathbf{X}(ST) =$ _____; that is, first
perform _____, and then perform _____.
    If A is the usual matrix for S and if B is the usual matrix for T, then the
matrix corresponding to ST is _____.

5 The inverse of a linear transformation T is the linear transformation
T^{-1} such that _____ $= I$ and _____ $= I$, where I is the linear transformation that has the rule $\mathbf{X}I =$ _____.
    The inverse of an $n \times n$ matrix A is the matrix A^{-1} such that
_____ $= I$ and _____ $= I$, where $I = \begin{bmatrix} & \\ & \end{bmatrix}$.

    To find the matrix A^{-1}:

(a) Form the matrix $[\ \vdots\ ]$.
(b) Reduce this matrix to $[\ \vdots\ ]$ and read off A^{-1}.
If A does not reduce to _____, then A has no _____.

6 If $T : \mathcal{R}^n \rightarrow \mathcal{R}^n$ is a linear transformation, the following are equivalent:

(a) T^{-1} exists.
(b) T is _____ $-$ _____.
(c) The only solution to $\mathbf{X}T = \mathbf{0}$ is $\mathbf{X} =$ _____.
(d) T carries the usual basis for $\mathcal{R}^n$ to a _____ for $\mathcal{R}^n$.

(e) T has rank = ____; that is, the image of T is ____-dimensional.

(f)   If A is the usual matrix for T, then det $A \neq$ ____.

7   If A is an $n \times n$ matrix, the following are equivalent:

(a) A^{-1} exists.

(b)   det $A \neq$ ____.

(c)   The rows of A are _____.

(d) A has rank = ____; that is, the reduced form for A is ____.

8   One of the most efficient methods for finding det A is to reduce A to

_____   _____ form. If A is in upper triangular

form, then det A is equal to the product of the elements on the ____

_____.

If A and B are $n \times n$ matrices, then det $AB =$ _____.

$|\det A|$ can be interpreted geometrically as _____

_____.

Review Exercises

1   Determine which of the following are linear transformations:

(a) $T: \mathcal{R}^2 \to \mathcal{R}^2$, where $(x, y)T = (x - y, x + 2y)$.

(b) $T: \mathcal{R}^2 \to \mathcal{R}^2$, where $(x, y)T = (x^2, 0)$.

(c) $T: \mathcal{R}^3 \to \mathcal{R}^3$, where $(x, y, z)T = (2x, y, z) + (1, 0, 0)$.

(d) $T: \mathcal{R}^3 \to \mathcal{R}^2$, where $(x, y, z)T = (0, x + y + z)$.

(e) $T: \mathcal{R}^2 \to \mathcal{R}^3$, where $(x, y)T = (y, 0, x)$.

(f) $T: \mathcal{R}^2 \to \mathcal{R}^2$, where $(x, y)T = (3x, -y)$.

(g) $T: \mathcal{R}^3 \to \mathcal{R}^3$, where $(x, y, z)T = (3x - 5y, x - y, 2x - 4y)$.

2   For each of the functions in Exercise 1 that is a linear transformation, find its usual matrix.

3(a)   For the functions in Exercise 1 that are linear transformations, which are invertible?

(b)   Find a numerical rule for each of the invertible linear transformations in Exercise 1.

4   Suppose that $T: \mathcal{R}^2 \to \mathcal{R}^2$ is the perpendicular projection onto the line $y = -x$. Find the usual matrix for T.

5   For each of these linear transformations T, find $(1, 2, -3)T$ and $(1, 2, -3)T^{-1}$:

(a)   The $45°$ clockwise rotation about the x axis in $\mathcal{R}^3$, as viewed from the positive side of the x axis.

(b)   The reflection in the plane $y = x$, followed by a counterclockwise rotation by $\pi/3$ radians about the y axis (as viewed from the positive side of the y axis).

(c) $(x, y, z)T = (3y + 4z, 2x - 3y, y - 4z)$.

6   Find the rule for ST where:

(a) $S = T = $ perpendicular projection onto the plane $y = x$ in $\mathcal{R}^3$.
(b) $(x, y, z)S = (\frac{1}{2}x, 3y - z, 2z)$, and T is the linear transformation defined in part (c) of Exercise 5.

7   Find the inverse (if it exists) for each of these matrices:

(a) $\begin{bmatrix} 1 & 1 & -3 & 4 \\ 2 & 3 & 0 & 1 \\ 1 & 0 & 2 & 0 \\ 3 & 7 & -6 & -9 \end{bmatrix}$ (b) $\begin{bmatrix} 1 & 2 & 4 \\ 0 & 1 & 3 \\ 2 & 3 & 5 \end{bmatrix}$

(c) $\begin{bmatrix} \frac{1}{2} & 1 & -2 \\ 1 & 0 & 3 \\ 2 & \frac{1}{4} & 0 \end{bmatrix}$ (d) $\begin{bmatrix} 1 & 5 & 0 \\ 3 & 1 & 2 \end{bmatrix}$

8   Let

$$A = \begin{bmatrix} 1 & 2 & 3 \\ 2 & 4 & 6 \\ 1 & 2 & 3 \end{bmatrix}$$

(a)   Find a nonzero matrix B such that

$$AB = \begin{bmatrix} 0 & 0 & 0 \\ 0 & 0 & 0 \\ 0 & 0 & 0 \end{bmatrix}$$

(b)   Is it possible to find a matrix C such that $AC = I$?

9   Solve for C:

$$C \begin{bmatrix} 2 & 4 & 0 \\ 0 & 4 & -2 \\ 0 & 2 & 2 \end{bmatrix} = \begin{bmatrix} 1 & 0 & 4 \\ 2 & 3 & 0 \end{bmatrix}$$

10   Suppose A, B, and C are three matrices such that $AB = AC$.

(a)   Give an example to show that B need not equal C.
(b)   If A^{-1} exists, is it true that $B = C$?

11   Prove or disprove that $(AB)^2 = A^2B^2$ for all $n \times n$ matrices A and B.

12 Let

$$A = \begin{bmatrix} 1 & 0 & 3 \\ 0 & 2 & 0 \\ 0 & 0 & -2 \end{bmatrix} \quad \text{and} \quad B = \begin{bmatrix} 2 & 4 & 5 \\ 4 & 0 & 5 \\ 2 & 12 & 10 \end{bmatrix}$$

Find det A, det B, and det AB.

13 Let

$$A = \begin{bmatrix} 1 & 2 & 0 \\ 0 & 3 & 5 \\ -1 & 1 & 5 \end{bmatrix}$$

Find det (A^5).

14 Let A and B be $n \times n$ matrices. Which of the following are always true:

(a) det A + det B = det $(A + B)$
(b) det (AB) = (det A)(det B)
(c) det (rA) = r(det A) for all scalars r
(d) det (A^k) = (det $A)^k$ for all positive integers k
(e) det (rA) = r^n(det A) for all scalars r

15 Find a number x so that

$$\det \begin{bmatrix} 5 & 1 & 0 \\ 1 & 0 & 2 \\ -3 & x & 1 \end{bmatrix} = 0$$

16 Suppose $T : \mathcal{R}^3 \to \mathcal{R}^3$ has its usual matrix

$$\begin{bmatrix} 1 & 1 & 2 \\ -2 & -2 & -4 \\ 1 & 1 & 2 \end{bmatrix}$$

Find the range of T.

17 Suppose $T : \mathcal{R}^3 \to \mathcal{R}^3$ has its usual matrix

$$\begin{bmatrix} \dfrac{\sqrt{2}}{2} & \dfrac{\sqrt{2}}{2} & 0 \\ -\dfrac{\sqrt{2}}{2} & \dfrac{\sqrt{2}}{2} & 0 \\ 0 & 0 & 1 \end{bmatrix}$$

Determine the geometric effect of T.

CHAPTER 5

THE SIMPLIFICATION OF MATRICES; EIGENVALUES AND EIGENVECTORS

In this chapter we consider the general problem of finding a *simple* matrix for a linear transformation T. The applications at the end of this chapter suggest some of the reasons for wanting such a matrix; in brief, simple matrices simplify problems.

Two types of problems will be considered:

(a) Given a geometric transformation T, to find a simple matrix B that allows us to perform numerical computations with T.

(b) Given a more or less complicated usual matrix A for T, to find a simpler matrix B that allows us to analyze what T does.

Both types of problems require finding a new basis, one on which the transformation T has a simple effect. The rows of the matrix B will tell us what T does to the new basis. If T's effect on the new basis is simple, B will also be simple.†

Problems of type (b) are dealt with by numerical computations involving the given matrix A; these are treated beginning in Sec. 29. To solve

---

† It is worth noting that this is yet another example of the transform-solve-invert process. We *transform* to a new basis, *solve* the transformed problem (easy, since the matrix B is simple), and *invert* the results back to obtain a solution of the original problem.

problems of type (a) we use the geometry of T to locate a basis on which T has a simple effect. Since, initially at least, problems of type (a) require no heavy computations, we begin with them.

26. MATRICES RELATIVE TO DIFFERENT BASES

Sometimes the usual matrix A for a geometric transformation is quite simple. For example, consider the linear transformation $T: \mathcal{R}^3 \to \mathcal{R}^3$ defined by the rule:

$\mathbf{X}T$ equals the perpendicular projection of $\mathbf{X}$ onto the xy plane.

Applying T to the usual basis vectors (Fig. 26-1), we have

$$(1, 0, 0)T = (1, 0, 0)$$
$$(0, 1, 0)T = (0, 1, 0)$$
$$(0, 0, 1)T = (0, 0, 0)$$

and thus the usual matrix for T is

$$A = \begin{bmatrix} 1 & 0 & 0 \\ 0 & 1 & 0 \\ 0 & 0 & 0 \end{bmatrix}$$

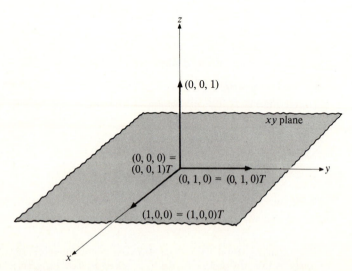

Fig. 26-1   The effect on the usual basis of the perpendicular projection onto the xy plane.

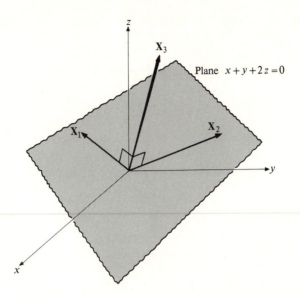

Fig. 26-2 X_1 and X_2 lie on the plane; X_3 is perpendicular to the plane.

For other geometric transformations, however, the usual matrix A might not be simple; also, it might be rather difficult to find directly. (We still need to find *some* matrix for the transformation, at the very least for efficient computation of its values.) For example, let $U : \mathcal{R}^3 \to \mathcal{R}^3$ be defined by the rule:

> XU equals the perpendicular projection of X onto the plane with equation $x + y + 2z = 0$.

The reader should convince himself of the difficulties in trying to find $(1, 0, 0)U$, $(0, 1, 0)U$, and $(0, 0, 1)U$. We will therefore postpone finding the usual matrix for U. Instead, we will look for a different matrix B for U. The rows of B will tell what U does, but not in terms of its effect on the usual basis; and B will actually be simpler than the usual matrix A.

Finding a matrix for the linear transformation T above was easy because we could readily see the geometric effect of T on the usual basis. This is the central idea in finding a simple matrix B for U: try to find a basis for $\mathcal{R}^3$ such that the effect of U on that basis is simple and is easy to see.

Because U is a projection onto the plane $x + y + 2z = 0$, the easiest basis to work with consists of two vectors X_1 and X_2 lying on that plane and a third vector X_3 perpendicular to the plane, as in Fig. 26-2. It is clear that X_1 and X_2 are not changed by U (since they already lie on the plane) and that $\mathbf{X}_3$ is projected to $\mathbf{0}$ (since $\mathbf{X}_3$ is perpendicular to the plane). That is,

$$
\begin{aligned}
\mathbf{X}_1 U &= \mathbf{X}_1 \\
\mathbf{X}_2 U &= \mathbf{X}_2 \\
\mathbf{X}_3 U &= \mathbf{0}
\end{aligned}
\tag{1}
$$

To form a matrix using the information in the system (1), we need to introduce numbers into the equations of (1) to use as entries of the matrix. To obtain these numbers, we write $\mathbf{X}_1 U$, $\mathbf{X}_2 U$, and $\mathbf{X}_3 U$ in terms of *coordinates* with respect to the basis $\mathbf{X}_1, \mathbf{X}_2, \mathbf{X}_3$. That is, we rewrite (1) as

$$
\begin{aligned}
\mathbf{X}_1 U &= 1\mathbf{X}_1 + 0\mathbf{X}_2 + 0\mathbf{X}_3 \\
\mathbf{X}_2 U &= 0\mathbf{X}_1 + 1\mathbf{X}_2 + 0\mathbf{X}_3 \\
\mathbf{X}_3 U &= 0\mathbf{X}_1 + 0\mathbf{X}_2 + 0\mathbf{X}_3
\end{aligned}
$$

in order to exhibit the coordinates for $\mathbf{X}_1 U$, $\mathbf{X}_2 U$, and $\mathbf{X}_3 U$. We then use these coordinates to form the rows of a matrix. Thus, the matrix B for U, relative to the basis $\mathbf{X}_1, \mathbf{X}_2, \mathbf{X}_3$, is

$$
B = \begin{bmatrix} 1 & 0 & 0 \\ 0 & 1 & 0 \\ 0 & 0 & 0 \end{bmatrix}
$$

We are not yet able to use this matrix in any computational way. For example, we are still unable to compute $(1, 2, 7)U$. In Sec. 28 we will learn how to use B to obtain the usual matrix A for U, at which time we will be able to compute $(1, 2, 7)U$ via the matrix product $(1, 2, 7)A$.

As in this last example, many problems are greatly simplified by changing to a new basis and then using that new basis to find a simple matrix. We now note the formal description of how to obtain a matrix for a linear transformation, using any basis.

Definition

Let $T : \mathcal{U} \to \mathcal{U}$ be a linear transformation and let $\mathbf{X}_1, \mathbf{X}_2, \ldots, \mathbf{X}_n$ be a basis for $\mathcal{U}$. Suppose that

$$
\begin{aligned}
\mathbf{X}_1 T &= a_{11}\mathbf{X}_1 + a_{12}\mathbf{X}_2 + \cdots + a_{1n}\mathbf{X}_n \\
\mathbf{X}_2 T &= a_{21}\mathbf{X}_1 + a_{22}\mathbf{X}_2 + \cdots + a_{2n}\mathbf{X}_n \\
&\ \ \vdots \\
\mathbf{X}_n T &= a_{n1}\mathbf{X}_1 + a_{n2}\mathbf{X}_2 + \cdots + a_{nn}\mathbf{X}_n
\end{aligned}
$$

The **matrix for** T **relative to the basis** $\mathbf{X}_1, \mathbf{X}_2, \ldots, \mathbf{X}_n$ is

$$\begin{bmatrix} a_{11} & a_{12} & \cdots & a_{1n} \\ a_{21} & a_{22} & \cdots & a_{2n} \\ \cdots\cdots\cdots\cdots\cdots\cdots\cdots \\ a_{n1} & a_{n2} & \cdots & a_{nn} \end{bmatrix}$$

Note that this method for obtaining a matrix relative to a new basis is analogous to the method used for obtaining the usual matrix:

(a)   We apply the transformation to the basis vectors $\mathbf{X}_1, \mathbf{X}_2, \ldots, \mathbf{X}_n$.

(b)   We use the *coordinates* (with respect to the basis $\mathbf{X}_1, \mathbf{X}_2, \ldots, \mathbf{X}_n$) of the results to form the rows of the matrix.

It has been mentioned before that coordinates provide the means of attaching numerical labels to vectors. Here coordinates provide the numerical entries of a matrix for a linear transformation. We now return to some examples of geometric transformations and to the problem of finding simple matrices B relative to carefully chosen new bases. The letter A will be consistently used for the usual matrix of a linear transformation, and the letter B for a (presumably simpler) matrix relative to a new basis.

Example 26-1

Let $T : \mathcal{R}^3 \to \mathcal{R}^3$ be the reflection in the plane $-3x + 2y - 4z = 0$. For our basis for $\mathcal{R}^3$ we choose any two vectors $\mathbf{X}_1$ and $\mathbf{X}_2$ that form a basis for the plane and we choose a third vector $\mathbf{X}_3$ perpendicular to the plane. (See Fig. 26-3.) We have

$$\mathbf{X}_1 T = \mathbf{X}_1$$
$$\mathbf{X}_2 T = \mathbf{X}_2$$
$$\mathbf{X}_3 T = -\mathbf{X}_3$$

In order to obtain a matrix, we supply all the coordinates:

$$\mathbf{X}_1 T = 1\mathbf{X}_1 + 0\mathbf{X}_2 + 0\mathbf{X}_3$$
$$\mathbf{X}_2 T = 0\mathbf{X}_1 + 1\mathbf{X}_2 + 0\mathbf{X}_3$$
$$\mathbf{X}_3 T = 0\mathbf{X}_1 + 0\mathbf{X}_2 + (-1)\mathbf{X}_3$$

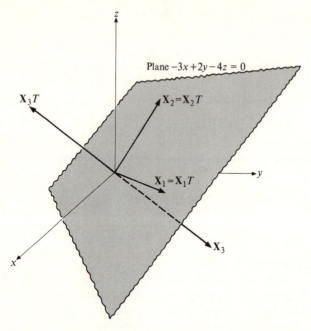

Fig. 26-3 Effect of the reflection in the plane $-3x + 2y - 4z = 0$ on a suitable basis.

We then use these coordinates as rows of the matrix for T with respect to the basis $\mathbf{X}_1, \mathbf{X}_2, \mathbf{X}_3$:

$$
B = \begin{bmatrix} 1 & 0 & 0 \\ 0 & 1 & 0 \\ 0 & 0 & -1 \end{bmatrix}
$$

One possible choice for a new basis is $\mathbf{X}_1 = (\frac{2}{3}, 1, 0)$, $\mathbf{X}_2 = (-\frac{4}{3}, 0, 1)$, $\mathbf{X}_3 = (-3, 2, -4)$.† A method for obtaining the basis vectors specifically is outlined in the exercises at the end of this section.   ●

Example 26-2

Let $U : \mathcal{R}^3 \to \mathcal{R}^3$ be the 90° counterclockwise rotation‡ about the line $\{a(1, 2, -2) : a \in \mathcal{R}\}$. As our basis for $\mathcal{R}^3$ we choose $\mathbf{X}_1$ lying on the axis of rotation and we choose $\mathbf{X}_2$ and $\mathbf{X}_3$ perpendicular both to $\mathbf{X}_1$ and to

---

† The vector $(-3, 2, -4)$ is perpendicular to the plane $-3x + 2y - 4z = 0$. In general, (a, b, c) is perpendicular to $ax + by + cz = d$. See Sec. 33 for a proof. A similar result is true in $\mathcal{R}^2$: (a, b) is perpendicular to the line $ax + by = c$.
‡ Whenever we deal with a rotation about an axis in $\mathcal{R}^3$, *counterclockwise* and *clockwise* mean as viewed looking toward the origin from a specified point on the axis of rotation. In this case the specified point is $(1, 2, -2)$.

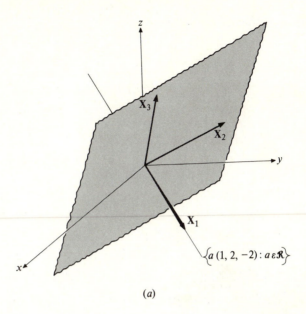

(a)

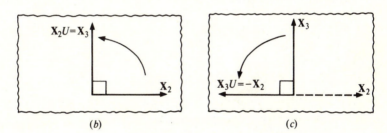

(b)   (c)

Fig. 26-4   Effect of the 90° rotation U on a suitable basis.

each other. In addition, it is desirable that $\mathbf{X}_2$ and $\mathbf{X}_3$ have the same length (Fig. 26-4a). Figures 26-4b and 26-4c illustrate the effect of U on the basis vectors $\mathbf{X}_2$ and $\mathbf{X}_3$, respectively. Thus we have

$$\mathbf{X}_1 U = \mathbf{X}_1$$
$$\mathbf{X}_2 U = \mathbf{X}_3$$
$$\mathbf{X}_3 U = -\mathbf{X}_2$$

Supplying coordinates again, we have

$$\mathbf{X}_1 U = 1\mathbf{X}_1 + 0\mathbf{X}_2 + 0\mathbf{X}_3$$
$$\mathbf{X}_2 U = 0\mathbf{X}_1 + 0\mathbf{X}_2 + 1\mathbf{X}_3$$
$$\mathbf{X}_3 U = 0\mathbf{X}_1 + (-1)\mathbf{X}_2 + 0\mathbf{X}_3$$

and we thus obtain the matrix

$$B = \begin{bmatrix} 1 & 0 & 0 \\ 0 & 0 & 1 \\ 0 & -1 & 0 \end{bmatrix}$$

for U with respect to the basis $\mathbf{X}_1$, $\mathbf{X}_2$, $\mathbf{X}_3$. Again, one possible choice for the new basis is $\mathbf{X}_1 = (1, 2, -2)$, $\mathbf{X}_2 = (-2, 1, 0)$, $\mathbf{X}_3 = (\frac{2}{3}, \frac{4}{3}, \frac{5}{3})$.   ●

Exercises

26-1 Let $T : \mathcal{R}^2 \to \mathcal{R}^2$ be the reflection in the line $\{a(1, 5) : a \in \mathcal{R}\}$.

(a) Sketch a suitable new basis for $\mathcal{R}^2$ (one that T has a simple effect on) and then find a simple matrix B for T with respect to this new basis. (Hint: In your sketch let $\mathbf{X}_1$ be any vector on the line and let $\mathbf{X}_2$ be any vector perpendicular to the line. Deduce what $\mathbf{X}_1 T$ and $\mathbf{X}_2 T$ are in terms of $\mathbf{X}_1$ and $\mathbf{X}_2$.)
(b) Find such a basis specifically. (Hint: The given line has the equation $5x - y = 0$.)

26-2 Let $S : \mathcal{R}^2 \to \mathcal{R}^2$ be the reflection in the line $y = x/3$.

(a) Sketch a suitable new basis for $\mathcal{R}^2$, and then find a simple matrix B for S with respect to this new basis.
(b) Find such a basis specifically.

26-3 Let $U : \mathcal{R}^3 \to \mathcal{R}^3$ be the perpendicular projection onto the plane $x - 3y + 2z = 0$.

(a) Sketch a suitable new basis for $\mathcal{R}^3$ and then find a simple matrix B for U with respect to this new basis.
(b) Find such a basis specifically. [Hint: To find two basis vectors on the plane, write the plane as the set of points of the form $(3a - 2b, a, b)$ and then separate the a and b parts.]

26-4 Let $T : \mathcal{R}^2 \to \mathcal{R}^2$ be the perpendicular projection onto the line $\{a(2, -1) : a \in \mathcal{R}\}$.

(a) Sketch a suitable new basis for $\mathcal{R}^2$ and then find a simple matrix B for T with respect to this basis.
(b) Find such a basis specifically.

26-5 Let $S : \mathcal{R}^3 \to \mathcal{R}^3$ be the $90°$ counterclockwise rotation about the line $\{a(1, 1, 1) : a \in \mathcal{R}\}$. Sketch a suitable new basis for $\mathcal{R}^3$ and then find a simple matrix B for S with respect to this basis.

26-6 Suppose that $T : \mathcal{V} \to \mathcal{V}$ is a linear transformation that satisfies

$$\begin{aligned}
\mathbf{B}_1 T &= \mathbf{B}_2 \\
\mathbf{B}_2 T &= \mathbf{B}_3 - 7\mathbf{B}_4 \\
\mathbf{B}_3 T &= 5\mathbf{B}_1 - 2\mathbf{B}_2 \\
\mathbf{B}_4 T &= \mathbf{B}_4 - \mathbf{B}_3
\end{aligned}$$

where $\mathbf{B}_1, \mathbf{B}_2, \mathbf{B}_3, \mathbf{B}_4$ is a basis for $\mathcal{V}$. Find the matrix B for T with respect to this basis.

26-7 Suppose that $T : \mathcal{R}^2 \to \mathcal{R}^2$ has the matrix

$$B = \begin{bmatrix} 1 & 0 \\ 0 & 0 \end{bmatrix}$$

relative to the basis $\mathbf{B}_1, \mathbf{B}_2$, where $\mathbf{B}_1$ is on the line $y = x$ and $\mathbf{B}_2$ is perpendicular to this line. Describe T geometrically. (Hint: First sketch $\mathbf{B}_1$ and $\mathbf{B}_2$ and their images $\mathbf{B}_1 T$ and $\mathbf{B}_2 T$. Then make a reasonable guess as to what T does.)

26-8 Suppose that $U : \mathcal{R}^3 \to \mathcal{R}^3$ has the matrix

$$B = \begin{bmatrix} 0 & 0 & 0 \\ 0 & 1 & 0 \\ 0 & 0 & 1 \end{bmatrix}$$

with respect to the basis $\mathbf{B}_1, \mathbf{B}_2, \mathbf{B}_3$, where $\mathbf{B}_1$ lies on the line $\{a(1, 3, 3) : a \in \mathcal{R}\}$ and $\mathbf{B}_2$ and $\mathbf{B}_3$ are perpendicular to this line. Describe U geometrically.

26-9 Suppose that the linear transformation $T : \mathcal{R}^3 \to \mathcal{R}^3$ has the rule $(x, y, z)T = (2x, 2y, 7y - 5z)$.

(a) Find the usual matrix A for T.
(b) Find the matrix B for T relative to the basis $(1, 0, 0)$, $(0, 1, 1)$, $(0, 0, 1)$. (Note: *Coordinates*, not actual vectors, go into the rows of B.)
(c) Use the matrix B to describe T geometrically.

26-10 Let $D : \mathcal{P}^3 \to \mathcal{P}^3$ be the derivative transformation; that is, $p(x)D$ equals the derivative of $p(x)$. Find the matrix for D relative to the basis x^3, x^2, x, 1.

27.   COORDINATES

Since it is the coordinates of $X_1 T$, $X_2 T$, ..., $X_n T$ that appear as the entries of the new matrix B for T (with respect to the basis X_1, X_2, ..., X_n), it is appropriate that we recall some of the details concerning coordinates, first dealt with in Sec. 16.

Recall that, by definition, the coordinates of a vector X with respect to a basis X_1, X_2, ..., X_n are the unique numbers a_1, a_2, ..., a_n such that

$$X = a_1 X_1 + a_2 X_2 + \cdots + a_n X_n$$

For example, relative to the basis $(2, 1)$, $(-1, 3)$ for $\mathcal{R}^2$, the coordinates of the vector $(6, 10)$ are 4 and 2, since

$$(6, 10) = 4(2, 1) + 2(-1, 3)$$

If we interpret these vectors as arrows in the plane, the coordinates can be viewed as "stretch factors." That is, if we stretch the vector $(2, 1)$ by a factor of 4 and stretch the vector $(-1, 3)$ by a factor of 2, and then add the results, we obtain $(6, 10)$, as in Fig. 27-1.

Relative to the basis $(-2, 6)$, $(1, 4)$, the same vector $(6, 10)$ has coordinates -1 and 4, since

$$(6, 10) = (-1)(-2, 6) + 4(1, 4)$$

Figure 27-2 illustrates this.

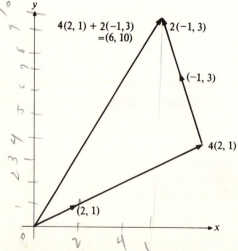

Fig. 27-1   Coordinates of $(6, 10)$ relative to the basis $(2, 1)$, $(-1, 3)$.

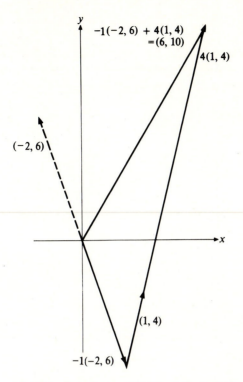

Fig. 27-2 Coordinates of $(6, 10)$ relative to the basis $(-2, 6), (1, 4)$.

Relative to the usual basis $(1, 0), (0, 1)$, the vector $(6, 10)$ has coordinates 6 and 10, since

$$(6, 10) = 6(1, 0) + 10(0, 1)$$

as shown in Fig. 27-3. In this case the coordinates of $(6, 10)$ turn out to be the entries of the vector itself. It is *only* relative to the usual basis for $\mathcal{R}^n$ that the coordinates of a vector equal the entries of that vector. The reader should be sure to note the difference between the coordinates of a vector and its entries.

To distinguish between vectors that are ordered *n*-tuples, such as (x, y), $(1, 2, 3)$, etc., and their coordinates with respect to a given basis, a new notation is introduced. If the vector **X** has coordinates a, b, c relative to a given basis, we call $[a \quad b \quad c]$ the *coordinate matrix* of **X** relative to the basis. Note that $[a \quad b \quad c]$ is a matrix and not a vector in $\mathcal{R}^3$. The square brackets and the lack of commas are a gentle reminder of this.

Referring to the previous example, the vector $(6, 10)$ has coordinate matrix $[4 \quad 2]$ relative to the basis $(2, 1), (-1, 3)$; it has coordinate matrix

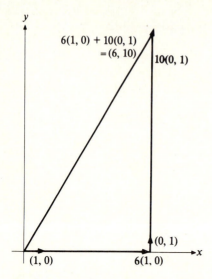

Fig. 27-3 Coordinates of (6, 10) relative to the usual basis.

$[-1 \quad 4]$ relative to the basis $(-2, 6), (1, 4)$; and it has coordinate matrix $[6 \quad 10]$ relative to the usual basis.

The notation $[\mathbf{X}]$ will denote the coordinate matrix for a vector $\mathbf{X}$ relative to a given basis. On occasion we will need to use coordinate matrices for the same vector $\mathbf{X}$ relative to an old basis as well as relative to a new basis. In this case the notation $[\mathbf{X}]_{\text{new}}$ will be used for the coordinate matrix of $\mathbf{X}$ relative to the new basis and $[\mathbf{X}]_{\text{old}}$ for the coordinate matrix of $\mathbf{X}$ relative to the old basis.

For example, if $\mathbf{X} = (6, 10)$, relative to the new basis $(2, 1), (-1, 3)$, we have $[\mathbf{X}]_{\text{new}} = [4 \quad 2]$. Relative to the usual basis we have $[\mathbf{X}]_{\text{old}} = [6 \quad 10]$.

We now develop a matrix method for finding coordinates relative to new bases. Finding the coordinates of $\mathbf{X}$ relative to the basis $\mathbf{X}_1, \mathbf{X}_2, \ldots,$ $\mathbf{X}_n$ requires solving the equation $\mathbf{X} = a_1\mathbf{X}_1 + a_2\mathbf{X}_2 + \cdots + a_n\mathbf{X}_n$ for the unknown coordinates $a_1, a_2, \ldots, a_n$. The following example illustrates the process.

Example 27-1

Let $\mathbf{X} = (a, b, c) \in \mathcal{R}^3$. To find $[\mathbf{X}]_{\text{new}}$ relative to the basis $(1, 2, 3),$ $(0, 1, 2), (-2, 3, 1)$, we must find $x, y,$ and z such that

$$(a, b, c) = x(1, 2, 3) + y(0, 1, 2) + z(-2, 3, 1)$$

This yields the system of equations (with x, y, and z as unknowns)

$$\begin{aligned} x \qquad\ -\ 2z &= a \\ 2x +\ y + 3z &= b \\ 3x + 2y +\ z &= c \end{aligned}$$

just as in Sec. 16. The key to an effective method is to introduce matrix notation, including $[x \quad y \quad z]$ for the matrix of coordinates. Using the definition of matrix multiplication, we can write this system as

$$[x \quad y \quad z]\begin{bmatrix} 1 & 2 & 3 \\ 0 & 1 & 2 \\ -2 & 3 & 1 \end{bmatrix} = [a \quad b \quad c]$$

Note that the new basis vectors appear as *rows* in the 3×3 matrix. If we let

$$P = \begin{bmatrix} 1 & 2 & 3 \\ 0 & 1 & 2 \\ -2 & 3 & 1 \end{bmatrix}$$

we can rewrite the system once more as

$$[x \quad y \quad z]P = [a \quad b \quad c]$$

That is,

$$[\mathbf{X}]_{new} P = [\mathbf{X}]_{old} \qquad\qquad (2)$$

We also have

$$[\mathbf{X}]_{new} = [\mathbf{X}]_{old} P^{-1}$$

if we multiply both sides of equation (2) on the right by P^{-1}.

To complete the solution, we need only find P^{-1} by reducing $[P \vdots I]$ to $[I \vdots P^{-1}]$ (as in Sec. 23) and compute the matrix product:

$$\begin{aligned} [\mathbf{X}]_{new} &= [a \quad b \quad c]P^{-1} \\ &= [a \quad b \quad c]\begin{bmatrix} \frac{5}{7} & -1 & -\frac{1}{7} \\ \frac{4}{7} & -1 & \frac{2}{7} \\ -\frac{2}{7} & 1 & -\frac{1}{7} \end{bmatrix} \\ &= [\tfrac{5}{7}a + \tfrac{4}{7}b - \tfrac{2}{7}c \quad -a - b + c \quad -\tfrac{1}{7}a + \tfrac{2}{7}b - \tfrac{1}{7}c] \quad \bullet \end{aligned}$$

The preceding example illustrates the following theorem (and its proof).

Theorem 27-1   For any vector $\mathbf{X}$ in $\mathcal{R}^n$,

1
$$[\mathbf{X}]_{new} = [\mathbf{X}]_{old}\, P^{-1}$$

2
$$[\mathbf{X}]_{old} = [\mathbf{X}]_{new}\, P$$

where the new basis vectors form the rows of P.

P takes coord of x wrt the basis $(x_1 - x_n)$ to coord wrt the usual basis

Let us consider a second example illustrating the use of this theorem.

Example 27-2

Suppose that we wish to compute the coordinates of the vector $(1, 2, 3)$ relative to the basis $(1, 2, 1)$, $(1, 0, 1)$, $(0, 1, 1)$ for $\mathcal{R}^3$. We use the basis vectors to form the rows of the matrix

$(1,2,3)[P] = (3,5,4)$

$$P = \begin{bmatrix} 1 & 2 & 1 \\ 1 & 0 & 1 \\ 0 & 1 & 1 \end{bmatrix}$$

We then compute

$$P^{-1} = \begin{bmatrix} \frac{1}{2} & \frac{1}{2} & -1 \\ \frac{1}{2} & -\frac{1}{2} & 0 \\ -\frac{1}{2} & \frac{1}{2} & 1 \end{bmatrix}$$

Using part (1) of the above theorem, we obtain the new coordinates of the vector $(1, 2, 3)$ by multiplying by P^{-1}:

$$[1 \quad 2 \quad 3]P^{-1} = [1 \quad 2 \quad 3]\begin{bmatrix} \frac{1}{2} & \frac{1}{2} & -1 \\ \frac{1}{2} & -\frac{1}{2} & 0 \\ -\frac{1}{2} & \frac{1}{2} & 1 \end{bmatrix} = [0 \quad 1 \quad 2]$$

Thus the new coordinate matrix of $(1, 2, 3)$ is $[0 \quad 1 \quad 2]$.   ●

Recall that when we dealt with the usual matrix A for a linear transformation T, we obtained the equation

$$XT = XA$$

In other words, we could compute XT by the simple matrix product XA. When dealing with a new basis, and hence dealing with new coordinates, we have an analogous situation. Only now we must use the new matrix B for T, relative to the new basis. That is, the coordinates of $\mathbf{X}T$ relative to the new basis can be computed by the simple matrix product $[\mathbf{X}]_{new} B$. For use in the next section, we state this fact formally.

Theorem 27-2

Let $T : \mathcal{R}^n \to \mathcal{R}^n$ be a linear transformation. Let B be the matrix for T relative to the basis $\mathbf{X}_1, \mathbf{X}_2, \ldots, \mathbf{X}_n$. Then

$$[\mathbf{X}T]_{new} = [\mathbf{X}]_{new} B$$

for all vectors $\mathbf{X}$ in $\mathcal{R}^n$.

The proof for $T : \mathcal{R}^2 \to \mathcal{R}^2$ is outlined in Exercise 27-8.

$2(1,0,0,1) + \frac{3}{2}(0,2,2,0) + (-1)(0,0,4,0) + (1)(0,0,1,2)$

$(2,3,0,4)$

Exercises

In Exercises 27-1 to 27-4, begin by finding the matrix P, and proceed as in Example 27-2.

27-1 Let $\mathbf{X} = (1, -2)$.

(a)   Find $[\mathbf{X}]$ relative to the basis $(2, 3)$, $(-1, 4)$.
(b)   Find $[\mathbf{X}]$ relative to the basis $(1, 2)$, $(5, 0)$.

27-2 Let $\mathbf{X} = (1, -5, -10)$. Find $[\mathbf{X}]$ relative to the basis $(3, 0, 1)$, $(2, -1, 0)$, $(1, 2, 4)$. $(2, -1, 3)$

27-3 Relative to the new basis $(1, 0, 0, 1)$, $(0, 2, 2, 0)$, $(0, 0, 4, 0)$, $(0, 0, 1, 2)$ for $\mathcal{R}^4$, find the coordinates of the vectors $(2, 3, 0, 4)$, $(1, 1, 1, -1)$, and (a, b, c, d). $(1, 1, 3, -10)$

27-4(a)   Find the coordinates of $(5, 3, 0)$, $(1, 1, 1)$, and (x, y, z) with respect to the new basis $\mathbf{X}_1 = (1, 0, 2)$, $\mathbf{X}_2 = (1, 1, 2)$, and $\mathbf{X}_3 = (2, 1, 5)$ for $\mathcal{R}^3$.
(b)   Find the old coordinates of $2\mathbf{X}_1 - \mathbf{X}_2$ by means of the matrix P.

27-5 In Theorem 27-1, how do we know that P^{-1} always exists?

27-6 Referring to the T and $\mathbf{X}_1, \mathbf{X}_2, \mathbf{X}_3$ of Example 26-1, find $[\mathbf{X}_1 T]_{new}$, $[\mathbf{X}_2 T]_{new}$, and $[\mathbf{X}_3 T]_{new}$.

27-7 Suppose that X has coordinate matrix $[a_1 \quad a_2 \quad \cdots \quad a_n]$ and $\dot{Y}$ has coordinate matrix $[b_1 \quad b_2 \quad \cdots \quad b_n]$ relative to a basis $X_1, X_2, \ldots, X_n$. Show that:

(a) $X + Y$ has coordinate matrix $[a_1 + b_1 \quad a_2 + b_2 \quad \cdots \quad a_n + b_n]$.
(b) rX has coordinate matrix $[ra_1 \quad ra_2 \quad \cdots \quad ra_n]$.

27-8 This exercise outlines a proof of Theorem 27-2 for the case $n = 2$. Let X_1, X_2 be a basis for $\mathcal{R}^2$ and let $T : \mathcal{R}^2 \to \mathcal{R}^2$ satisfy

$$X_1 T = aX_1 + bX_2$$
$$X_2 T = cX_1 + dX_2 \qquad \begin{bmatrix} a & b \\ c & d \end{bmatrix}$$

(a) Find the matrix B for T relative to the basis X_1, X_2.
(b) An arbitrary vector X in $\mathcal{R}^2$ can be written as $rX_1 + sX_2$. Explain why.
(c) Suppose that $X = rX_1 + sX_2$. Show that $XT = (ra + sc)X_1 + (rb + sd)X_2$.
(d) Using part (c) of this exercise, find $[XT]_{\text{new}}$.
(e) Using part (b), find $[X]_{\text{new}}$.
(f) Find $[X]_{\text{new}} B$. Your answers to parts (d) and (f) should be equal. This proves that $[XT]_{\text{new}} = [X]_{\text{new}} B$.

*27-9 Write a computer program that computes new coordinates, using the method of Theorem 27-1.

28. THE RELATIONSHIP BETWEEN THE USUAL MATRIX AND A NEW MATRIX FOR A LINEAR TRANSFORMATION

In Sec. 26, we considered the problem of finding a simple matrix B for a geometric transformation T. In this section we derive a formula relating the usual matrix A to the new matrix B, and we use it to find the usual matrix for geometric transformations such as projections, reflections, and rotations in $\mathcal{R}^2$ and $\mathcal{R}^3$.

Since B is the matrix used to perform computations with *new* coordinates (Theorem 27-2), it is not surprising that the matrix P used in changing coordinates (Theorem 27-1) enters into the formula.

Theorem 28-1   Let $T : \mathcal{R}^n \to \mathcal{R}^n$ be a linear transformation. Let A be the usual matrix for T and let B be the matrix for T relative to a new basis. Then

$$A = P^{-1}BP \qquad\qquad (1)$$

or, equivalently,

$$B = PAP^{-1} \tag{2}$$

where P is the matrix whose rows are the new basis vectors.

Proof   We know that $\mathbf{X}A = \mathbf{X}T$ for all vectors $\mathbf{X}$. On the other hand, using B, we can compute $\mathbf{X}T$ for any vector $\mathbf{X}$ by the following process, according to the results of Sec. 27:

(a)   Find $[\mathbf{X}]_{\text{new}}$, by means of the product $\mathbf{X}P^{-1}$ (Theorem 27-1).

(b)   Multiply by B on the right to obtain the new coordinates of $\mathbf{X}T$ (Theorem 27-2), yielding $\mathbf{X}P^{-1}B$.

(c)   Multiply the new coordinates by P to find the old coordinates (Theorem 27-1), yielding $\mathbf{X}P^{-1}BP$.

Since the result of this three-step process is $\mathbf{X}T$, we have

$$\mathbf{X}(P^{-1}BP) = \mathbf{X}T$$

Since we also have

$$\mathbf{X}A = \mathbf{X}T$$

it can be seen that

$$\mathbf{X}A = \mathbf{X}(P^{-1}BP)$$

for all vectors $\mathbf{X}$. Therefore, we must have

$$A = P^{-1}BP$$

This proves formula (1).
   If we solve $A = P^{-1}BP$ for B, we obtain formula (2).   ■

This theorem is especially important. It has many uses, including finding the usual matrix A for certain geometric linear transformations. Whenever the relationship $A = P^{-1}BP$ is used, we are simplifying a problem by *transforming* to a new basis (by means of P^{-1}), *solving* (by means of the simpler matrix B), and *inverting* (by means of P) to find a solution to the original problem. The following examples illustrate this problem-solving technique.

Example 28-1

Let $T : \mathcal{R}^2 \to \mathcal{R}^2$ be the reflection in the line $\{a(1, 6) : a \in \mathcal{R}\}$. We want to find the usual matrix A for T. We will first find a simple matrix B for T, and then use Theorem 28-1 to find the usual matrix A. The easiest basis to work with consists of a vector $\mathbf{X}_1$ on the given line and a vector $\mathbf{X}_2$ perpendicular to the line. Then $\mathbf{X}_1 T = \mathbf{X}_1$ and $\mathbf{X}_2 T = -\mathbf{X}_2$, so that we have

$$B = \begin{bmatrix} 1 & 0 \\ 0 & -1 \end{bmatrix}$$

In order to be able to use Theorem 28-1, we must have specific vectors $\mathbf{X}_1$ and $\mathbf{X}_2$, since they will form the rows of P. The vector $\mathbf{X}_1 = (1, 6)$ lies on the line and $\mathbf{X}_2 = (-6, 1)$ is perpendicular to the line.† Using these vectors as rows, we have

$$P = \begin{bmatrix} 1 & 6 \\ -6 & 1 \end{bmatrix}$$

The reader can verify, by reducing $[P \mid I]$ to $[I \mid P^{-1}]$, that

$$P^{-1} = \begin{bmatrix} \frac{1}{37} & -\frac{6}{37} \\ \frac{6}{37} & \frac{1}{37} \end{bmatrix}$$

Lastly, we compute the usual matrix A for T by means of formula (1) in Theorem 28-1:

$$A = P^{-1}BP$$
$$= (P^{-1}B)P$$
$$= \begin{bmatrix} \frac{1}{37} & \frac{6}{37} \\ \frac{6}{37} & -\frac{1}{37} \end{bmatrix} \begin{bmatrix} 1 & 6 \\ -6 & 1 \end{bmatrix}$$
$$= \begin{bmatrix} -\frac{35}{37} & \frac{12}{37} \\ \frac{12}{37} & \frac{35}{37} \end{bmatrix}$$

●

Example 28-2

Let U be the perpendicular projection of $\mathcal{R}^3$ onto the plane $x + y + 2z = 0$. (This is the linear transformation discussed in Sec. 26.) We want to find its usual matrix A. We know that relative to the basis $\mathbf{X}_1, \mathbf{X}_2, \mathbf{X}_3$,

† In general, $(-b, a)$ is perpendicular to (a, b). (See Sec. 33.)

where $\mathbf{X}_1$ and $\mathbf{X}_2$ lie on the plane and $\mathbf{X}_3$ is perpendicular to the plane, its matrix is

$$B = \begin{bmatrix} 1 & 0 & 0 \\ 0 & 1 & 0 \\ 0 & 0 & 0 \end{bmatrix}$$

Since $A = P^{-1}BP$, we must find P and P^{-1}, and therefore we must know $\mathbf{X}_1$, $\mathbf{X}_2$, and $\mathbf{X}_3$ specifically since these form the rows of P. Writing the plane as the set of points of the form $(-a - 2b, a, b)$ and separating the a and the b parts, we have $a(-1, 1, 0) + b(-2, 0, 1)$. Thus $\mathbf{X}_1$ and $\mathbf{X}_2$ may be chosen as $(-1, 1, 0)$ and $(-2, 0, 1)$. We may choose $(1, 1, 2)$ for $\mathbf{X}_3$. These give the three rows of P:

$$P = \begin{bmatrix} -1 & 1 & 0 \\ -2 & 0 & 1 \\ 1 & 1 & 2 \end{bmatrix},$$

Reducing $[P \mid I]$ to $[I \mid P^{-1}]$, we find

$$P^{-1} = \begin{bmatrix} -\frac{1}{6} & -\frac{1}{3} & \frac{1}{6} \\ \frac{5}{6} & -\frac{1}{3} & \frac{1}{6} \\ -\frac{1}{3} & \frac{1}{3} & \frac{1}{3} \end{bmatrix}$$

Therefore, the usual matrix for U is

$$A = P^{-1}BP$$
$$= \begin{bmatrix} \frac{5}{6} & -\frac{1}{6} & -\frac{1}{3} \\ -\frac{1}{6} & \frac{5}{6} & -\frac{1}{3} \\ -\frac{1}{3} & -\frac{1}{3} & \frac{1}{3} \end{bmatrix}$$

To compute $\mathbf{X}U$ for any vector $\mathbf{X}$ we form the matrix product $\mathbf{X}A$. For example, $(1, 1, 0)U = (1, 1, 0)A = (\frac{2}{3}, \frac{2}{3}, -\frac{2}{3})$. Thus the product $\mathbf{X}A$ allows us to find the perpendicular projection of any vector onto the plane $x + y + 2z = 0$. ●

It should be noted that the matrices A and B represent the same linear transformation, relative to different bases. In fact, any linear transformation will have many matrix representations, one for each choice of basis.†

† This is analogous to the familiar fact that any rational number has many representations as a fraction. For example $\frac{9}{15}, \frac{3}{5}, \frac{6}{10}, \frac{33}{55}$, etc., all represent the same number.

Definition

Two $n \times n$ matrices A and B are said to be **similar** provided each represents the same linear transformation T relative to some basis. Equivalently, A and B are similar provided $B = PAP^{-1}$ for some $n \times n$ matrix P.

In general, we seek a simple matrix B that is similar to the usual matrix A for a given linear transformation. We hope that the simple matrix B will reduce the complexity of computations. The following theorem points out certain relationships between A and B, each of which indicates a potential benefit of working with a simple matrix B.

Theorem 28-2

Suppose that A and B are similar $n \times n$ matrices; that is, $B = PAP^{-1}$ for some matrix P. Then

1   $\det A = \det B$.

2   A is invertible $\Leftrightarrow B$ is invertible.

3   $A^k = P^{-1}B^k P$ for $k = 1, 2, 3, \ldots$.

Proof

We leave the verifications as exercises.   ∎

The following example illustrates the use of this theorem.

Example 28-3

Let

$$A = \begin{bmatrix} 6 & 6 & 4 \\ -2 & -1 & -2 \\ -5 & -6 & -3 \end{bmatrix}$$

Suppose we wish to find A^{10}. Although we could find A^{10} directly, we know from Theorem 28-2 that $A^{10} = P^{-1}B^{10}P$. It can be shown that A is similar to

$$B = \begin{bmatrix} 1 & 0 & 0 \\ 0 & 2 & 0 \\ 0 & 0 & -1 \end{bmatrix}$$

with

$$P = \begin{bmatrix} 1 & 0 & 1 \\ 1 & 2 & 0 \\ 1 & 1 & 1 \end{bmatrix}$$

It is easily determined that

$$P^{-1} = \begin{bmatrix} 2 & 1 & -2 \\ -1 & 0 & 1 \\ -1 & -1 & 2 \end{bmatrix}$$

Whereas A^{10} would be relatively difficult to compute directly, by inspection

$$B^{10} = \begin{bmatrix} 1^{10} & 0 & 0 \\ 0 & 2^{10} & 0 \\ 0 & 0 & (-1)^{10} \end{bmatrix} = \begin{bmatrix} 1 & 0 & 0 \\ 0 & 1024 & 0 \\ 0 & 0 & 1 \end{bmatrix}$$

Therefore,

$$A^{10} = P^{-1}B^{10}P$$

$$= \begin{bmatrix} 2 & 1 & -2 \\ -1 & 0 & 1 \\ -1 & -1 & 2 \end{bmatrix} \begin{bmatrix} 1 & 0 & 0 \\ 0 & 1024 & 0 \\ 0 & 0 & 1 \end{bmatrix} \begin{bmatrix} 1 & 0 & 1 \\ 1 & 2 & 0 \\ 1 & 1 & 1 \end{bmatrix}$$

$$= \begin{bmatrix} 2 & 1024 & -2 \\ -1 & 0 & 1 \\ -1 & -1024 & 2 \end{bmatrix} \begin{bmatrix} 1 & 0 & 1 \\ 1 & 2 & 0 \\ 1 & 1 & 1 \end{bmatrix}$$

$$= \begin{bmatrix} 1024 & 2046 & 0 \\ 0 & 1 & 0 \\ -1023 & -2046 & 1 \end{bmatrix}$$

We can also immediately determine from B that $\det A = -2$ (since $\det B = -2$) and therefore that A is invertible.   ●

Exercises

28-1(a) Find the usual matrix for the reflection in the plane $-3x + 2y - 4z = 0$ of Example 26-1. (Hint: Example 26-1 gives a simple matrix B and spells out the basis vectors to use as the rows of P. Then use Theorem 28-1.)
(b) Compute the image of $(1, 2, 0)$.

28-2(a) Find the usual matrix for the $90°$ counterclockwise rotation about the line $\{a(1, 2, -2) : a \in \mathcal{R}\}$. (See Example 26-2.)
(b) Compute the image of the vector $(2, 2, 2)$.

28-3(a) Find the usual matrix for the perpendicular projection of $\mathcal{R}^3$ onto the plane $x - y = 0$.

(b) Find the projection of (2, 1, 1) onto the plane.

28-4(a) Find the usual matrix for the 90° counterclockwise rotation about the line $\{a(1, 0, -2) : a \in \mathcal{R}\}$ in $\mathcal{R}^3$. [Hint: $(0, \sqrt{5}, 0)$ and $(2, 0, 1)$ are perpendicular vectors of the same length and are both perpendicular to $(1, 0, -2)$.]

(b) Find the image of the vector (1, 1, 0).

28-5 Suppose that $T : \mathcal{R}^2 \to \mathcal{R}^2$ is the reflection in the line $y = -2x$. Find $(1, 5)T$.

28-6 A linear transformation T has the usual matrix

$$A = \begin{bmatrix} 3 & -1 \\ 2 & 0 \end{bmatrix}$$

Find its matrix relative to the basis $(1, -1)$, $(2, -1)$. The new matrix should be simpler than A. Use this new matrix to analyze the geometric effect of T.

28-7 The usual matrix

$$A = \begin{bmatrix} 7 & 10 & 0 \\ -4 & -6 & 0 \\ 2 & -2 & 0 \end{bmatrix}$$

for a linear transformation T is similar to

$$B = \begin{bmatrix} 2 & 0 & 0 \\ 0 & -1 & 0 \\ 0 & 0 & 0 \end{bmatrix}$$

where B is the matrix for the linear transformation relative to the basis $(4, 5, 0)$, $(1, 2, 0)$, $(10, 17, -1)$. Using Theorem 28-2:

(a) Compute A^6.
(b) Compute det A.
(c) Determine if A is invertible.
(d) Find the effect the linear transformation T has on volumes.
(e) Find the effect T^{12} has on volumes.

28-8 Prove Theorem 28-2.

*28-9(a)   Write a computer program for computing a new matrix B where the input is the usual matrix A and a new basis.
(b)   Write a computer program for computing the usual matrix A where the input is a new matrix B and the new basis.

29.   EIGENVALUES AND EIGENVECTORS

We will now consider problems of the second type mentioned in the introduction to this chapter: given the usual matrix A for a linear transformation T, to find a simpler matrix B. As before, we need a new basis that is affected in a simple way by T. In searching for simple matrices for projections, reflections, and rotations, it proved advantageous to seek basis vectors that had properties such as $\mathbf{X}_1 T = \mathbf{X}_1$ or $\mathbf{X}_2 T = -\mathbf{X}_2$ or $\mathbf{X}_3 T = \mathbf{0}$. In each case the resulting matrix B was simple because properties such as these gave rise to rows with many zeros. For our purposes, the simplest way in which T can affect a basis vector $\mathbf{X}_i$ is to "stretch" it, that is, to have $\mathbf{X}_i T = \lambda \mathbf{X}_i$ for some scalar λ (the Greek letter lambda). If we have such a vector, it gives rise to a simple ith row in the matrix B: $[0 \cdots 0 \; \lambda \; 0 \cdots 0]$, with λ as the ith entry. For example, the vector $\mathbf{X}_1$ above is stretched by a factor 1 and gives rise to a row $[1 \quad 0 \quad 0]$. Similarly, $\mathbf{X}_2 T = -\mathbf{X}_2$ leads to a row $[0 \quad -1 \quad 0]$, and $\mathbf{X}_3 T = 0$ leads to a row $[0 \quad 0 \quad 0]$.

The reader may have noticed that we are using the word *stretch* rather broadly: if the stretch factor is *negative*, the vector's direction is being reversed, and if the stretch factor is $\frac{1}{2}$ or $\frac{1}{3}$, etc., the vector is actually being shortened.

Nonzero vectors that are stretched are called eigenvectors, and the amount of stretch is called an eigenvalue.†

Definition

Let $T : \mathfrak{V} \to \mathfrak{V}$ be a linear transformation. A nonzero vector $\mathbf{X} \in \mathfrak{V}$ is an **eigenvector** if $\mathbf{X}T = \lambda \mathbf{X}$ for some number λ.

A number λ is called an **eigenvalue** if there is a nonzero vector $\mathbf{X} \in \mathfrak{V}$ such that $\mathbf{X}T = \lambda \mathbf{X}$.

Note that $\mathbf{0}$ is never an eigenvector (this is built into the definition of eigenvector). This is deliberate, since our primary use for eigenvectors will be to form a basis, and we cannot allow $\mathbf{0}$ as a basis vector.

---

† The words eigenvector and eigenvalue both stem from the German word *eigen*, meaning "own" or "self." The terminology is appropriate, since an eigenvector is its *own* image, except for a stretch factor. The terminology *characteristic vector* and *characteristic value* are also used.

Example 29-1

$$(1,5) \begin{bmatrix} -4 & -5 \\ 1 & 2 \end{bmatrix} = (1,5)$$

Let $T : \mathcal{R}^2 \to \mathcal{R}^2$ be the linear transformation with the rule $(x, y)T = (-4x + y, -5x + 2y)$. The vector $(1, 5)$ is an eigenvector, since

$$(1, 0)T = \begin{bmatrix} -4 & -5 \\ 1 & 2 \end{bmatrix}$$
$$(0, 1)T =$$

$$(1, 5)T = (1, 5) = 1(1, 5)$$

The vector $(1, 1)$ is also an eigenvector, since

$$(1, 1)T = -3(1, 1)$$

The number 1 is the eigenvalue corresponding to the eigenvector $(1, 5)$, and the number -3 is the eigenvalue corresponding to the eigenvector $(1, 1)$.

The two eigenvectors $(1, 5)$ and $(1, 1)$ form a basis for $\mathcal{R}^2$. The matrix we obtain relative to this basis is

$$B = \begin{bmatrix} 1 & 0 \\ 0 & -3 \end{bmatrix}$$

Compare this matrix with the usual matrix

$$A = \begin{bmatrix} -4 & -5 \\ 1 & 2 \end{bmatrix}$$

Both A and B represent T, but B shows more clearly the geometric effect of T; for example, the first row of B reveals that $(1, 5)$ is stretched by a factor of 1. Other information about T can be obtained easily from B: since det $B \neq 0$, T is invertible; since det $B = -3$, T magnifies areas by a factor of 3.   ●

The matrix

$$B = \begin{bmatrix} 1 & 0 \\ 0 & -3 \end{bmatrix}$$

of Example 29-1 is called a diagonal matrix.

A *diagonal matrix* is an $n \times n$ matrix in which every entry not on the main diagonal (upper left to lower right) is zero.

For example,

$$\begin{bmatrix} 1 & 0 & 0 \\ 0 & 2 & 0 \\ 0 & 0 & 7 \end{bmatrix} \qquad \begin{bmatrix} -3 & 0 & 0 \\ 0 & 1 & 0 \\ 0 & 0 & 0 \end{bmatrix} \qquad \begin{bmatrix} 6 & 0 & 0 & 0 \\ 0 & -3 & 0 & 0 \\ 0 & 0 & 0 & 0 \\ 0 & 0 & 0 & \frac{1}{2} \end{bmatrix}$$

are diagonal matrices, whereas

$$\begin{bmatrix} 2 & 0 & 0 \\ 0 & 3 & 1 \\ 0 & 0 & -1 \end{bmatrix} \qquad \begin{bmatrix} 0 & 0 & 1 \\ 0 & 1 & 0 \\ 1 & 0 & 0 \end{bmatrix}$$

are not.

Not all linear transformations can have such simple matrices. Note, however, that the perpendicular projections and the reflections treated earlier in this chapter have diagonal matrices such as

$$\begin{bmatrix} 1 & 0 & 0 \\ 0 & 1 & 0 \\ 0 & 0 & 0 \end{bmatrix} \qquad \begin{bmatrix} 1 & 0 & 0 \\ 0 & 1 & 0 \\ 0 & 0 & -1 \end{bmatrix}$$

with respect to carefully chosen bases. In each case we chose basis vectors that were stretched; that is, we chose a basis of eigenvectors. This is the key to obtaining a diagonal matrix.

Theorem 29-1

Let $T : \mathcal{V} \to \mathcal{V}$ be a linear transformation. T has a diagonal matrix $\Leftrightarrow \mathcal{V}$ has a basis consisting of eigenvectors for T.

When T has a diagonal matrix, the diagonal entries are the eigenvalues.

Proof

$\Rightarrow$: This is left as an exercise.

$\Leftarrow$: Suppose that T has $\mathbf{X}_1, \mathbf{X}_2, \ldots, \mathbf{X}_n$ as a basis of eigenvectors. Then $\mathbf{X}_1 T = \lambda_1 \mathbf{X}_1, \mathbf{X}_2 T = \lambda_2 \mathbf{X}_2, \ldots, \mathbf{X}_n T = \lambda_n \mathbf{X}_n$ for some numbers $\lambda_1, \lambda_2, \ldots, \lambda_n$ (the λ's are the corresponding eigenvalues). Therefore, the matrix corresponding to T relative to this basis is the diagonal matrix

$$\begin{bmatrix} \lambda_1 & 0 & \cdots & 0 \\ 0 & \lambda_2 & \cdots & 0 \\ \cdots & \cdots & \cdots & \cdots \\ 0 & 0 & \cdots & \lambda_n \end{bmatrix}$$

The main diagonal entries are the eigenvalues.   ∎

Each different eigenvalue gives rise to at least one eigenvector to use in the new basis, and each eigenvector in the basis yields a simple row for B. Thus we have another test for a linear transformation having a diagonal matrix.

Theorem 29-2

Let $T : \mathcal{V} \to \mathcal{V}$ be a linear transformation and suppose that dim $\mathcal{V} = n$. If T has n *distinct* eigenvalues, then the corresponding eigenvectors form a basis for $\mathcal{V}$, and T has a diagonal matrix, whose main diagonal entries are the eigenvalues.

Proof   We will prove this theorem only for the case where $n = 2$.

Suppose that λ_1 and λ_2 are two distinct eigenvalues (hence $\lambda_1 \neq \lambda_2$). Suppose that X_1 is an eigenvector for λ_1 and X_2 is an eigenvector for λ_2. Then X_1 and any multiple of X_1 is stretched by a factor of λ_1. But since X_2 is stretched by a different factor λ_2, X_2 cannot be a multiple of X_1. Therefore, X_1 and X_2 are independent. Since dim $\mathcal{V} = 2$, then X_1, X_2 is a basis for $\mathcal{V}$. Therefore, $\mathcal{V}$ has a basis of eigenvectors, namely X_1, X_2, and by Theorem 29-1, T has a diagonal matrix.   ∎

The concepts of eigenvectors and eigenvalues are defined above in terms of linear transformations. But since any $n \times n$ matrix A represents a linear transformation, these concepts can also be defined in terms of matrices.

Definition

Let A be an $n \times n$ matrix. A nonzero vector $X \in \mathcal{R}^n$ is said to be an **eigenvector** for A, provided $XA = \lambda X$ for some number λ, and in that case the number λ is called an **eigenvalue** for A.

Theorems 29-1 and 29-2 have analogs for matrices, extremely useful in applications.

Theorem 29-3   Let A be an $n \times n$ matrix. A is similar to a diagonal matrix $\Leftrightarrow \mathcal{R}^n$ has a basis consisting of eigenvectors for A.

When A is similar to a diagonal matrix B, the diagonal entries of B are the eigenvalues.

Theorem 29-4   Let A be an $n \times n$ matrix. If A has n *distinct* eigenvalues, then the corresponding eigenvectors form a basis for $\mathcal{R}^n$, and A is similar to a diagonal matrix B whose main diagonal entries are the eigenvalues.

The proof of these theorems rests on the idea of interpreting an $n \times n$ matrix, however it arises, as the matrix for a linear transformation $T : \mathcal{R}^n \to \mathcal{R}^n$. For example,

$$\begin{bmatrix} 2 & 1 \\ 1 & 3 \end{bmatrix}$$

can be interpreted as the matrix for the unique linear transformation $T : \mathcal{R}^2 \to \mathcal{R}^2$ satisfying

$$(1, 0)T = (2, 1)$$
$$(0, 1)T = (1, 3)$$

Since $XT = \lambda X$ precisely when $XA = \lambda X$, Theorems 29-1 and 29-2 give us Theorems 29-3 and 29-4 automatically.

All four theorems of this section hold because each eigenvector that can be found yields a simple row for B, and if n simple rows are found, B will be simple from top to bottom; that is, B will be diagonal. We are not always able to obtain a diagonal matrix B, unfortunately, but the more eigenvectors we can find, the simpler the matrix B becomes.

Thus far, any eigenvectors we have found have been found geometrically. In the next section we study a numerical method for finding eigenvalues and eigenvectors, given an $n \times n$ matrix A. The method is of great importance since it leads to simplification of many problems.

Exercises

29-1 Verify that each of the following vectors is an eigenvector for the given linear transformation, and find the corresponding eigenvalue:

(a) $X = (1, 7)$, where $(x, y)T = (-5x + y, 2y)$
(b) $X = (4, 3)$, where $(x, y)U = (-2x - 4y, -4x + y/3)$
(c) $X_1 = (2, 1, 0)$ and $X_2 = (1, 3, 1)$, where $(x, y, z)T = (x - 2y, 3x - 6y, -5z)$

29-2 Verify that each of the following vectors is an eigenvector for the given matrix, and find the corresponding eigenvalue:

(a) $X_1 = (1, 0)$ and $X_2 = (-1, 1)$ for $A = \begin{bmatrix} 2 & 0 \\ 1 & 1 \end{bmatrix}$

(b) $X_1 = (1, 1)$ and $X_2 = (1, -1)$ for $A = \begin{bmatrix} 3 & 1 \\ 1 & 3 \end{bmatrix}$

(c) $X_1 = (0, 0, 1)$ and $X_2 = (-2, 2, -1)$ for

$$A = \begin{bmatrix} 0 & 1 & 0 \\ 1 & 0 & 1 \\ 0 & 0 & 1 \end{bmatrix}$$

29-3 Let $T : \mathcal{R}^2 \to \mathcal{R}^2$ be the reflection in the y axis. By drawing a sketch, find two independent eigenvectors for T, and find their corresponding eigenvalues.

29-4   Let S be the perpendicular projection of $\mathcal{R}^3$ onto the xy plane. Find three independent eigenvectors for S, and find their corresponding eigenvalues.

29-5   Let T be the $90°$ clockwise rotation of $\mathcal{R}^2$ about the origin. Explain geometrically why T has no eigenvectors.

29-6   Let U be the $90°$ clockwise rotation of $\mathcal{R}^3$ about the x axis, as viewed from the positive side of the x axis. How many independent eigenvectors does U have?

29-7   Let $T : \mathcal{R}^3 \to \mathcal{R}^3$ be the perpendicular projection onto a plane.

(a)   Explain why T has three independent eigenvectors.
(b)   Does T have a diagonal matrix? Explain.

29-8   Let $U : \mathcal{R}^3 \to \mathcal{R}^3$ be a reflection in a plane.

(a)   Explain why U has three independent eigenvectors.
(b)   Does U have a diagonal matrix? Explain.

29-9   Suppose that $\mathbf{X}$ is an eigenvector for T and that λ is its corresponding eigenvalue. Assuming that T^{-1} exists:

(a)   Show that $\mathbf{X}$ is an eigenvector for T^{-1}.
(b)   Find its corresponding eigenvalue.

29-10   Prove the implication " $\Rightarrow$ " of Theorem 29-1.

29-11   Suppose that $\mathbf{X}$ is an eigenvector for T. Prove that every nonzero multiple of $\mathbf{X}$ is also an eigenvector for T.

*29-12   Write a computer program that computes A^k using the formula $A^k = P^{-1}B^kP$. The input should be a new basis and a diagonal matrix B relative to this basis.

30.   FINDING EIGENVALUES AND EIGENVECTORS

As indicated in prior sections, to find a simple matrix B for a linear transformation T we need basis vectors $\mathbf{X}$ that are stretched by T (eigenvectors.) That is, we must find numbers λ and nonzero vectors $\mathbf{X} = (x_1, x_2, \ldots, x_n)$ such that

$$XT = \lambda X$$

A test for the existence of nonzero solutions **X** to this equation will now be developed.

In matrix notation, we must solve

$$XA = \lambda X$$

that is,

$$XA - \lambda X = 0$$

Since $\lambda X = X(\lambda I)$, this last equation is equivalent to

$$XA - X(\lambda I) = 0$$

or, by the distributive law for matrices,

$$X(A - \lambda I) = 0$$

This last equation has a nonzero solution **X** provided that the transformation represented by $A - \lambda I$ is noninvertible (Theorem 23-3). This will be so if and only if the determinant of $A - \lambda I$ is zero (Theorem 24-2). Thus we obtain the following test:

$$XT = \lambda X \text{ has a nonzero solution} \Leftrightarrow \det (A - \lambda I) = 0.$$

The equation $\det (A - \lambda I) = 0$ is called the *characteristic equation* for A. Solving it for λ gives us the eigenvalues; for each eigenvalue λ we must then solve $X(A - \lambda I) = 0$ to obtain eigenvectors **X**.

This is summarized in the following theorem.

(A ~ can be ~ any matrix usual most often)

Theorem 30-1

Let A be an $n \times n$ matrix.

1   To find eigenvalues for A: solve $\det (A - \lambda I) = 0$ for λ.

2   To find eigenvectors for A: for each eigenvalue λ found in (1), solve the system $X(A - \lambda I) = 0$ for **X**.

Theorem 30-1 also enables us to find eigenvalues and eigenvectors for linear transformations $T : \mathcal{R}^n \to \mathcal{R}^n$. First find T's usual matrix A; the eigenvalues and eigenvectors for A found by means of Theorem 30-1 are those of T, as well. The following examples illustrate this technique.

Example 30-1

Suppose that T is given by the rule $(x, y)T = (-4x + y, -5x + 2y)$ (as in Example 29-1). We will find the eigenvalues and eigenvectors for T, using Theorem 30-1. The usual matrix A for T is

$$A = \begin{bmatrix} -4 & -5 \\ 1 & 2 \end{bmatrix}$$

Hence

$$A - \lambda I = \begin{bmatrix} -4 & -5 \\ 1 & 2 \end{bmatrix} - \lambda \begin{bmatrix} 1 & 0 \\ 0 & 1 \end{bmatrix} = \begin{bmatrix} -4 - \lambda & -5 \\ 1 & 2 - \lambda \end{bmatrix} \qquad (3)$$

Therefore, the characteristic equation $\det (A - \lambda I) = 0$ gives tne equation

$$\lambda^2 + 2\lambda - 3 = 0$$

This factors as $(\lambda + 3)(\lambda - 1) = 0$, and $\lambda_1 = -3$ and $\lambda_2 = 1$ are the two solutions. Hence -3 and 1 are eigenvalues.

We now find an eigenvector corresponding to $\lambda_1 = -3$. To do this we solve $\mathbf{X}(A - \lambda I) = \mathbf{0}$ for $\mathbf{X}$, with $\lambda = -3$. Using $\lambda = -3$ in the expression (3) for $A - \lambda I$, we have

$$\mathbf{X} \begin{bmatrix} -4 + 3 & -5 \\ 1 & 2 + 3 \end{bmatrix} = \mathbf{0}$$

that is,

$$\mathbf{X} \begin{bmatrix} -1 & -5 \\ 1 & 5 \end{bmatrix} = \mathbf{0}$$

or

$$(x_1, x_2) \begin{bmatrix} -1 & -5 \\ 1 & 5 \end{bmatrix} = (0, 0) \qquad (4)$$

This is the system of equations

$$-1x_1 + 1x_2 = 0$$
$$-5x_1 + 5x_2 = 0$$

whose matrix is

$$\begin{bmatrix} -1 & 1 \\ -5 & 5 \end{bmatrix}$$

[Note that the rows of this matrix are the columns of the matrix in equation (4).] This latter matrix reduces to

$$\begin{bmatrix} 1 & -1 \\ 0 & 0 \end{bmatrix}$$

which yields $\{(a, a) : a \in \mathcal{R}\}$ as the solution set. Therefore (taking $a = 1$), $(1, 1)$ is an eigenvector corresponding to $\lambda_1 = -3$. So are $(2, 2)$, $(4, 4)$, $(-\frac{2}{5}, -\frac{2}{5})$, etc.

Similarly, to find an eigenvector corresponding to $\lambda_2 = 1$, we solve $\mathbf{X}(A -- 1I) = \mathbf{0}$ for $\mathbf{X}$. Using $\lambda = 1$ in (3) yields

$$\mathbf{X}\begin{bmatrix} -5 & -5 \\ 1 & 1 \end{bmatrix} = \mathbf{0}$$

This gives a system of equations with matrix

$$\begin{bmatrix} -5 & 1 \\ -5 & 1 \end{bmatrix}$$

(Note that the rows become columns again.) This matrix reduces to

$$\begin{bmatrix} 1 & -\frac{1}{5} \\ 0 & 0 \end{bmatrix}$$

Thus we have $\{(a/5, a) : a \in \mathcal{R}\}$ as the solution set. Therefore (taking $a = 5$), $(1, 5)$ is an eigenvector corresponding to $\lambda_2 = 1$. So are $(\frac{1}{3}, \frac{5}{3})$, $(-2, -10)$, etc.

We choose one eigenvector for each of the two eigenvalues, say $\mathbf{X}_1 = (1, 1)$ and $\mathbf{X}_2 = (1, 5)$. Since we have two distinct eigenvalues and a two-dimensional vector space, the eigenvectors form a basis for $\mathcal{R}^2$, and relative to this basis, T has the diagonal matrix

$$B = \begin{bmatrix} -3 & 0 \\ 0 & 1 \end{bmatrix}$$

whose diagonal entries are the eigenvalues (Theorem 29-2).   ●

Example 30-2

Let $(x, y)T = (y, x)$. Here

$$A = \begin{bmatrix} 0 & 1 \\ 1 & 0 \end{bmatrix}$$

and

$$A - \lambda I = \begin{bmatrix} -\lambda & 1 \\ 1 & -\lambda \end{bmatrix}$$

(When forming $A - \lambda I$, we simply subtract λ from each main diagonal entry of A.) Therefore, the characteristic equation $\det (A - \lambda I) = 0$ is $\lambda^2 - 1 = 0$. This has two solutions; the two eigenvectors are $\lambda_1 = 1$ and $\lambda_2 = -1$.

To find an eigenvector corresponding to $\lambda_1 = 1$, we solve $\mathbf{X}(A - 1I) = \mathbf{0}$ for $\mathbf{X}$; that is, we solve

$$\mathbf{X} \begin{bmatrix} -1 & 1 \\ 1 & -1 \end{bmatrix} = \mathbf{0}$$

This gives us $\{a(1, 1) : a \in \mathcal{R}\}$ as solution set. Therefore, $(1, 1)$ [or any nonzero multiple of $(1, 1)$] is an eigenvector.

To find an eigenvector corresponding to $\lambda_2 = -1$, we solve $\mathbf{X}(A - (-1)I) = \mathbf{0}$ for $\mathbf{X}$; that is, we solve

$$\mathbf{X} \begin{bmatrix} 1 & 1 \\ 1 & 1 \end{bmatrix} = \mathbf{0}$$

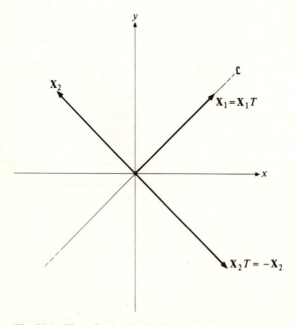

Fig. 30-1 The reflection in the line $\mathcal{L}$. $\mathbf{X}_1$ and $\mathbf{X}_2$ are eigenvectors, with corresponding eigenvalues 1 and -1.

This yields $\{a(-1, 1) : a \in \mathcal{R}\}$ as solution set. Therefore, $(-1, 1)$ [or any nonzero multiple of $(-1, 1)$] is an eigenvector.

Again, by Theorem 29-2, the vectors $X_1 = (1, 1)$ and $X_2 = (-1, 1)$ are a basis for $\mathcal{R}^2$, and T has

$$B = \begin{bmatrix} 1 & 0 \\ 0 & -1 \end{bmatrix}$$

as a diagonal matrix relative to this basis. T is now seen to be the reflection in the line $\mathcal{L} = \{a(1, 1) : a \in \mathcal{R}\}$ (Fig. 30-1). ●

Example 30-3

Let

$$A = \begin{bmatrix} -2 & -2 & -4 \\ 2 & 3 & 2 \\ 3 & 2 & 5 \end{bmatrix}$$

To find eigenvalues for A, we must solve

$$\det \begin{bmatrix} -2-\lambda & -2 & -4 \\ 2 & 3-\lambda & 2 \\ 3 & 2 & 5-\lambda \end{bmatrix} = 0$$

Taking the determinant, we obtain $-\lambda^3 + 6\lambda^2 - 11\lambda + 6 = 0$, which must be solved for λ. The left side factors as $-(\lambda - 1)(\lambda - 2)(\lambda - 3)$. (One way to see this is to find by trial and error that $\lambda = 1$ is a solution to the equation and then to divide the left side by $\lambda - 1$.†) Therefore, 1, 2, and 3 are eigenvalues and A is similar to the diagonal matrix

$$B = \begin{bmatrix} 1 & 0 & 0 \\ 0 & 2 & 0 \\ 0 & 0 & 3 \end{bmatrix}$$

Since A is similar to B, $B = PAP^{-1}$. In many applications we must find the matrix P. Therefore, we must find the corresponding basis of eigenvectors. The reader can check that they turn out to be the following:

For $\lambda = 1$, any vector of the form $(a, 0, a) = a(1, 0, 1)$

For $\lambda = 2$, any vector of the form $(a, 2a, 0) = a(1, 2, 0)$

For $\lambda = 3$, any vector of the form $(a, a, a) = a(1, 1, 1)$

† The characteristic equation can often be difficult to solve. A vast literature exists on approximate numerical methods for finding eigenvalues. Computers are used to find approximate roots and then the eigenvectors are found approximately.

Taking $a = 1$ reveals $(1, 0, 1)$, $(1, 2, 0)$, and $(1, 1, 1)$ as a basis of eigenvectors. Hence

$$P = \begin{bmatrix} 1 & 0 & 1 \\ 1 & 2 & 0 \\ 1 & 1 & 1 \end{bmatrix}$$

●

Example 30-4

(Rotations have no eigenvectors.)

Suppose that T is the 90° counterclockwise rotation of $\mathcal{R}^2$ around the origin. Then T has

$$\begin{bmatrix} 0 & 1 \\ -1 & 0 \end{bmatrix}$$

as its usual matrix. To find eigenvalues we must solve

$$\det \begin{bmatrix} -\lambda & 1 \\ -1 & -\lambda \end{bmatrix} = 0$$

We obtain $\lambda^2 + 1 = 0$, which has no real solutions. Hence T has no eigenvalues (and therefore no eigenvectors). This fact can be seen geometrically—T does not stretch any vector, but rather T rotates them. Therefore, we conclude that T has no diagonal matrix.   ●

Example 30-5

Suppose that $T : \mathcal{R}^3 \rightarrow \mathcal{R}^3$ has the rule $(x, y, z)T = (2x, 2y, 7y - 5z)$. To find eigenvalues we must solve

$$\det \begin{bmatrix} 2-\lambda & 0 & 0 \\ 0 & 2-\lambda & 7 \\ 0 & 0 & -5-\lambda \end{bmatrix} = 0$$

The characteristic equation is $(2-\lambda)^2(-5-\lambda) = 0$, and so the eigenvalues are $\lambda_1 = 2$ and $\lambda_2 = -5$. We have only two eigenvalues and therefore can guarantee only two simple rows in B. That is, we know we will find a basis vector that is stretched by 2 and another that is stretched by -5, but we have no assurance that we will find a third eigenvector to form a complete *basis*, and thus no assurance that T will have a diagonal matrix. To complete the problem, we *must* look for eigenvectors.

To find eigenvectors corresponding to $\lambda_1 = 2$, we solve

$$\mathbf{X} \begin{bmatrix} 0 & 0 & 0 \\ 0 & 0 & 7 \\ 0 & 0 & -7 \end{bmatrix} = \mathbf{0}$$

This system of equations has the solution set $\{(a, b, b) : a, b \in \mathcal{R}\}$. Hence the eigenvectors will be of the form (a, b, b). Separating the a and the b parts yields $a(1, 0, 0) + b(0, 1, 1)$. In this case we have *two* natural choices for eigenvectors corresponding to the eigenvalue 2 : $(1, 0, 0)$ and $(0, 1, 1)$. Since they are independent, they can both be put in a new basis.

Corresponding to $\lambda_2 = -5$ we obtain $(0, 0, a) = a(0, 0, 1)$ as eigenvectors. If we take $(0, 0, 1)$ as a particular one, then $\mathbf{X}_1 = (1, 0, 0)$, $\mathbf{X}_2 = (0, 1, 1)$, $\mathbf{X}_3 = (0, 0, 1)$ form a basis of eigenvectors, and relative to this basis, T has diagonal matrix

$$B = \begin{bmatrix} 2 & 0 & 0 \\ 0 & 2 & 0 \\ 0 & 0 & -5 \end{bmatrix}$$

The first two rows arise since both $\mathbf{X}_1 T = 2\mathbf{X}_1$ and $\mathbf{X}_2 T = 2\mathbf{X}_2$. We were fortunate to get two basis vectors for $\lambda_1 = 2$. If we had not, there would have been no diagonal matrix for T.   ●

Example 30-6

Let

$$A = \begin{bmatrix} 3 & 1 & 0 \\ 0 & 2 & 1 \\ 0 & 1 & 2 \end{bmatrix}$$

be the usual matrix for a linear transformation T. The characteristic equation is

$$\det \begin{bmatrix} 3 - \lambda & 1 & 0 \\ 0 & 2 - \lambda & 1 \\ 0 & 1 & 2 - \lambda \end{bmatrix} = 0$$

Evaluating the determinant by means of the first column yields $(3 - \lambda) \times (\lambda^2 - 4\lambda + 4 - 1) = 0$, which factors as $(\lambda - 3)^2(\lambda - 1) = 0$. Again we have two eigenvalues, $\lambda_1 = 3$ and $\lambda_2 = 1$. We will get at least two simple rows for B, since there must be eigenvectors $\mathbf{X}_1$ and $\mathbf{X}_2$ satisfying $\mathbf{X}_1 T = 3\mathbf{X}_1$ and $\mathbf{X}_2 T = 1\mathbf{X}_2$.

To find eigenvectors for $\lambda_1 = 3$, we solve

$$\mathbf{X} \begin{bmatrix} 0 & 1 & 0 \\ 0 & -1 & 1 \\ 0 & 1 & -1 \end{bmatrix} = \mathbf{0}$$

The solutions are of the form $a(0, 1, 1)$. Thus only one basis vector $\mathbf{X}_1 = (0, 1, 1)$ is an eigenvector for $\lambda_1 = 3$.

Similarly, eigenvectors for $\lambda_2 = 1$ are of the form $a(0, 1, -1)$, and thus we obtain one more basis vector $\mathbf{X}_2 = (0, 1, -1)$.

We have $\mathbf{X}_1 T = 3\mathbf{X}_1$ and $\mathbf{X}_2 T = 1\mathbf{X}_2$, and thus two rows of B are

$$\begin{bmatrix} 3 & 0 & 0 \\ 0 & 1 & 0 \\ - & - & - \end{bmatrix}$$

But B cannot be diagonal, since we cannot find a third eigenvector for use in the basis.   ●

Let us note one final fact regarding diagonal matrices. Not every transformation has a diagonal matrix. Examples 30-4 and 30-6 illustrate this point. In Example 30-4 the characteristic equation was $\lambda^2 + 1 = 0$, which has no real solution. In Example 30-6, only two eigenvectors could be found for a new basis, and hence only two simple rows could be found for B.

We can guarantee solutions to the characteristic equation (that is, we can guarantee that we will find eigenvalues) by using complex numbers for scalars and thereby permitting complex eigenvalues. This is true since every polynomial with complex coefficients factors completely. But even then, examples such as Example 30-6 show that we cannot always obtain a diagonal matrix B.

For the sake of completeness we now describe Jordan form matrices. Jordan form is the simplest form that we are assured of obtaining. We assume for these few paragraphs only that we are using complex numbers $c + di$ for scalars (and thus also for entries in matrices, eigenvalues, etc.).

First we define a Jordan block. A *Jordan block* is an $n \times n$ matrix of the form

$$\begin{bmatrix} \lambda & 1 & 0 & 0 & \cdots & 0 \\ 0 & \lambda & 1 & 0 & \cdots & 0 \\ 0 & 0 & \lambda & 1 & \cdots & 0 \\ \multicolumn{6}{c}{\dotfill} \\ 0 & 0 & 0 & 0 & \cdots & 1 \\ 0 & 0 & 0 & 0 & \cdots & \lambda \end{bmatrix}$$

having equal entries on the main diagonal, 1s just above the main diagonal, and zeros everywhere else. For example,

$$\begin{bmatrix} 3i & 1 \\ 0 & 3i \end{bmatrix} \qquad \begin{bmatrix} 3 & 1 & 0 & 0 \\ 0 & 3 & 1 & 0 \\ 0 & 0 & 3 & 1 \\ 0 & 0 & 0 & 3 \end{bmatrix} \qquad \begin{bmatrix} 5-2i & 1 & 0 \\ 0 & 5-2i & 1 \\ 0 & 0 & 5-2i \end{bmatrix}$$

are Jordan blocks. Even the 1×1 matrix $[7]$ is a Jordan block.

A *Jordan form matrix* is an $n \times n$ matrix made up of Jordan blocks strung along its main diagonal. All entries not in these blocks are zeros. For example,

$$\begin{bmatrix} i & 1 & 0 & 0 & 0 \\ 0 & i & 0 & 0 & 0 \\ 0 & 0 & 4 & 1 & 0 \\ 0 & 0 & 0 & 4 & 1 \\ 0 & 0 & 0 & 0 & 4 \end{bmatrix} \qquad \begin{bmatrix} 4 & 0 & 0 & 0 & 0 & 0 \\ 0 & 3 & 1 & 0 & 0 & 0 \\ 0 & 0 & 3 & 1 & 0 & 0 \\ 0 & 0 & 0 & 3 & 0 & 0 \\ 0 & 0 & 0 & 0 & 2-i & 1 \\ 0 & 0 & 0 & 0 & 0 & 2-i \end{bmatrix}$$

are Jordan form matrices. Using this terminology, we have the following theorem (whose proof will be omitted).

Theorem 30-2   Let A be an $n \times n$ matrix with complex numbers as entries. Then A is similar to a Jordan form matrix B. Further, the diagonal entries of B are the eigenvalues of A, and they are repeated in B as often as they occur as roots of the characteristic equation.

In Example 30-6 we obtained the characteristic equation that factored as $(\lambda - 3)^2(\lambda - 1) = 0$, giving rise to two eigenvalues. Although A is not similar to a diagonal matrix, Theorem 30-2 guarantees a Jordan form matrix similar to A. Since the eigenvalues are 3 and 1, and since the eigenvalue 3 is repeated, the Jordan form matrix is

$$\begin{bmatrix} 3 & 1 & 0 \\ 0 & 3 & 0 \\ 0 & 0 & 1 \end{bmatrix}$$

Exercises

For each of the linear transformations or matrices in Exercises 30-1 to 30-12:

(a) Find the characteristic equation.
(b) Find all the eigenvalues.
(c) Find the corresponding eigenvectors.
(d) Find a basis of eigenvectors (if such a basis exists) and a diagonal matrix relative to this basis.

30-1 $(x, y)T = (5x, 4x + 2y)$

30-2 $\begin{bmatrix} 5 & 1 \\ 4 & 5 \end{bmatrix}$

30-3 $(x, y)T = (x, x + y)$

30-4 $\begin{bmatrix} 2 & 0 \\ 3 & 2 \end{bmatrix}$

30-5 $(x, y, z)T = (x, 3x + 2y + z, x + 4z)$

30-6 $(x, y, z)T = (4x, -7y, 3z)$

30-7 $(x, y, z)T = (z, 0, x)$

30-8 $(x, y, z)T = (-y + \frac{1}{2}x, x + 3y, -2y + z)$

30-9 $\begin{bmatrix} 1 & 1 & 0 \\ 2 & 0 & 0 \\ 1 & 0 & 2 \end{bmatrix}$

30-10 $\begin{bmatrix} 1 & 0 & 1 \\ 0 & 1 & 0 \\ 1 & 0 & 1 \end{bmatrix}$

30-11 The linear transformation whose usual matrix is

$$\begin{bmatrix} 1 & 1 & 3 \\ 0 & 0 & -3 \\ 0 & 1 & 4 \end{bmatrix}$$

30-12 $$\begin{bmatrix} 0 & 0 & 4 \\ 2 & 4 & -2 \\ 1 & 0 & 3 \end{bmatrix}$$

30-13 Suppose a linear transformation $T : \mathcal{R}^3 \to \mathcal{R}^3$ has its usual matrix in upper triangular form

$$\begin{bmatrix} a & b & c \\ 0 & d & e \\ 0 & 0 & f \end{bmatrix}$$

What can be said about eigenvalues for T?

30-14 Suppose A is a Jordan *block* matrix of size at least 2×2. Show that A cannot be similar to a diagonal matrix B. (Hint: How many eigenvalues does A have? How many independent eigenvectors for use in a new basis?)

30-15 Show that a 3×3 upper triangular matrix with two or more equal entries on its main diagonal and no zeros above the main diagonal cannot be similar to a diagonal matrix.

30-16 Find as many independent eigenvectors as possible for

$$A = \begin{bmatrix} 2 & 1 & 4 & 2 & 5 \\ 0 & 3 & 1 & 0 & -1 \\ 0 & 0 & 2 & 5 & 1 \\ 0 & 0 & 0 & 2 & 1 \\ 0 & 0 & 0 & 0 & 3 \end{bmatrix}$$

What will its Jordan form matrix B be?

31. APPLICATIONS

Example 31-1

Growth Patterns and Recursion Formulas

Assume that rabbits do not reproduce during the first month of their lives but that beginning with the second month each pair of rabbits has one pair of offspring per month. Assuming that none of the rabbits die, if we begin with one pair of newborn rabbits, how many pairs of rabbits are alive after n months?

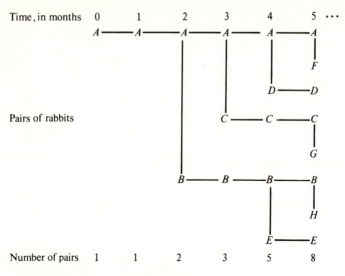

Fig. 31-1 Growth of a hypothetical rabbit population.

The growth pattern is shown in Fig. 31-1. At time 0 we have one pair of rabbits—the original pair A. At time 1 (1 month later) we still have only the original pair A. At time 2 we have two pairs—the original pair A and their new offspring B. At time 3 we have three pairs—A, the pair B born at time 2, and a new pair of offspring C born to A. At time 4 we have five pairs—A, B, C, the new pair D produced by A, and the new pair E produced by B.

If we let x_n equal the number of pairs of rabbits at the end of n months, then $x_0 = 1$, $x_1 = 1$, $x_2 = 2$, $x_3 = 3$, $x_4 = 5$, $x_5 = 8$, ... This sequence of numbers is called a Fibonacci sequence (after Leonardo Fibonacci, who in 1202 posed this breeding problem). We seek a formula that gives x_n.

At any given time the number of pairs of rabbits is equal to the number of pairs alive the previous month plus the number of newborn pairs; that is,

$$x_n = x_{n-1} + \text{number of pairs born in month } n$$

Since one pair of rabbits is born to each pair alive 2 months previously, the number of newborn pairs is x_{n-2}. Therefore, we have the following *recursion formula*:

$$x_n = x_{n-1} + x_{n-2}$$

for $n = 2, 3, 4, \ldots$. We can use this equation to compute x_n, although it is not practical if n is large. For some purposes it is advantageous to have a formula that allows us to compute x_n directly. We now derive such a formula.

The recursion formula $x_n = x_{n-1} + x_{n-2}$ can be written as the matrix equation

$$(x_n, x_{n-1}) = (x_{n-1}, x_{n-2})\begin{bmatrix} 1 & 1 \\ 1 & 0 \end{bmatrix}$$

If we let

$$A = \begin{bmatrix} 1 & 1 \\ 1 & 0 \end{bmatrix}$$

then we have

$$(x_2, x_1) = (x_1, x_0)A = (1, 1)A$$
$$(x_3, x_2) = (x_2, x_1)A = (1, 1)AA = (1, 1)A^2$$
$$(x_4, x_3) = (x_3, x_2)A = (1, 1)A^2A = (1, 1)A^3$$

and, in general,

$$(x_{n+1}, x_n) = (1, 1)A^n$$

Thus to find x_n we need only find A^n and then read off x_n from this equation. However, even though A is a relatively simple matrix, it is not easy to find A^n directly if n is large.

Fortunately, Theorem 28-2 provides an easy way to compute A^n. We first find a diagonal matrix B similar to A; by Theorem 28-2, $A^n = P^{-1}B^nP$. Since B^n is easily computed (it is diagonal), so is A^n.

For the matrix A we have the characteristic equation $\lambda^2 - \lambda - 1 = 0$, and from this we obtain the eigenvalues $\lambda_1 = (1 + \sqrt{5})/2$ and $\lambda_2 = (1 - \sqrt{5})/2$. Therefore,

$$B = \begin{bmatrix} \dfrac{1 + \sqrt{5}}{2} & 0 \\ 0 & \dfrac{1 - \sqrt{5}}{2} \end{bmatrix}$$

Since B is diagonal, B^n is obtained by raising the diagonal entries of B to the nth power, and we have

$$B^n = \begin{bmatrix} \left(\dfrac{1 + \sqrt{5}}{2}\right)^n & 0 \\ 0 & \left(\dfrac{1 - \sqrt{5}}{2}\right)^n \end{bmatrix}$$

Corresponding to λ_1 we have an eigenvector $(1 + \sqrt{5}, 2)$, and corresponding to λ_2 we have an eigenvector $(1 - \sqrt{5}, 2)$. Thus

$$P = \begin{bmatrix} 1 + \sqrt{5} & 2 \\ 1 - \sqrt{5} & 2 \end{bmatrix}$$

and it follows that

$$P^{-1} = \begin{bmatrix} \dfrac{1}{2\sqrt{5}} & \dfrac{-1}{2\sqrt{5}} \\ \dfrac{-1 + \sqrt{5}}{4\sqrt{5}} & \dfrac{1 + \sqrt{5}}{4\sqrt{5}} \end{bmatrix}$$

Therefore, since

$$(x_{n+1}, x_n) = (1, 1)A^n = (1, 1)P^{-1}B^nP$$

we need only find $P^{-1}B^nP$ and multiply by $(1, 1)$ to obtain the formula

$$x_n = \frac{1}{\sqrt{5}}\left(\frac{1 + \sqrt{5}}{2}\right)^{n+1} - \frac{1}{\sqrt{5}}\left(\frac{1 - \sqrt{5}}{2}\right)^{n+1}$$

Using a calculator, we find that at the end of 3 years $(n = 36)$ we have approximately 24 million pairs of rabbits, and after 5 years $(n = 60)$ about $2\frac{1}{2}$ trillion pairs of rabbits. ●

Example 31-2

Systems of Differential Equations

We wish to solve the system of differential equations

$$\begin{align} f' &= 3f + g \\ g' &= 2f + 2g \end{align} \tag{5}$$

A solution to this system is any ordered pair of functions $(f(x), g(x))$ that satisfies both equations.

If we introduce vector and matrix notation

$$\mathbf{X} = (f, g)$$

$$\mathbf{X}' = (f', g')$$

$$A = \begin{bmatrix} 3 & 2 \\ 1 & 2 \end{bmatrix}$$

the system (5) can be rewritten as

$$\mathbf{X}' = \mathbf{X}A$$

Our aim is to transform this system to a simpler system (one with a diagonal matrix), solve the simpler system, and then invert to obtain the solution to the original system.

It is easily verified that 4 and 1 are eigenvalues for A, and $(1, 1)$ and $(-1, 2)$ are corresponding eigenvectors. Thus we have the diagonal matrix

$$B = \begin{bmatrix} 4 & 0 \\ 0 & 1 \end{bmatrix}$$

relative to the basis $(1, 1)$ and $(-1, 2)$. By Theorem 28-1, $A = P^{-1}BP$, where

$$P = \begin{bmatrix} 1 & 1 \\ -1 & 2 \end{bmatrix}$$

Therefore,

$$\mathbf{X}' = \mathbf{X}A$$
$$= \mathbf{X}P^{-1}BP$$

Multiplying both members of this equation on the right by P^{-1}, we have

$$\mathbf{X}'P^{-1} = \mathbf{X}P^{-1}B \tag{6}$$

If we let $\mathbf{Y} = (y_1, y_2) = \mathbf{X}P^{-1}$ (and hence $\mathbf{Y}' = (y_1', y_2') = \mathbf{X}'P^{-1}$), we obtain from (6) the equation

$$(y_1', y_2') = \mathbf{Y}' = \mathbf{Y}B$$

$$= (y_1, y_2) \begin{bmatrix} 4 & 0 \\ 0 & 1 \end{bmatrix}$$

$$= (4y_1, y_2)$$

Thus the new, simpler system is

$$y_1' = 4y_1$$
$$y_2' = y_2$$

This system is easily solved, and its solutions are $y_1 = ae^{4x}$ and $y_2 = be^x$, where a and b are arbitrary constants. That is,

$$\mathbf{Y} = (y_1, y_2) = (ae^{4x}, be^x)$$

To find the solution $\mathbf{X}$ to the original system, we note that $\mathbf{X} = \mathbf{Y}P$ (since $\mathbf{Y} = \mathbf{X}P^{-1}$). Therefore, the original system (5) has solutions

$$\mathbf{X} = \mathbf{Y}P$$

$$= (ae^{4x}, be^x)\begin{bmatrix} 1 & 1 \\ -1 & 2 \end{bmatrix}$$

$$= (ae^{4x} - be^x, ae^{4x} + 2be^x)$$

that is, $f(x) = ae^{4x} - be^x$ and $g(x) = ae^{4x} + 2be^x$.   ●

Example 31-3

Stability of Population Age Groups

Assume that we are studying a certain type of living being (human, animal, or insect) that has a maximum life-span of n years. Its population may be divided into age groups. We begin by setting up equations concerning the sizes of these groups. Let

$$x_i = \text{No. of beings of age } i \text{ alive now}$$
$$y_i = \text{No. of beings of age } i \text{ alive 1 year from now}$$

where $i = 0, 1, 2, \ldots, n$. Let

$p_i = $ probability that a being of age i will be alive 1 year from now
$b_i = $ the number of beings born during the year to each being of age i

We have the following equations:

$$\begin{aligned} y_1 &= p_0 x_0 \\ y_2 &= p_1 x_1 \\ &\ \ \vdots \\ y_n &= p_{n-1} x_{n-1} \end{aligned} \tag{7}$$

(These equations show the number of beings in each age group that survive to become members of the next age group.)

$$y_0 = b_0 x_0 + b_1 x_1 + \cdots + b_n x_n \tag{8}$$

(This equation gives the number of births.)

We introduce the vector and matrix notation:

$$\mathbf{X} = (x_0, x_1, \ldots, x_n)$$
$$\mathbf{Y} = (y_0, y_1, \ldots, y_n)$$

$$A = \begin{bmatrix} b_0 & p_0 & 0 & \cdots & 0 \\ b_1 & 0 & p_1 & \cdots & 0 \\ \hdotsfor{5} \\ b_{n-1} & 0 & 0 & \cdots & p_{n-1} \\ b_n & 0 & 0 & \cdots & 0 \end{bmatrix}$$

Then the equations in (7) and (8) can be rewritten as

$$\mathbf{Y} = \mathbf{X}A \tag{9}$$

We can use equation (9) to find a population distribution by age groups such that the percentage in each group remains the same 1 year from now. Such a population distribution can be represented by a vector $\mathbf{X}$ that in 1 year becomes a *multiple* $\lambda\mathbf{X}$ of itself. What we seek then is a vector $\mathbf{X}$ and a scalar λ such that

$$\mathbf{Y} = \mathbf{X}A = \lambda\mathbf{X}$$

Hence this population problem reduces to one of finding eigenvalues and their eigenvectors.

For example, if it happened that $\lambda = 2$ were an eigenvalue, then the corresponding eigenvector $\mathbf{X}$ would give a breakdown of population such that the number in each age group would double each year. The *relative* sizes of the age groups would remain the same. If $\lambda = 1$ were an eigenvalue, then the corresponding eigenvector would give a population distribution such that each age group would *remain the same size* from year to year (such a vector is called a *stable vector*). ●

Example 31-4

Markov Chains

Suppose that we perform the same experiment over and over and that each time we repeat this experiment one of two outcomes occurs. (These possible outcomes, denoted a_1 and a_2, are called *states*.) Also assume that the probability that outcome a_j follows outcome a_i in succession remains the same regardless of how often we repeat the experiment;

call this number p_{ij}. Such a sequence of experiments is called a *Markov chain*, and the matrix

$$Q = \begin{bmatrix} p_{11} & p_{12} \\ p_{21} & p_{22} \end{bmatrix}$$

is called a *transition matrix*.

For example, the experiment might consist of determining whether people in a certain office are late for work each day. We take the states to be $a_1 = $ "on time" and $a_2 = $ "late." Suppose that (on the basis of observations) we know the transition matrix is

$$Q = \begin{bmatrix} \frac{7}{10} & \frac{3}{10} \\ \frac{9}{10} & \frac{1}{10} \end{bmatrix}$$

The entry $p_{11} = \frac{7}{10}$ means that if a person is on time one day there is a $\frac{7}{10}$ probability that the person will be on time the next day; the entry $p_{12} = \frac{3}{10}$ means that if a person is on time one day there is a $\frac{3}{10}$ probability that the person will be late the next day; etc.

The transition matrix Q can be used to predict the number of people who will arrive on time or late on future days. Suppose that 50 people work in the office and that today 40 were on time and 10 were late. Form the vector $X_0 = (40, 10)$. The product $X_1 = X_0 Q$ gives the number of people we expect to be on time or late tomorrow:

$$X_1 = X_0 Q = (40, 10) \begin{bmatrix} \frac{7}{10} & \frac{3}{10} \\ \frac{9}{10} & \frac{1}{10} \end{bmatrix} = (37, 13)$$

Thus we expect 37 to be on time tomorrow and 13 to be late tomorrow. [To see this, note that in computing $37 = 40(\frac{7}{10}) + 10(\frac{9}{10})$ the first summand means that $\frac{7}{10}$ of the 40 people who were on time today will be on time tomorrow and the second summand means that $\frac{9}{10}$ of the 10 people who were late today will be on time tomorrow. These two numbers added together give the grand total of 37 on time tomorrow. The number 13 has a similar interpretation.]

If $X_2 = X_1 Q$, then similar reasoning shows that X_2 gives the number we expect to arrive on time or late the day after tomorrow. In this case we have

$$X_2 = X_1 Q = (X_0 Q)Q = X_0 Q^2 = (40, 10) \begin{bmatrix} \frac{76}{100} & \frac{24}{100} \\ \frac{72}{100} & \frac{28}{100} \end{bmatrix} = (37.6, 12.4)$$

In general, if $\mathbf{X}_n$ gives the numbers arriving on time and late n days from now,

$$\mathbf{X}_n = \mathbf{X}_0 Q^n$$

If we use *probabilities* as the entries of $\mathbf{X}_0$, then the entries of $\mathbf{X}_n$ give the probabilities of being on time and late on the nth day. In our example, 40 out of 50 were on time and 10 out of 50 were late. Using the probabilities $0.8(= \frac{40}{50})$ and $0.2(= \frac{10}{50})$ as the entries of $\mathbf{X}_0$, we have $\mathbf{X}_0 = (0.8, 0.2)$. The vector $\mathbf{X}_0$ is called the *initial probability vector*. Then

$$\mathbf{X}_1 = \mathbf{X}_0 Q = (0.8, 0.2)\begin{bmatrix} 0.7 & 0.3 \\ 0.9 & 0.1 \end{bmatrix} = (0.74, 0.26)$$

gives the probabilities for tomorrow: we expect 74 percent to arrive on time tomorrow and 26 percent to arrive late. (Since our population is 50, these percentages give 37 on time and 13 late, as above.)

What happens in this office in the long run? That is, what are the entries of the vector $\mathbf{X}_n$ as n gets larger and larger? Since $\mathbf{X}_n = \mathbf{X}_0 Q^n$, we are faced with the problem of computing high powers of the matrix Q. To do this, we will find a diagonal matrix B similar to Q, and then use the fact that $Q^n = P^{-1}B^nP$ (Theorem 28-2).

It can be verified that Q has eigenvalues 1 and $-\frac{1}{5}$, with corresponding eigenvectors $(3, 1)$ and $(-1, 1)$. Thus Q is similar to the matrix

$$B = \begin{bmatrix} 1 & 0 \\ 0 & -\frac{1}{5} \end{bmatrix}, \text{ where } P = \begin{bmatrix} 3 & 1 \\ -1 & 1 \end{bmatrix} \quad \text{and} \quad P^{-1} = \begin{bmatrix} \frac{1}{4} & -\frac{1}{4} \\ \frac{1}{4} & \frac{3}{4} \end{bmatrix}.$$

Therefore,

$$Q^n = P^{-1}B^nP$$

$$= P^{-1}\begin{bmatrix} 1^n & 0 \\ 0 & (-\frac{1}{5})^n \end{bmatrix}P$$

$$= \begin{bmatrix} \frac{3}{4} + \frac{1}{4}(-\frac{1}{5})^n & \frac{1}{4} - \frac{1}{4}(-\frac{1}{5})^n \\ \frac{3}{4} - \frac{3}{4}(-\frac{1}{5})^n & \frac{1}{4} + \frac{3}{4}(-\frac{1}{5})^n \end{bmatrix}$$

We have the distribution on the nth day given by

$$\mathbf{X}_n = \mathbf{X}_0 Q^n$$
$$= (0.8, 0.2)Q^n$$
$$= (\tfrac{3}{4} + (0.8)(\tfrac{1}{4})(-\tfrac{1}{5})^n - (0.2)(\tfrac{3}{4})(-\tfrac{1}{5})^n, \ \tfrac{1}{4} - (0.8)(\tfrac{1}{4})(-\tfrac{1}{5})^n + (0.2)(\tfrac{3}{4})(-\tfrac{1}{5})^n)$$

As n increases, the first entry approaches $\frac{3}{4}$ and the second entry approaches $\frac{1}{4}$. Thus, in the long run we can expect $\frac{3}{4}$ of the office workers to arrive on time and $\frac{1}{4}$ to be late.

The preceding computations were based on the fact that $\mathbf{X}_0 = (0.8, 0.2)$. The same long-run probabilities $\frac{3}{4}$ and $\frac{1}{4}$ will be obtained regardless of the entries in the initial probability vector $\mathbf{X}_0$. To see this, note that wherever the numbers 0.8 and 0.2 appear in $\mathbf{X}_n$ they are multiplied by $(-\frac{1}{5})^n$. Thus, as n increases, these terms approach zero. Hence any initial probability vector $\mathbf{X}_0$ will yield $\frac{3}{4}$ and $\frac{1}{4}$. ●

Exercises

31-1   Suppose that the rabbits multiply as in Example 31-1 except that each pair reproduces *two* pairs of rabbits (rather than one pair as in the example). Let x_n equal the number of pairs at time n.

(a)   Write a recursion formula that gives x_n in terms of x_{n-1} and x_{n-2}.
(b)   Find a formula for x_n.

31-2   Suppose that the maximum life-span of a certain species of animal is 3 years. The probability that a newborn animal lives to be 1 year old is $\frac{2}{3}$ and the probability that a 1-year-old animal lives to be 2 years old is $\frac{1}{2}$. The animal does not reproduce until it is 2 years old; during its last year it produces (on the average) three offspring.

(a)   Set up a matrix A as in Example 31-3.
(b)   Find a distribution of the population into age groups that remains stable.

31-3   Suppose that $\frac{1}{2}$ the people in a certain town are conservative and $\frac{1}{2}$ are liberal. Suppose further that the following table gives the probabilities that a person keeps or changes his political tendency during any 1-year period.

Tendency now	Tendency 1 year from now	
	Conservative	Liberal
Conservative	$\dfrac{8}{10}$	$\dfrac{2}{10}$
Liberal	$\dfrac{1}{10}$	$\dfrac{9}{10}$

Using the notation of Example 31-4:

(a) Find X_0 and Q.
(b) Find the eigenvalues and corresponding eigenvectors for Q.
(c) Fid a formula for Q^n.
(d) Find X_n.
(e) Find what happens to X_n in the long run.

31-4 In an experiment, white mice are kept in a cage with two connecting compartments: a feeding compartment and an exercise compartment.
The mice can freely move back and forth between the two compartments,
and it is known that in any 1-min period the probability of a mouse's
changing compartments is given by the following table.

Now	One minute from now	
	Feeding	Exercise
Feeding	$\dfrac{3}{5}$	$\dfrac{2}{5}$
Exercise	$\dfrac{1}{2}$	$\dfrac{1}{2}$

At the start of the experiment, the experimenter wishes to place a supply
of mice in the cage so that the percentage of mice in each of the two
compartments does not change from minute to minute. What percentage
of the mice should be placed in each of the two compartments initially?

31-5 Solve the system of differential equations

$$f' = 2f + g$$
$$g' = 3f + 4g$$

31-6 Solve the system of differential equations

$$f' = f + g$$
$$g' = 2g$$
$$h' = 2f - h$$

*31-7 Write a computer program that generates Fibonacci numbers
using the recursion formula $x_n = x_{n-1} + x_{n-2}$. Use the computer to compare x_{20} with the value for x_{20} computed by the formula derived in
Example 31-1.

*31-8   Write a computer program that permits one to input birth rates b_i and survival rates p_i and an initial population distribution $(x_0, x_1, \ldots, x_n)$, and then computes population figures into the future, year by year. (Such a program with $n \approx 90$ could be used to project population figures for the United States, for example, in order to plan for the future.)

Review of Chapter 5

1   If $X = (a_1, a_2, \ldots, a_n)$ is a vector in $\mathcal{R}^n$, then the coordinate matrix of X relative to the usual basis is $[X] = [$ $]$. If $X_1, X_2, \ldots,$ X_n is a different basis for $\mathcal{R}^n$ and if $X = b_1 X_1 + b_2 X_2 + \cdots + b_n X_n$, then $[X]_{new} = [$ $]$.

$[X]_{old}$ and $[X]_{new}$ are related by the formula $[X]_{new} = [X]_{old} P^{-1}$, where P is the matrix whose rows are _____

_____. Equivalently, $[X]_{old} =$ _____.

2   A linear transformation T has many matrices that correspond to it. The usual matrix A for T is obtained by applying T to the usual basis and using the usual coordinates as rows of the matrix.

If $X_1, X_2, \ldots, X_n$ is a new basis, we apply T to these basis vectors and use the *new* coordinates of the results as rows of the matrix B. That is, if

$$X_1 T = a_{11} X_1 + a_{12} X_2 + \cdots + a_{1n} X_n$$
$$X_2 T = a_{21} X_1 + a_{22} X_2 + \cdots + a_{2n} X_n$$
$$\vdots$$
$$X_n T = a_{n1} X_1 + a_{n2} X_2 + \cdots + a_{nn} X_n$$

then

$$B = \begin{bmatrix} & & \\ & & \\ & & \end{bmatrix}$$

is the matrix for T relative to the basis $X_1, X_2, \ldots, X_n$.

3   If A is the usual matrix for T and if B is a new matrix for T, then

$$A = \underline{\quad\quad}$$

or, equivalently,

$$B = \underline{\qquad\qquad}$$

where the _____ of P are formed by using _____. If A and B are both matrices for T, then A and B are said to be similar.

4 Given a linear transformation T, we seek a simple matrix for T to aid in performing computations and to help in analyzing the geometric effect of T. The simplest matrices (for many purposes) are the diagonal matrices. A linear transformation $T : \mho \rightarrow \mho$ has a diagonal matrix

$$B = \begin{bmatrix} a_1 & 0 & \cdots & 0 \\ 0 & a_2 & \cdots & 0 \\ \cdots & \cdots & \cdots & \cdots \\ 0 & 0 & \cdots & a_n \end{bmatrix}$$

relative to a basis $\mathbf{X}_1$, $\mathbf{X}_2$, ..., $\mathbf{X}_n$ exactly when $\mathbf{X}_1 T = $ _____, $\mathbf{X}_2 T = $ _____, ..., $\mathbf{X}_n T = $ _____. The numbers $a_1, a_2, \ldots, a_n$ are called _____ for T, and the vectors $\mathbf{X}_1$, $\mathbf{X}_2$, ..., $\mathbf{X}_n$ are called _____ for T.

5 To find eigenvalues for A, where A is an $n \times n$ matrix, solve the characteristic equation det ($\underline{\qquad}$) $= 0$ for λ.

  To find eigenvectors corresponding to an eigenvalue λ, solve the system of equations _____ $=$ _____ for $\mathbf{X}$.

6 The eigenvectors and eigenvalues for a geometric linear transformation T can often be found by inspection. For example, if a plane is left unchanged by T, then for any vector $\mathbf{X}$ in the plane, $\mathbf{X}T = $ __$\mathbf{X}$; if T is the reflection in a plane and if $\mathbf{X}$ is perpendicular to this plane, then $\mathbf{X}T = $ __$\mathbf{X}$; if T is the perpendicular projection onto a plane and if $\mathbf{X}$ is perpendicular to this plane, then $\mathbf{X}T = 0 = $ __$\mathbf{X}$.

  For other kinds of transformations, to find eigenvalues and eigenvectors, we first find the usual matrix A, and then find A's eigenvalues and eigenvectors as in 5.

7 Let $\mho$ be an n-dimensional vector space and let $T : \mho \rightarrow \mho$ be a linear transformation. If T has __ distinct eigenvalues, then T has a diagonal matrix B. Even if T has fewer than __ eigenvalues, T will have a diagonal matrix B if there are enough independent eigenvectors to make up a complete _____ of eigenvectors.

Review Exercises

Each of the matrices in Exercises 1 to 7 is the usual matrix for a linear transformation T.

(a) Find (if possible) a diagonal matrix B similar to A.
(b) Find a basis of eigenvectors relative to which T has B as diagonal matrix.

1 $A = \begin{bmatrix} 2 & 3 \\ -1 & -2 \end{bmatrix}$ **2** $A = \begin{bmatrix} 2 & -3 \\ 3 & 2 \end{bmatrix}$

3 $A = \begin{bmatrix} 2 & 1 & 0 \\ 0 & 2 & 0 \\ 0 & 0 & 3 \end{bmatrix}$ **4** $A = \begin{bmatrix} 3 & 2 & 0 \\ 2 & 0 & 0 \\ 0 & 0 & 4 \end{bmatrix}$

5 $A = \begin{bmatrix} 2 & 6 & 0 \\ -2 & -5 & 0 \\ 1 & 3 & -1 \end{bmatrix}$ **6** $A = \begin{bmatrix} 0 & 2 & 1 \\ 0 & 1 & 0 \\ 4 & 0 & 0 \end{bmatrix}$

7 $A = \begin{bmatrix} -\frac{1}{2} & 0 & -\frac{3}{2} \\ \frac{3}{2} & 1 & \frac{3}{2} \\ -\frac{3}{2} & 0 & -\frac{1}{2} \end{bmatrix}$

8 Let T be the reflection in the line $y = 3x$ in $\mathcal{R}^2$.

(a) Find a diagonal matrix for T.
(b) Find the usual matrix for T.
(c) Find $(1, -2)T$.

9 Let T be the perpendicular projection of $\mathcal{R}^2$ onto the line $y = -4x$.

(a) Find a diagonal matrix for T.
(b) Find the usual matrix for T.
(c) Find $(1, 0)T$.

10 Let T be the reflection of $\mathcal{R}^3$ in the plane $x - 5y + z = 0$.

(a) Find a diagonal matrix for T.
(b) Find the usual matrix for T.
(c) Find $(0, 1, 0)T$.

11 Let T be the perpendicular projection of R^3 onto the plane $x - z = 0$.

(a) Find a diagonal matrix for T.
(b) Find the usual matrix for T.
(c) Find $(3, -2, 3)T$.

12 Let $T: R^5 \to R^5$ be a linear transformation.

(a) Suppose that T has eigenvalues $1, 2, -2, 3, -3$. Does T have a diagonal matrix? If so, what?
(b) Suppose that T has eigenvalues $1, 2, -2$, and that corresponding to the eigenvalue 1, T has three independent eigenvectors. Does T have a diagonal matrix? If so, what?
(c) Suppose that T has $2, 3$ as eigenvalues and that corresponding to the eigenvalue 2, it has two independent eigenvectors, and corresponding to the eigenvalue 3, it has three independent eigenvectors. Does T have a diagonal matrix? If so, what?
(d) Suppose that T has only the eigenvalue 4 and that the set of solutions to $\mathbf{X}T = 4\mathbf{X}$ is $\{a(1, 0, 1, 0, 0) + b(0, -2, 0, 1, 0) + c(1, 3, 0, 0, 1) : a, b, c \in R\}$. Does T have a diagonal matrix? If so, what?

13 Let $T: R^3 \to R^3$. Suppose that T has eigenvalues 2 and 3 and that the set of solutions to $\mathbf{X}(A - 2I) = \mathbf{0}$ is $\langle(1, 1, 0), (0, 1, 1)\rangle$ and the set of solutions to $\mathbf{X}(A - 3I) = \mathbf{0}$ is $\langle(1, 2, 4)\rangle$.

(a) Find a diagonal matrix for T.
(b) Find a matrix P such that $PAP^{-1} = B$.
(c) Find the usual matrix A for T.

14 Let $T: R^3 \to R^3$ have the rule $(x, y, z)T = (x + 3z, 3y + z, z)$. Explain why T does not have a diagonal matrix.

15 Suppose $(x, y, z)T = (z, x + 2z, 3z)$.

(a) Find all the eigenvalues and eigenvectors for T.
(b) If possible, find a diagonal matrix for T.

16 Suppose $(x, y, z)T = (x + 2y + 3z, x + 2y + 3z, x + 2y + 3z)$.

(a) Find all the eigenvalues and eigenvectors for T.
(b) If possible, find a diagonal matrix for T.

17 Let A be the matrix in Exercise 7 above. Using the formula $A^n = P^{-1}B^nP$:

(a) Find A^7.
(b) Find A^{-1}.

18 Suppose that $T : \mathcal{R}^2 \rightarrow \mathcal{R}^2$ has the following rule: $\mathbf{X}T$ is obtained by first rotating $\mathbf{X}$ by $90°$ counterclockwise and then reflecting this vector in the y axis. Find a diagonal matrix for T.

19 Suppose that $T : \mathcal{R}^2 \rightarrow \mathcal{R}^2$ has the following rule: $\mathbf{X}T$ is obtained by first rotating $\mathbf{X}$ by $45°$ counterclockwise and then reflecting this vector in the line $y = x$. Find a diagonal matrix for T.

CHAPTER 6

EUCLIDEAN GEOMETRY
IN VECTOR SPACES

One reason for studying vector spaces is that they provide suitable ve-hicles for developing geometry. In any vector space one can define many of the basic geometric objects (such as lines, planes, segments, rays, and triangles) and some of the basic geometric concepts (such as parallelism and dimension). In fact, any geometric concept that depends only on vector addition and scalar multiplication can be defined in *any* vector space. For example, in any vector space with real scalars we can define a line in accordance with the theorems of Sec. 4: a line is a set of points of the form $\{X + aY : a \in \mathcal{R}\}$. Parallel lines can be defined to be coplanar lines having no point of intersection.

However, euclidean geometry involves some other concepts—to be specific, *length* and *angle*—that cannot be defined solely on the basis of vector addition and scalar multiplication. Length and angle are the vital concepts that give euclidean geometry its unique flavor. Other familiar concepts of euclidean geometry—for example, congruence, similarity, and area—are defined in terms of length and angle. It should be noted that while the ideas of length and angle have not been *formally* introduced thus far, we have touched upon transformations—rotations, perpendicu-lar projections, and reflections—whose definitions really depend on the concepts of length and angle. It is impossible to discuss $\mathcal{R}^2$ and $\mathcal{R}^3$ in any depth without touching upon euclidean ideas since $\mathcal{R}^2$ is the vector space model for euclidean plane geometry and $\mathcal{R}^3$ is the model for euclidean space geometry.

We will now formally study the euclidean concepts of length and angle, beginning with their most familiar form in $\mathcal{R}^2$ and $\mathcal{R}^3$. We will then show how to generalize these ideas, first to any $\mathcal{R}^n$, later to other vector spaces. In doing so, euclidean geometry becomes much more widely applicable.

It happens that in $\mathcal{R}^n$ the concepts of both length and angle can be based on the single fundamental idea of the *dot product* of vectors. We will therefore begin with a discussion of the dot product, and then use this to develop various geometric concepts in $\mathcal{R}^n$. Later, in Sec. 36, we go beyond $\mathcal{R}^n$, to a class of vector spaces called inner product spaces.

32. LENGTH AND DISTANCE

In Sec. 21 we used the dot product of vectors in $\mathcal{R}^n$ to formulate the definition of matrix multiplication. In this chapter the dot product will be used as the starting point for our discussion of geometry in vector spaces. Before defining length of vectors and distance between vectors, we review the definition of dot product and discuss some of its basic properties.

Definition

Let $\mathbf{X} = (x_1, x_2, \ldots, x_n)$ and $\mathbf{Y} = (y_1, y_2, \ldots, y_n)$ be vectors in $\mathcal{R}^n$. The **dot product** $\mathbf{X} \cdot \mathbf{Y}$ of $\mathbf{X}$ and $\mathbf{Y}$ is defined by the rule

$$\mathbf{X} \cdot \mathbf{Y} = x_1 y_1 + x_2 y_2 + \cdots + x_n y_n$$

Note that the dot product of vectors is a *number*. For example,

$$(1, 2, -1) \cdot (4, -5, 2) = 1(4) + 2(-5) + (-1)(2) = -8$$
$$(2, 3, 1, 1) \cdot (3, -2, 2, 0) = 2(3) + 3(-2) + 1(2) + 1(0) = 2$$

The following theorem lists the four fundamental properties of the dot product, properties that permit us to use the dot product in defining length and angle.

Theorem 32-1   Let $\mathbf{X}$, $\mathbf{Y}$, and $\mathbf{Z}$ be arbitrary vectors in $\mathcal{R}^n$ and let r be an arbitrary scalar. The dot product has the following properties:

1   $\mathbf{X} \cdot \mathbf{Y} = \mathbf{Y} \cdot \mathbf{X}$

2   $(\mathbf{X} + \mathbf{Y}) \cdot \mathbf{Z} = \mathbf{X} \cdot \mathbf{Z} + \mathbf{Y} \cdot \mathbf{Z}$

3   $(r\mathbf{X}) \cdot \mathbf{Y} = r(\mathbf{X} \cdot \mathbf{Y})$

4   $\mathbf{X} \cdot \mathbf{X} > 0$ if and only if $\mathbf{X} \neq \mathbf{0}$

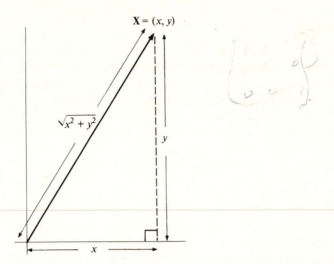

Fig. 32-1   The Pythagorean theorem gives the length of a vector (x, y).

Each of these four properties is easily verified, and their proofs are left to the reader as an exercise.

We will now use the dot product to define the concepts of length and distance for vectors in $\mathcal{R}^n$. The definition of length is suggested by the situation in $\mathcal{R}^2$. The length of a vector $\mathbf{X} = (x, y)$, considered to be the arrow from the origin to the point (x, y), is $\sqrt{x^2 + y^2}$, by the Pythagorean theorem (Fig. 32-1). In terms of the dot product, since $x^2 + y^2 = (x, y) \cdot (x, y)$, we have $x^2 + y^2 = \mathbf{X} \cdot \mathbf{X}$. Therefore, $\mathbf{X}$ has length $\sqrt{\mathbf{X} \cdot \mathbf{X}}$.

The situation in $\mathcal{R}^3$ is similar: if $\mathbf{X} = (x, y, z)$, then the length of $\mathbf{X} = \sqrt{x^2 + y^2 + z^2} = \sqrt{(x, y, z) \cdot (x, y, z)} = \sqrt{\mathbf{X} \cdot \mathbf{X}}$. Using $\mathcal{R}^2$ and $\mathcal{R}^3$ as our models, we define length in any $\mathcal{R}^n$ as follows.

Definition

The **length** of a vector $\mathbf{X}$ in $\mathcal{R}^n$ is defined to be $\sqrt{\mathbf{X} \cdot \mathbf{X}}$. We denote the length of $\mathbf{X}$ by $|\mathbf{X}|$.

Example 32-1

The length of $(3, 4)$ is

$$\sqrt{(3, 4) \cdot (3, 4)} = \sqrt{3(3) + 4(4)} = \sqrt{25} = 5$$

The length $|(1, 0, -3)|$ is

$$\sqrt{(1, 0, -3) \cdot (1, 0, -3)} = \sqrt{1(1) + 0(0) + (-3)(-3)} = \sqrt{10}$$

As a final example, even though we cannot visualize the length of a vector in $\mathcal{R}^4$, the definition applies; we have, for example,

$$|(1, 1, 2, 1)| = \sqrt{(1, 1, 2, 1) \cdot (1, 1, 2, 1)} = \sqrt{7} \qquad \bullet$$

If a vector $\mathbf{X}$ in $\mathcal{R}^2$ or $\mathcal{R}^3$ is represented by an arrow, then multiplying $\mathbf{X}$ by a scalar r stretches the arrow by a factor $|r|$. (The reader is advised to prove this fact, as he now can, using the definition $|\mathbf{X}| = \sqrt{\mathbf{X} \cdot \mathbf{X}}$. See Exercise 32-5.) This property allows us to convert vectors so that they have length 1. Vectors of length 1 are called *unit vectors*. We can adjust the length of a nonzero vector down (or up) to 1 by multiplying the vector by a suitable scalar. In fact, if $\mathbf{X}$ has length $r \neq 0$, then $(1/r)\mathbf{X}$ has length 1.

Example 32-2

The length of the vector $(3, 1)$ is $\sqrt{(3, 1) \cdot (3, 1)} = \sqrt{10}$. Therefore, $(1/\sqrt{10})(3, 1) = (3/\sqrt{10}, 1/\sqrt{10})$ has length 1 and the same direction as $(3, 1)$. (See Fig. 32-2.) $\qquad \bullet$

The distance between two vectors in $\mathcal{R}^n$ can also be defined by means of the dot product. The source of our definition is the formula $\sqrt{(c - a)^2 + (d - b)^2}$ for the distance between two points $\mathbf{X} = (a, b)$ and $\mathbf{Y} = (c, d)$ in the plane. (See Fig. 32-3.) In terms of the dot product we have

$$\begin{aligned}
\sqrt{(c - a)^2 + (d - b)^2} &= \sqrt{(c - a, d - b) \cdot (c - a, d - b)} \\
&= \sqrt{((c, d) - (a, b)) \cdot ((c, d) - (a, b))} \\
&= \sqrt{(\mathbf{Y} - \mathbf{X}) \cdot (\mathbf{Y} - \mathbf{X})} \\
&= |\mathbf{Y} - \mathbf{X}|
\end{aligned}$$

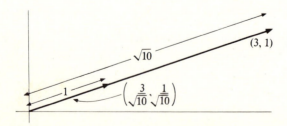

Fig. 32-2 $(3/\sqrt{10}, 1/\sqrt{10})$ is the unit vector having the same direction as $(3, 1)$.

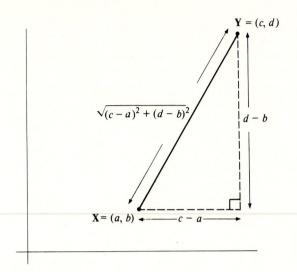

Fig. 32-3

Therefore, the distance between **X** and **Y** is the length of the vector **Y** − **X**.

The above derivation can be reformulated in terms of vectors drawn as arrows. In Fig. 32-4 arrows represent the points **X** and **Y**, and the vector representing the segment from **X** to **Y** is labeled **Z**. By the head-to-tail method for addition of vectors we have **X** + **Z** = **Y**, and hence **Z** = **Y** − **X**. Thus the length of **Z** is equal to $|\mathbf{Y} - \mathbf{X}|$. Since the length of **Z** is the distance between **X** and **Y**, we again have $|\mathbf{Y} - \mathbf{X}|$ as that distance. We use this idea of taking the length of **Y** − **X** to define distance in any $\mathcal{R}^n$.

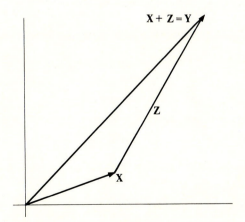

Fig. 32-4 The distance between **X** and **Y** is given by **Z** = **Y** − **X**.

Definition

The **distance** between any two vectors X and Y in $\mathcal{R}^n$ is $|Y - X|$.

Example 32-3

The distance between $(1, 2)$ and $(4, -3)$ is found by first obtaining their difference $(3, -5)$ and then computing the length of this difference:

$$|(4, -3) - (1, 2)| = |(3, -5)| = \sqrt{9 + 25} = \sqrt{34}$$

The distance between $(1, 2, 3, 0)$ and $(2, 1, 3, 2)$ is

$$|(1, -1, 0, 2)| = \sqrt{1 + 1 + 0 + 4} = \sqrt{6} \qquad \bullet$$

Exercises

32-1 Find the length of each vector.

(a) $(1, 1)$ (b) $(1, 2, 5)$
(c) $(\frac{3}{5}, \frac{4}{5})$ (d) $(-1, 0, 3, 2, 1)$
(e) $(1/\sqrt{3}, 1/\sqrt{3}, 1/\sqrt{3})$
(f) $-86(2, 3, 2)$ (Hint: Find $|(2, 3, 2)|$ first.)
(g) $(2, 0, 0) + (0, 0, 3)$ (Hint: The answer is *not* 5.)

32-2 Find the distance between each pair of vectors.

(a) $(1, 1)$ and $(2, 2)$
(b) $(3, 5)$ and $(-3, -5)$
(c) $(4, 5, 6)$ and $(-1, -1, 5)$
(d) $(1, 2, 3)$ and $(1, 2, 3)$
(e) $(1, 2, 0, 4)$ and $(2, 2, 2, 2)$
(f) $(0, -3, 2, 1, 4)$ and $(2, -3, 2, 0, 2)$

32-3 Let $X = (2, 1)$.

(a) Find the length of X.
(b) Find the length of $(1/|X|)X$.
(c) Explain why X and $(1/|X|)X$ lie on the same line. (Hint: Verify that they both lie on the line with equation $y = \frac{1}{2}x$.)

32-4 For each of the following vectors, find a unit vector (vector of length 1) that has the same direction.

(a) (3, 4)   (b) $(-3, -2)$
(c) (1, 2, 5)   (d) $(4, 0, -2)$
(e) (1, 1, 1, 1)   (f) (0.3, 0.4, 0)

32-5 Let $X = (x_1, x_2, \ldots, x_n)$ be an arbitrary vector in $\mathcal{R}^n$. Show that for any scalar r, $|rX| = |r||X|$.

32-6 Prove Theorem 32-1.

32-7 Let X be any vector in $\mathcal{R}^n$. Show that $0 \cdot X = 0$.

*32-8 Write a computer program that computes lengths of vectors in $\mathcal{R}^3$, or, more generally, in any $\mathcal{R}^n$.

*32-9 Write a computer program that computes the distance between any two given vectors in $\mathcal{R}^n$.

*32-10 The vector (1, 2, 2) has length 3. Write a computer program that finds other vectors in $\mathcal{R}^3$ whose entries are all positive integers and whose length is an integer.

33. PERPENDICULARITY AND ANGLES

In this section we will see how the geometric ideas of perpendicularity and angle can be generalized from $\mathcal{R}^2$ and $\mathcal{R}^3$ to any $\mathcal{R}^n$.

The fundamental relationship between length and angle in $\mathcal{R}^2$ and $\mathcal{R}^3$ is given by the law of cosines *a opposite side c*

$$c^2 = a^2 + b^2 - 2ab \cos \theta \tag{1}$$

where a, b, and c are the lengths of the three sides of a triangle and θ is the angle opposite the side of length c (Fig. 33-1). We will derive a formula for $\cos \theta$ from (1).

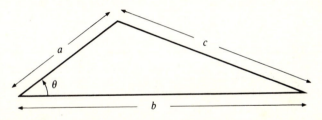

Fig. 33-1

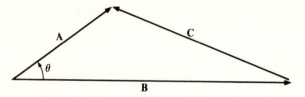

Fig. 33-2 $A = B + C$, and thus $C = A - B$.

First we redraw the triangle of Fig. 33-1, using arrows, or vectors, for its sides, as in Fig. 33-2. The length of C is the distance between A and B, which equals $|A - B|$, as worked out in the previous section. We can now restate the law of cosines in terms of the lengths of the vectors A, B, and C.

$$|A - B|^2 = |A|^2 + |B|^2 - 2|A||B| \cos \theta$$

In terms of the dot product (the underlying concept in defining length), this equation becomes

$$(A - B) \cdot (A - B) = A \cdot A + B \cdot B - 2|A||B| \cos \theta$$

Working on the left side of this equation using properties of the dot product, we obtain

$$A \cdot A - 2A \cdot B + B \cdot B = A \cdot A + B \cdot B - 2|A||B| \cos \theta$$

which simplifies to

$$-2A \cdot B = -2|A||B| \cos \theta$$

Therefore,

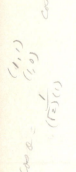

$$\cos \theta = \frac{A \cdot B}{|A||B|}$$

a formula that gives the angle between two vectors in $\mathcal{R}^2$ (or $\mathcal{R}^3$). Although we used $\mathcal{R}^2$ as a motivating force to obtain this formula, we will adopt this formula to define the angle between two vectors in any $\mathcal{R}^n$.

Definition

The **angle** (in radians) between any two nonzero vectors X and Y in $\mathcal{R}^n$ is the unique number θ between zero and π such that

$$\cos \theta = \frac{X \cdot Y}{|X||Y|}$$

In giving this definition, we are relying on a fact to be proved in Sec. 34, namely that

$$-1 \le \frac{\mathbf{X} \cdot \mathbf{Y}}{|\mathbf{X}||\mathbf{Y}|} \le 1$$

This inequality (the Cauchy-Schwarz inequality) assures us that our expression for $\cos \theta$ will always lie between -1 and 1, and thus that the angle θ will always be defined.

Example 33-1

Let $\mathbf{X} = (1, 1, 0)$ and $\mathbf{Y} = (1, 0, 1)$ be two vectors in $\mathcal{R}^3$. The angle θ between them is determined by the formula

$$\cos \theta = \frac{\mathbf{X} \cdot \mathbf{Y}}{|\mathbf{X}||\mathbf{Y}|} = \frac{(1, 1, 0) \cdot (1, 0, 1)}{|(1, 1, 0)||(1, 0, 1)|} = \frac{1}{\sqrt{2}\sqrt{2}} = \frac{1}{2}$$

Since $\cos \theta = \frac{1}{2}$, we have $\theta = \pi/3$ radians, or $60°$.   ●

Example 33-2

Let $\mathbf{X} = (1, 2)$ and $\mathbf{Y} = (-6, 3)$. The angle between them is determined by

$$\cos \theta = \frac{(1, 2) \cdot (-6, 3)}{|(1, 2)||(-6, 3)|} = \frac{0}{\sqrt{5}\sqrt{45}} = 0$$

Since $\cos \theta = 0$, we have $\theta = \pi/2$ radians, or $90°$. Hence $\mathbf{X}$ and $\mathbf{Y}$ are perpendicular.   ●

Example 33-3

Let $\mathbf{X} = (1, 1)$ and $\mathbf{Y} = (4, \sqrt{2})$. We have

$$\cos \theta = \frac{4 + \sqrt{2}}{\sqrt{2}\sqrt{18}} = \frac{4 + \sqrt{2}}{6} \approx 0.902$$

In this case it is not easy to compute θ. However, we may use tables for the cosine function to find the angle θ (whose cosine is 0.902) approximately: $\theta \approx 0.45$ radians, or $26°$.   ●

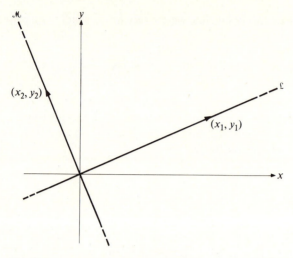

Fig. 33-3 Perpendicular lines $\mathcal{L}$ and $\mathcal{M}$.

In $\mathcal{R}^2$ and $\mathcal{R}^3$, perpendicularity of **X** and **Y** requires that $\theta = \pi/2$, and thus that $\cos \theta = 0$. In terms of the formula

$$\cos \theta = \frac{\mathbf{X} \cdot \mathbf{Y}}{|\mathbf{X}| \, |\mathbf{Y}|}$$

the only way we can have $\cos \theta = 0$ is to have $\mathbf{X} \cdot \mathbf{Y} = 0$. We will adopt this condition as our general definition of perpendicularity.

Definition

Let **X** and **Y** be two vectors in $\mathcal{R}^n$. We say that **X** and **Y** are **perpendicular** or **orthogonal** if $\mathbf{X} \cdot \mathbf{Y} = 0$.

For example, the vectors $(1, 3, -2)$ and $(4, 0, 2)$ are perpendicular since $(1, 3, -2) \cdot (4, 0, 2) = 0$.

We can use the above condition for perpendicularity to prove the familiar fact that two nonvertical lines in $\mathcal{R}^2$ are perpendicular if and only if their slopes are negative reciprocals of each other. In Fig. 33-3 we have drawn arrows on two lines $\mathcal{L}$ and $\mathcal{M}$ meeting at the origin. The slope of $\mathcal{L}$ is y_1/x_1 and the slope of $\mathcal{M}$ is y_2/x_2. By the above condition for perpendicularity, $\mathcal{L}$ is perpendicular to $\mathcal{M}$ if and only if $(x_1, y_1) \cdot (x_2, y_2) = 0$. That is, if and only if $x_1 x_2 + y_1 y_2 = 0$. Rearranging terms, this becomes $y_2/x_2 = -x_1/y_1$. Therefore, we have perpendicularity if and only if $y_2/x_2 = -1/(y_1/x_1)$, that is, if and only if the slope of $\mathcal{M}$ is the negative reciprocal of the slope of $\mathcal{L}$. If $\mathcal{L}$ and $\mathcal{M}$ meet in a point other than the origin, we take lines $\mathcal{L}'$ parallel to $\mathcal{L}$ and $\mathcal{M}'$ parallel to $\mathcal{M}$ that do meet at the origin. Since parallel lines have the same slope, the above argument applied to $\mathcal{L}'$ and $\mathcal{M}'$ also yields the desired result.

We can use our condition for perpendicularity to find vectors perpendicular to planes in $\mathcal{R}^3$. We asserted in Chap. 5 that the vector (a, b, c) is perpendicular to the plane $ax + by + cz = d$. We now prove this.

Theorem 33-1

The vector (a, b, c) is perpendicular to every vector in the plane $ax + by + cz = 0$.

Proof

When we say that a vector is perpendicular to a plane, we mean that the *line* $\mathcal{L}$ determined by the vector is perpendicular to every line segment that lies in the plane and intersects $\mathcal{L}$. Thus, in the present situation it suffices to show that (a, b, c) is perpendicular to every segment that lies in the plane and passes through the origin. Each such segment is determined by a vector (arrow) lying in the plane. Hence we must show that (a, b, c) is perpendicular to each vector (x_0, y_0, z_0) in the plane.

Since (x_0, y_0, z_0) lies in the plane, it must satisfy the equation of the plane. That is, we must have

$$ax_0 + by_0 + cz_0 = 0$$

Rewriting the left side in terms of the dot product, we obtain

$$(a, b, c) \cdot (x_0, y_0, z_0) = 0$$

Therefore, (a, b, c) is perpendicular to (x_0, y_0, z_0). ∎

Corollary 33-2

The vector (a, b, c) is also perpendicular to the plane $ax + by + cz = d$, for any $d \neq 0$.

Proof

The plane $ax + by + cz = d$ is parallel to the plane $ax + by + cz = 0$, since the system consisting of these two equations has no solution. Since (a, b, c) is perpendicular to one of two parallel planes, it is perpendicular to the other. ∎

One advantage of working with perpendicular vectors is that they are automatically independent, and hence they can be used as part of a basis.

Theorem 33-3

Let $\mathbf{X}_1, \mathbf{X}_2, \ldots, \mathbf{X}_k$ be nonzero vectors in $\mathcal{R}^n$, each of them perpendicular to the others. Then they are independent.

Proof   To show that these vectors are independent, we must show that the equation

$$a_1X_1 + a_2 X_2 + \cdots + a_k X_k = 0 \tag{2}$$

has the unique solution $a_1 = 0, a_2 = 0, \ldots, a_k = 0$. We will show that for any i, $a_i = 0$. In order to do this, we take the dot product of X_i with both sides of (2):

$$(a_1X_1 + a_2 X_2 + \cdots + a_k X_k) \cdot X_i = 0 \cdot X_i$$

Using the properties of the dot product, we have

$$a_1(X_1 \cdot X_i) + a_2(X_2 \cdot X_i) + \cdots + a_k(X_k \cdot X_i) = 0$$

Since X_i is perpendicular to each of the other vectors, every dot product $X_j \cdot X_i$ is zero except $X_i \cdot X_i$. Thus we have

$$a_i(X_i \cdot X_i) = 0$$

Since $X_i \cdot X_i \neq 0$, we must have $a_i = 0$, as desired.   ∎

$(4, v) = 0 \qquad x - ay = v$

$a = \left(\dfrac{(x, 4)}{(4, 4)} \right)$

Exercises

33-1 Find the angle between each pair of vectors.

(a) $(1, 2), (-2, 1)$
(b) $(0, 1), (\sqrt{3}, -1)$
(c) $(\sqrt{2}/2, 0, \sqrt{2}/2), (\sqrt{2}/2, -\sqrt{2}/2, 0)$
(d) $(1, -2, 0), (2, 1, 3)$
(e) $(6, 0, 4), (8, \sqrt{39}, 1)$
(f) $(1, 1, 1, 1), (-2, -2, -2, -2)$

33-2(a) Find the angle between $(\sqrt{3}, 1)$ and $(1, \sqrt{3})$.
(b) Find the angle between $(5\sqrt{3}, 5)$ and $(2, 2\sqrt{3})$.
(c) Explain geometrically why the answers to (a) and (b) are the same.
(d) If X and Y are nonzero vectors in any $\mathscr{R}^n$ and r and s are positive scalars, prove that the angle between X and Y is equal to the angle between rX and sY.

33-3 Find a nonzero vector perpendicular to the given vector.

(a) $(1, 2, 3)$ [Hint: Use any nonzero solution to $(1, 2, 3) \cdot (x, y, z) = 0$]
(b) $(2, 1, 1, -4, 3)$
(c) (a, b)

33-4 Find a vector perpendicular to $(1, 2, 3)$ that has the same length as $(1, 2, 3)$. (Hint: Find any perpendicular vector, and then multiply by a suitable scalar.)

33-5(a) Find a vector perpendicular to the plane $2x - y + 4z = 0$. Then, by dividing by a suitable scalar, find a vector one unit long that is perpendicular to the plane.
(b) Find a vector one unit long that is perpendicular to the plane $3x + 2y + 3z = 8$.

33-6(a) Find a plane perpendicular to $(1, 2, -3)$.
(b) Find a plane perpendicular to $(1, 4, -2)$ that passes through the point $(2, -2, 2)$.

33-7 Find a vector perpendicular to both $(2, 0, 1)$ and $(1, 3, -2)$.

33-8 Prove that the diagonals of a square are perpendicular. [Hint: Let X and Y be vectors representing two adjacent sides of the square, as in Fig. 33-4. Find vectors (in terms of X and Y) that represent the two diagonals and show that they satisfy the condition for perpendicularity.]

33-9 Let X and Y be vectors in $\mathcal{R}^n$. Prove that $X + Y$ and $X - Y$ are perpendicular if and only if $|X| = |Y|$.

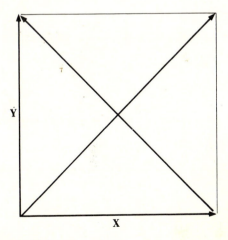

Fig. 33-4

*33-10   Write a computer program that finds the cosine of the angle between any two vectors in $\mathcal{R}^n$.

*33-11   Write a computer program that finds a unit vector pointing in the same direction as a given vector.

*33-12   Write a computer program that finds a unit vector perpendicular to a given vector $\mathbf{X} = (a, b, c)$. (Hint: Have the computer first compute a solution to $ax + by + cz = 0$.)

34.   PROJECTIONS

In previous chapters we have seen many examples of perpendicular projections, especially projections onto planes in $\mathcal{R}^3$. Historically, the original impetus for studying projections was probably the desire to make a scientific study of map making and of perspective in art. Both of these involve projecting three-dimensional objects onto plane surfaces (maps or the canvas of paintings). Modern mathematics and physics have given rise to many further uses for projections. The physical applications are all based on the idea that the perpendicular projection of a vector onto a line $\mathcal{L}$ gives the portion of the vector that acts along $\mathcal{L}$. An example will clarify this point.

In Fig. 34-1a, the vector $\mathbf{F}$ represents the force of the wind pushing against the sail of a sailboat as the sailboat moves to the right along the

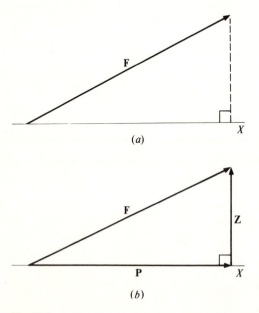

(a)

(b)

Fig. 34-1

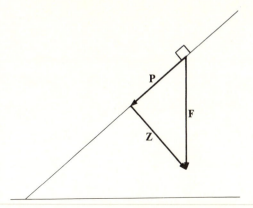

Fig. 34-2

line X. The perpendicular projection **P** of **F** on the line X (Fig. 34-1b) is that "part" of **F** that tends to act along X, that is, the component of the force that actually helps push the boat in the desired direction. The vector labeled **Z** in Fig. 34-1b is perpendicular to X, and represents the part of the force of the wind that is tending to tilt the sailboat over, not contributing to its motion along X. Note that by the head-to-tail addition of vectors, $\mathbf{F} = \mathbf{P} + \mathbf{Z}$, so that **P** and **Z** add up to **F**; that is, **P** and **Z** are "components" or "parts" of **F**.

As a second example, consider a box on an inclined plane undergoing the force of gravity **F** (Fig. 34-2). The projection **P** (obtained by dropping a perpendicular from **F** to the inclined plane) is the component of **F** that tends to pull the box down the plane; **Z** is the component of **F** that is wasted (insofar as motion down the plane is concerned) and contributes only to friction between the box and the plane.

In view of the many applications of projections, we now develop a method for computing projections. The dot product provides such a method.

Figure 34-3 shows two vectors **X** and **Y** meeting at an angle of θ radians (where $\theta < \pi/2$) and the perpendicular projection of **X** on **Y**, labeled **P**. We will now derive a formula for **P**. From the triangle formed by **X**, **P**, and the line perpendicular to **Y** we see that

$$\cos\theta = \frac{|\mathbf{P}|}{|\mathbf{X}|}$$

or

$$|\mathbf{P}| = |\mathbf{X}|\cos\theta$$

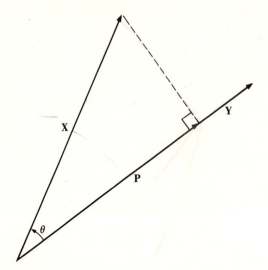

Fig. 34-3

From Sec. 33 we know that

$$\cos \theta = \frac{\mathbf{X} \cdot \mathbf{Y}}{|\mathbf{X}| \, |\mathbf{Y}|}$$

and hence the length of **P** is given by

$$|\mathbf{P}| = |\mathbf{X}| \frac{\mathbf{X} \cdot \mathbf{Y}}{|\mathbf{X}| \, |\mathbf{Y}|} = \frac{\mathbf{X} \cdot \mathbf{Y}}{|\mathbf{Y}|}$$

Since **P** lies on the same line as **Y** and has the same direction, $\mathbf{P} = r\mathbf{Y}$ for some scalar $r > 0$. Therefore, $|\mathbf{P}| = r|\mathbf{Y}|$; that is, $r = |\mathbf{P}|/|\mathbf{Y}|$. Thus

$$\mathbf{P} = r\mathbf{Y} = \left(\frac{|\mathbf{P}|}{|\mathbf{Y}|}\right)\mathbf{Y} = \left(\frac{\mathbf{X} \cdot \mathbf{Y}}{|\mathbf{Y}| \, |\mathbf{Y}|}\right)\mathbf{Y}$$

or, finally,

$$\mathbf{P} = \left(\frac{\mathbf{X} \cdot \mathbf{Y}}{\mathbf{Y} \cdot \mathbf{Y}}\right)\mathbf{Y} \tag{3}$$

If it happens that $\pi/2 \leq \theta \leq \pi$, then the same formula holds. This is left as an exercise.

We adopt (3) as the definition of projection in any $\mathcal{R}^n$.

Definition

Let **X** and **Y** be vectors in $\mathcal{R}^n$ with $\mathbf{Y} \neq \mathbf{0}$. The **projection** of **X** on **Y**, denoted $P_\mathbf{Y}(\mathbf{X})$, is given by

$$P_\mathbf{Y}(\mathbf{X}) = \left(\frac{\mathbf{X} \cdot \mathbf{Y}}{\mathbf{Y} \cdot \mathbf{Y}}\right)\mathbf{Y}$$

Note that $P_\mathbf{Y}(\mathbf{X})$ lies on the same line as **Y** and is a scalar multiple of **Y**, the scalar being $(\mathbf{X} \cdot \mathbf{Y})/(\mathbf{Y} \cdot \mathbf{Y})$.

Example 34-1

We will find the projection of $\mathbf{X} = (2, 5)$ on $\mathbf{Y} = (6, 3)$. We have

$$\frac{\mathbf{X} \cdot \mathbf{Y}}{\mathbf{Y} \cdot \mathbf{Y}} = \frac{(2, 5) \cdot (6, 3)}{(6, 3) \cdot (6, 3)} = \frac{27}{45} = \frac{3}{5}$$

Hence $P_\mathbf{Y}(\mathbf{X}) = \frac{3}{5}\mathbf{Y} = \frac{3}{5}(6, 3) = (\frac{18}{5}, \frac{9}{5})$. This is illustrated in Fig. 34-4. ●

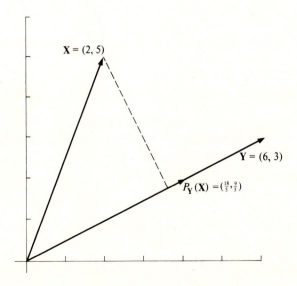

Fig. 34-4 The projection of $(2, 5)$ on $(6, 3)$.

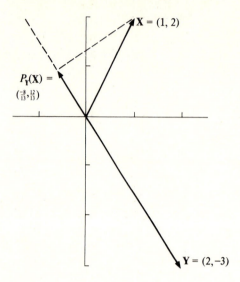

Fig. 34-5 The projection of $(1, 2)$ on $(2, -3)$.

Example 34-2

We will find the projection of $X = (1, 2)$ on $Y = (2, -3)$. We have

$$\frac{X \cdot Y}{Y \cdot Y} = \frac{(1, 2) \cdot (2, -3)}{(2, -3) \cdot (2, -3)} = \frac{-4}{13}$$

Hence $P_Y(X) = -\frac{4}{13}Y = -\frac{4}{13}(2, -3) = (-\frac{8}{13}, \frac{12}{13})$. This is illustrated in Fig. 34-5.   ●

Example 34-3

We will find the perpendicular projection of $X = (2, 1)$ on the line $y = 3x$. This is another application of the projection defined above. We first choose a *vector* that lies along the line $y = 3x$, and then project X onto that vector. This line consists of all vectors of the form $a(1, 3)$, and we select $Y = (1, 3)$. Thus the solution is

$$P_Y(X) = \left(\frac{X \cdot Y}{Y \cdot Y}\right)Y = \frac{5}{10}(1, 3) = \left(\frac{1}{2}, \frac{3}{2}\right)$$

(See Fig. 34-6.) In Exercise 34-19, the reader is asked to show that the result is independent of the choice of Y, as long as Y is a nonzero vector lying on the given line.   ●

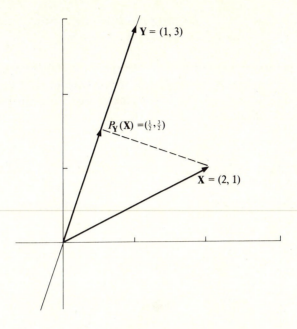

Fig. 34-6   The projection of $(2, 1)$ on the line $y = 3x$.

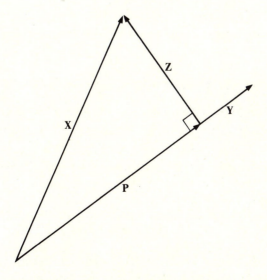

Fig. 34-7

The projection $P_Y(X)$ can also be used to find a vector Z perpendicular to Y. (In physical applications this is the component of X that is perpendicular to Y.)

Suppose that P is the projection of X on Y and Z is perpendicular to Y, as in Fig. 34-7. $P + Z = X$ according to the head-to-tail addition of vectors. Solving for Z yields $Z = X - P$, the desired vector perpendicular to Y.

Theorem 34-1   Let X and Y be nonzero vectors in any $\mathcal{R}^n$ and let P be the projection of X on Y. Then the vector $X - P$ is perpendicular to Y and lies in the plane determined by X and Y (that is, $X - P$ lies in the subspace spanned by X and Y).

Proof   This is left as an exercise.   ∎

Example 34-4

We wish to find a basis consisting of two perpendicular vectors for the plane $-x + 3y + 4z = 0$. Since the points on the plane satisfy $x = 3a + 4b$, $y = a$, $z = b$, they are all of the form $a(3, 1, 0) + b(4, 0, 1)$. Let $X_1 = (3, 1, 0)$ and $X_2 = (4, 0, 1)$. Then X_1 and X_2 are a basis for the plane, but are not perpendicular (Fig. 34-8). By Theorem 34-1, the vector

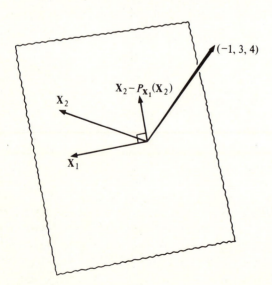

Fig. 34-8 X_1 and $X_2 - P_{X_1}(X_2)$ are perpendicular vectors that form a basis for the plane $-x + 3y + 4z = 0$.

$X_2 - P = X_2 - P_{X_1}(X_2)$ will be perpendicular to X_1 and will still lie in the given plane.

$$X_2 - P_{X_1}(X_2) = X_2 - \left(\frac{X_2 \cdot X_1}{X_1 \cdot X_1}\right)X_1$$

$$= (4, 0, 1) - \frac{(4, 0, 1) \cdot (3, 1, 0)}{(3, 1, 0) \cdot (3, 1, 0)}(3, 1, 0)$$

$$= (4, 0, 1) - \tfrac{12}{10}(3, 1, 0)$$

$$= (\tfrac{2}{5}, -\tfrac{6}{5}, 1)$$

Hence the two vectors $(3, 1, 0)$ and $(\tfrac{2}{5}, -\tfrac{6}{5}, 1)$ are perpendicular. Since they are perpendicular, they are independent, by Theorem 33-3. Since a plane is two-dimensional, these two vectors form the required basis. ●

Example 34-5

We will find a basis for $\mathcal{R}^3$ having two vectors on the plane $x + 2y - 3z = 0$ and having all three vectors in the basis perpendicular to each other. The points on the plane are all of the form $(-2a + 3b, a, b) = a(-2, 1, 0) + b(3, 0, 1)$. We let $X_1 = (-2, 1, 0)$ and $X_2 = (3, 0, 1)$. Then

$$X_2 - P_{X_1}(X_2) = (3, 0, 1) - \frac{(3, 0, 1) \cdot (-2, 1, 0)}{(-2, 1, 0) \cdot (-2, 1, 0)}(-2, 1, 0)$$

$$= (\tfrac{3}{5}, \tfrac{6}{5}, 1)$$

Therefore, $(-2, 1, 0)$ and $(\tfrac{3}{5}, \tfrac{6}{5}, 1)$ are two perpendicular vectors on the given plane. The third vector for the basis can be any vector perpendicular to the plane. By Theorem 33-1, $(1, 2, -3)$ will do. Thus our basis is $(-2, 1, 0)$, $(\tfrac{3}{5}, \tfrac{6}{5}, 1)$, $(1, 2, -3)$. ●

The chief application of Theorem 34-1 is in constructing entire new bases consisting of perpendicular vectors. Such a basis is called an *orthogonal basis*. We saw in Chap. 5 that for some geometric transformations, the simplest basis to use is orthogonal. There are other uses for orthogonal bases, a few of which will emerge in the theorems, examples, and applications of this chapter. We now consider an algorithm for finding orthogonal bases, called the Gram-Schmidt process, a generalization of Theorem 34-1.

We begin with any basis $X_1, X_2, \ldots, X_n$ for $\mathcal{R}^n$. We will use projections of these vectors to construct an orthogonal basis $Y_1, Y_2, \ldots, Y_n$. In what follows, remember that the **Y**'s are the new basis vectors, and will turn out to be perpendicular to each other. To get started we let

$$Y_1 = X_1$$

Then, to get a vector Y_2 perpendicular to Y_1, we take X_2 and subtract its projection on Y_1:

$$Y_2 = X_2 - P_{Y_1}(X_2)$$

(by Theorem 34-1, Y_2 is perpendicular to Y_1). Next we let

$$Y_3 = X_3 - P_{Y_1}(X_3) - P_{Y_2}(X_3)$$

that is, we take X_3 and subtract from it its projections on Y_1 and Y_2. The vector Y_3 is perpendicular to both Y_1 and Y_2 (Exercise 34-14). We continue this process, taking each X_i and subtracting its projections on all the Y_j's so far determined, until finally Y_n is defined. Then we will have n perpendicular vectors $Y_1, Y_2, \ldots, Y_n$. None of these can be **0** (Exercise 34-13), and since they are perpendicular they are independent (by Theorem 33-3). Thus the vectors are a basis (by Corollary 14-2). The following theorem summarizes the above process.

Theorem 34-2

The Gram-Schmidt process   Let $X_1, X_2, \ldots, X_n$ be a basis for $\mathscr{R}^n$. Then the n vectors $Y_1, Y_2, \ldots, Y_n$ defined by

$$Y_1 = X_1$$
$$Y_i = X_i - P_{Y_1}(X_i) - P_{Y_2}(X_i) - \cdots - P_{Y_{i-1}}(X_i)$$

where $i = 2, 3, \ldots, n$, is an orthogonal basis for $\mathscr{R}^n$.

Note once again that the ith vector, Y_i, is obtained by subtracting from X_i all of its projections on the previously obtained Y's.

Corollary 34-3

If we replace the vectors $Y_1, Y_2, \ldots, Y_n$ of Theorem 34-2 by $(1/|Y_1|)Y_1$, $(1/|Y_2|)Y_2, \ldots, (1/|Y_n|)Y_n$, we obtain an orthogonal basis consisting of *unit* vectors.

A basis consisting of perpendicular vectors each one unit long is called an *orthonormal* basis. Thus the vectors $(1/|Y_1|)Y_1$, $(1/|Y_2|)Y_2, \ldots$, $(1/|Y_n|)Y_n$ obtained in this corollary constitute an orthonormal basis. Orthonormal bases are a generalization of the *usual* basis, which also consists of perpendicular unit vectors.

Example 34-6

We will use $X_1 = (1, 2, 3)$, $X_2 = (1, 0, -1)$, $X_3 = (0, 1, 3)$ to construct an orthonormal basis for $\mathcal{R}^3$. We first let

$$Y_1 = X_1 = (1, 2, 3)$$

Next we have

$$
\begin{aligned}
Y_2 &= X_2 - P_{Y_1}(X_2) \\
&= (1, 0, -1) - (-\tfrac{2}{14})(1, 2, 3) \\
&= (\tfrac{8}{7}, \tfrac{2}{7}, -\tfrac{4}{7})
\end{aligned}
$$

Finally,

$$
\begin{aligned}
Y_3 &= X_3 - P_{Y_1}(X_3) - P_{Y_2}(X_3) \\
&= (0, 1, 3) - \frac{11}{14}(1, 2, 3) - \frac{-\tfrac{10}{7}}{\tfrac{84}{49}}\left(\tfrac{8}{7}, \tfrac{2}{7}, -\tfrac{4}{7}\right) \\
&= (0, 1, 3) - (\tfrac{11}{14}, \tfrac{11}{7}, \tfrac{33}{14}) - (-\tfrac{20}{21}, -\tfrac{5}{21}, \tfrac{10}{21}) \\
&= (\tfrac{1}{6}, -\tfrac{1}{3}, \tfrac{1}{6})
\end{aligned}
$$

Therefore, $Y_1 = (1, 2, 3)$, $Y_2 = (\tfrac{8}{7}, \tfrac{2}{7}, -\tfrac{4}{7})$, $Y_3 = (\tfrac{1}{6}, -\tfrac{1}{3}, \tfrac{1}{6})$ is an orthogonal basis. To make this basis orthonormal we divide each vector by its length, obtaining $(1/\sqrt{14})(1, 2, 3)$, $(1/\sqrt{\tfrac{12}{7}})(\tfrac{8}{7}, \tfrac{2}{7}, -\tfrac{4}{7})$, $(1/\sqrt{\tfrac{1}{6}})(\tfrac{1}{6}, -\tfrac{1}{3}, \tfrac{1}{6})$, a basis consisting of perpendicular vectors, each of length 1. ●

Projections also provide the starting point for one of the most important theorems in the theory of vector spaces, known as the Cauchy-Schwarz inequality.

Theorem 34-4   *Cauchy-Schwarz inequality*   In any $\mathcal{R}^n$, $|X \cdot Y| \le |X||Y|$ for all vectors X and Y.

Proof   If $Y = 0$, the inequality is true since we then have $|X \cdot 0| \le |X||0|$, or $0 \le 0$. Thus we will assume that $Y \ne 0$ for the remainder of the proof. Our starting point is the obvious fact that $0 \le |Z|^2$, where Z is the vector of Fig. 34-7. If we let P denote the projection of X on Y, then $Z = X - P$.

Using various definitions, and properties of the dot product, we have

$$0 \leq |\mathbf{Z}|^2 = \mathbf{Z} \cdot \mathbf{Z}$$

$$= (\mathbf{X} - \mathbf{P}) \cdot (\mathbf{X} - \mathbf{P})$$

$$= \left(\mathbf{X} - \left(\frac{\mathbf{X} \cdot \mathbf{Y}}{\mathbf{Y} \cdot \mathbf{Y}}\right)\mathbf{Y}\right) \cdot \left(\mathbf{X} - \left(\frac{\mathbf{X} \cdot \mathbf{Y}}{\mathbf{Y} \cdot \mathbf{Y}}\right)\mathbf{Y}\right)$$

$$= \mathbf{X} \cdot \mathbf{X} - 2\left(\frac{\mathbf{X} \cdot \mathbf{Y}}{\mathbf{Y} \cdot \mathbf{Y}}\right)\mathbf{Y} \cdot \mathbf{X} + \left(\frac{\mathbf{X} \cdot \mathbf{Y}}{\mathbf{Y} \cdot \mathbf{Y}}\right)^2 \mathbf{Y} \cdot \mathbf{Y}$$

$$= \mathbf{X} \cdot \mathbf{X} - 2\left(\frac{(\mathbf{X} \cdot \mathbf{Y})^2}{\mathbf{Y} \cdot \mathbf{Y}}\right) + \frac{(\mathbf{X} \cdot \mathbf{Y})^2}{\mathbf{Y} \cdot \mathbf{Y}}$$

$$= \mathbf{X} \cdot \mathbf{X} - \frac{(\mathbf{X} \cdot \mathbf{Y})^2}{\mathbf{Y} \cdot \mathbf{Y}}$$

Thus

$$0 \leq \mathbf{X} \cdot \mathbf{X} - \frac{(\mathbf{X} \cdot \mathbf{Y})^2}{\mathbf{Y} \cdot \mathbf{Y}}$$

and therefore

$$(\mathbf{X} \cdot \mathbf{Y})^2 \leq (\mathbf{X} \cdot \mathbf{X})(\mathbf{Y} \cdot \mathbf{Y})$$

That is (using the definition of length),

$$(\mathbf{X} \cdot \mathbf{Y})^2 \leq |\mathbf{X}|^2 |\mathbf{Y}|^2$$

Taking the positive square root of both sides, we obtain the desired inequality

$$|\mathbf{X} \cdot \mathbf{Y}| \leq |\mathbf{X}| |\mathbf{Y}| \qquad \blacksquare$$

Many important theorems in mathematics state that for two quantities A and B, we have $A \leq B$. The Cauchy-Schwarz inequality is one such theorem. It is often advantageous to know in what special cases we can replace the sign " $\leq$ " by " $=$ ". In the Cauchy-Schwarz inequality this amounts to determining which vectors $\mathbf{X}$ and $\mathbf{Y}$ satisfy $|\mathbf{X} \cdot \mathbf{Y}| = |\mathbf{X}| |\mathbf{Y}|$. In the first line of the proof, we began with

$$0 \leq |\mathbf{Z}|^2$$

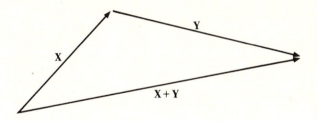

Fig. 34-9

Had we been able to begin with $0 = |\mathbf{Z}|^2$, we would have obtained

$$0 = \mathbf{X} \cdot \mathbf{X} - \frac{(\mathbf{X} \cdot \mathbf{Y})^2}{\mathbf{Y} \cdot \mathbf{Y}}$$

and hence we would have obtained the *equality*

$$|\mathbf{X} \cdot \mathbf{Y}| = |\mathbf{X}||\mathbf{Y}|$$

But in order to have $0 = |\mathbf{Z}|^2$, we must have $|\mathbf{Z}| = 0$; that is, we must have $\mathbf{Z} = \mathbf{0}$. By examining Fig. 34-7 one can see that $\mathbf{Z} = \mathbf{0}$ when $\mathbf{X}$ and $\mathbf{Y}$ lie on the same line, in which case $\mathbf{Y} = r\mathbf{X}$ for some scalar r. These remarks prove the following corollary.

Corollary 34-5 Let $\mathbf{X}$ and $\mathbf{Y}$ be vectors in $\mathcal{R}^n$. $|\mathbf{X} \cdot \mathbf{Y}| = |\mathbf{X}||\mathbf{Y}|$ if and only if $\mathbf{Y} = r\mathbf{X}$ for some scalar r.

The Cauchy-Schwarz inequality not only is important in its own right, but also can be used to prove the triangle inequality, a vital part of euclidean geometry. In geometric terms, the triangle inequality states that one side of a triangle cannot be longer than the sum of the lengths of the other two sides. In vector notation (see Fig. 34-9) the side $\mathbf{X} + \mathbf{Y}$ of the triangle is no longer than $|\mathbf{X}| + |\mathbf{Y}|$; that is, $|\mathbf{X} + \mathbf{Y}| \leq |\mathbf{X}| + |\mathbf{Y}|$. This result holds in any $\mathcal{R}^n$. Thus we have generalized a basic fact of euclidean *plane* geometry to n dimensional vector spaces.

Theorem 34-6 *Triangle inequality* If $\mathbf{X}$ and $\mathbf{Y}$ are vectors in $\mathcal{R}^n$, then $|\mathbf{X} + \mathbf{Y}| \leq |\mathbf{X}| + |\mathbf{Y}|$.

Proof
$$|X + Y|^2 = (X + Y) \cdot (X + Y)$$
$$= X \cdot X + 2(X \cdot Y) + Y \cdot Y$$
$$= |X|^2 + 2(X \cdot Y) + |Y|^2$$
$$\leq |X|^2 + 2|X \cdot Y| + |Y|^2$$
$$\leq |X|^2 + 2|X||Y| + |Y|^2 \quad \text{by Cauchy-Schwarz inequality}$$
$$= (|X| + |Y|)^2$$

Since we have $|X + Y|^2 \leq (|X| + |Y|)^2$ and since $|X + Y|$ and $|X| + |Y|$ are both nonnegative, taking the square root of both sides yields the desired result.   ∎

Again it is natural to ask when we can replace the sign $\leq$ by $=$. This is left as an exercise.

Exercises

34-1 Find $P_Y(X)$ and $P_X(Y)$ for the following vectors.

(a) $X = (3, 1)$, $Y = (3, 7)$
(b) $X = (-4, 2)$, $Y = (-2, -4)$
(c) $X = (0, 1, -5, 2)$, $Y = (1, 0, 2, -1)$

34-2   Find a vector perpendicular to Y by computing $Z = X - P_Y(X)$ (as in Theorem 34-1). Check by showing that $Z \cdot Y = 0$.

(a) $X = (1, 2)$, $Y = (-4, 1)$
(b) $X = (1, 2, 0)$, $Y = (2, 3, 1)$
(c) $X = (1, 2, 3)$, $Y = (-2, 1, 0)$
(d) $X = (0, 1, -2, -1)$, $Y = (1, 3, -2, 1)$

34-3   For each vector Y of Exercise 34-2 find a *unit* vector Z perpendicular to Y.

34-4   Find an orthonormal basis for $\mathcal{R}^2$ (that is, a basis consisting of perpendicular unit vectors) where one of the vectors is on the line $y = -3x$.

34-5(a)   Find an orthonormal basis for the plane $x - 5y - 2z = 0$.
(b)   Find a basis for $\mathcal{R}^3$ using the two vectors from part (a) and a third unit vector perpendicular to each of the first two vectors.

34-6   Find an orthonormal basis for $\mathcal{R}^3$ having two of its three vectors on the plane $2x - y - z = 0$.

34-7 Find an orthonormal basis for $\mathcal{R}^3$ having two of its three vectors on the plane $x - 2z = 0$.

34-8 Find an orthonormal basis for $\mathcal{R}^4$ having three of its four vectors lying in the subspace of solutions of $x_1 - 2x_2 - 3x_3 + x_4 = 0$.

34-9 Use the Gram-Schmidt process to convert the basis $(1, 1, 0)$, $(0, -2, 3)$, $(1, 2, 4)$ for $\mathcal{R}^3$ to an orthonormal basis.

34-10 Use the Gram-Schmidt process to convert the basis $(0, 1, 0, 0)$, $(2, 0, 1, 0)$, $(1, 1, 0, 0)$, $(3, 0, 0, -1)$ for $\mathcal{R}^4$ to an orthonormal basis.

34-11 Prove Theorem 34-1.

34-12 When we derived the formula

$$P_Y(X) = \frac{X \cdot Y}{Y \cdot Y} Y$$

for $\mathcal{R}^2$ (at the beginning of this section), we assumed that $0 \le \theta < \pi/2$. Prove that the same result is true for $\pi/2 \le \theta < \pi$.

34-13 In the Gram-Schmidt process (Theorem 34-2):

(a) Prove that the vector Y_2 cannot equal 0. (Hint: Express Y_2 as a linear combination of X_2 and X_1.)
(b) Prove that Y_3 cannot equal 0. (Hint: Express Y_3 as a linear combination of X_3, X_1, and X_2.)

34-14 Prove that the vector Y_3 defined in the Gram-Schmidt process is perpendicular to Y_1 and Y_2. (Hint: Take its dot product with Y_1 and Y_2 and use the definitions of the projections involved.)

34-15 Verify that the Cauchy-Schwarz inequality holds for each pair of vectors X, Y in Exercise 34-1.

34-16 Let $a_1, a_2, \ldots, a_n, b_1, b_2, \ldots, b_n$ be real numbers. Prove that $(a_1 b_1 + a_2 b_2 + \cdots + a_n b_n)^2 \le (a_1^2 + a_2^2 + \cdots + a_n^2)(b_1^2 + b_2^2 + \cdots + b_n^2)$.

34-17(a) Let a and b be positive real numbers. By applying the Cauchy-Schwarz inequality to the vectors $(\sqrt{a}, \sqrt{b})$, $(1/\sqrt{a}, 1/\sqrt{b})$, prove that $4 \le (a + b)(1/a + 1/b)$.
(b) Let $a_1, a_2, \ldots, a_n$ be n positive numbers. Prove that

$$n^2 \le (a_1 + a_2 + \cdots + a_n)(1/a_1 + 1/a_2 + \cdots + 1/a_n)$$

by using the Cauchy-Schwarz inequality. [Hint: Model a proof after (a).]

34-18   In the triangle inequality (Theorem 34-6), when can the sign $\leq$ be replaced by $=$? (Hint: In the proof of the triangle inequality, the $\leq$ sign was used twice. Determine what conditions on **X** and **Y** allow the equality sign to be used instead of the $\leq$ sign.)

34-19(a) Let $\mathbf{X} = (2, 1)$ and let $\mathcal{L} = \{a(1, 3) : a \in \mathcal{R}\}$ be a line through the origin. Show that for *any* choice of a nonzero vector **Y** on $\mathcal{L}$, we have $P_{\mathbf{Y}}(\mathbf{X}) = (\frac{1}{2}, \frac{3}{2})$. Thus the projection of **X** on $\mathcal{L}$ is independent of the vector chosen to represent $\mathcal{L}$.
(b)   Repeat part (a) for the general case where $\mathbf{X} = (x, y)$ and $\mathcal{L} = \{a(b, c) : a \in \mathcal{R}\}$.

*34-20   Write a computer program that finds the projection of a vector **X** on a vector **Y**.

*34-21   Write a computer program that uses the results of Theorem 34-1 to find a vector perpendicular to **Y**.

*34-22   Write a computer program that carries out the Gram-Schmidt process on any given basis to find an orthogonal basis for $\mathcal{R}^n$.

35.   ORTHONORMAL BASES

The usual basis for $\mathcal{R}^n$ is the easiest one to use in performing many computations. This is due, in part, to two features: The vectors are perpendicular and they all have unit length. These two features are the defining properties of an orthonormal basis, and consequently working with *any* orthonormal basis simplifies several types of computations, as we will see in this section. (This simplification accounts for our desire to obtain orthonormal bases and the great importance of the Gram-Schmidt process for obtaining them.)

Suppose, for instance, that $T : \mathcal{R}^n \to \mathcal{R}^n$ is a linear transformation whose usual matrix is A and whose new matrix is B. We know that A and B have the relationship $B = PAP^{-1}$, where the rows of P are the new basis vectors. Here we obtain our first benefit from using an orthonormal basis: If the new basis is orthonormal, we can compute P^{-1} effortlessly. For, as we will see, the columns of P^{-1} are the same as the rows of P. Thus, if the new basis were $(\sqrt{3}/2, 1/2), (-1/2, \sqrt{3}/2)$, which is orthonormal,

$$
P = \left[\begin{array}{c|c} \dfrac{\sqrt{3}}{2} & \dfrac{1}{2} \\ \hline -\dfrac{1}{2} & \dfrac{\sqrt{3}}{2} \end{array} \right]
$$

and we obtain instantly

$$P^{-1} = \begin{bmatrix} \dfrac{\sqrt{3}}{2} & -\dfrac{1}{2} \\ \dfrac{1}{2} & \dfrac{\sqrt{3}}{2} \end{bmatrix}$$

The reader should verify that the product of these two matrices is I.

This is the second time that we have had occasion to interchange the rows and columns of matrices, and it is not the last. Since this interchanging process occurs so often, it has a special name. The matrix obtained by interchanging the rows and columns of a matrix A is called the transpose of A.

Definition

The **transpose** of a matrix A is the matrix whose i, j entry is the j, i entry of A for all i and j. We denote the transpose of A by A^T, read "A transpose."

Thus A^T is the matrix whose columns are the rows of A, taken in the same order. For example, if

$$A = \begin{bmatrix} 1 & 2 & 3 & 4 \\ 5 & 6 & 7 & 8 \\ 9 & 10 & 11 & 12 \end{bmatrix}$$

then

$$A^T = \begin{bmatrix} 1 & 5 & 9 \\ 2 & 6 & 10 \\ 3 & 7 & 11 \\ 4 & 8 & 12 \end{bmatrix}$$

Using this terminology, we can rephrase the previous comment concerning the matrix P as follows.

Theorem 35-1

If P is the $n \times n$ matrix whose rows are the vectors in an orthonormal basis for $\mathcal{R}^n$, then $P^{-1} = P^T$.

Proof

Let $\mathbf{X}_1, \mathbf{X}_2, \ldots, \mathbf{X}_n$ be an orthonormal basis for $\mathcal{R}^n$. Let the rows of P be these basis vectors, so that

$$P = \begin{bmatrix} \mathbf{X}_1 \\ \hline \mathbf{X}_2 \\ \hline \vdots \\ \hline \mathbf{X}_n \end{bmatrix}$$

Then

$$P^T = \begin{bmatrix} \mathbf{X}_1 & \mathbf{X}_2 & \cdots & \mathbf{X}_n \end{bmatrix}$$

that is, the basis vectors form the columns of P^T. We will show that $P^T = P^{-1}$. When we form the product PP^T, the i, j entry is $\mathbf{X}_i \cdot \mathbf{X}_j$. Hence

$$PP^T = \begin{bmatrix} \mathbf{X}_1 \cdot \mathbf{X}_1 & \mathbf{X}_1 \cdot \mathbf{X}_2 & \cdots & \mathbf{X}_1 \cdot \mathbf{X}_n \\ \mathbf{X}_2 \cdot \mathbf{X}_1 & \mathbf{X}_2 \cdot \mathbf{X}_2 & \cdots & \mathbf{X}_2 \cdot \mathbf{X}_n \\ \cdots\cdots\cdots\cdots\cdots\cdots\cdots\cdots\cdots \\ \mathbf{X}_n \cdot \mathbf{X}_1 & \mathbf{X}_n \cdot \mathbf{X}_2 & \cdots & \mathbf{X}_n \cdot \mathbf{X}_n \end{bmatrix} = \begin{bmatrix} 1 & 0 & \cdots & 0 \\ 0 & 1 & \cdots & 0 \\ \cdots\cdots\cdots\cdots \\ 0 & 0 & \cdots & 1 \end{bmatrix}$$

The last equality is obtained by using the fact that for an orthonormal basis, $\mathbf{X}_i \cdot \mathbf{X}_i = 1$ and $\mathbf{X}_i \cdot \mathbf{X}_j = 0$ if $i \neq j$. Thus $PP^T = I$, and by Theorem 23-1 we also have $P^T P = I$. Hence $P^T = P^{-1}$.   ∎

Corollary 35-2   If $T : \mathcal{R}^n \to \mathcal{R}^n$ is a linear transformation whose usual matrix is A and whose new matrix relative to a new orthonormal basis is B, then $B = PAP^T$, and, equivalently, $A = P^T BP$.

Proof   We know that $B = PAP^{-1}$ (Theorem 28-1). Since $P^{-1} = P^T$ by the above theorem, $B = PAP^T$.   ∎

In trying to find simple matrices B for geometric transformations, we are often led to a new basis that is orthonormal, or at least orthogonal. For example, such bases arise for reflections in $\mathcal{R}^2$ and rotations in $\mathcal{R}^3$. The following examples show how Theorem 35-1 and its corollary simplify some of the calculations involved.

Example 35-1

In this example we obtain the usual matrix A for an arbitrary reflection in $\mathcal{R}^2$. Let $T : \mathcal{R}^2 \to \mathcal{R}^2$ be the reflection in a line $\mathcal{L}$ through the origin, where $\mathcal{L}$ meets the positive x axis in an angle of θ radians. Let $\mathbf{X}_1$ be a unit vector on $\mathcal{L}$ and let $\mathbf{X}_2$ be a unit vector perpendicular to $\mathcal{L}$ (Fig. 35-1). The matrix relative to this basis is easy to find: since T is a reflection in $\mathcal{L}$, the new matrix for T is

$$B = \begin{bmatrix} 1 & 0 \\ 0 & -1 \end{bmatrix}$$

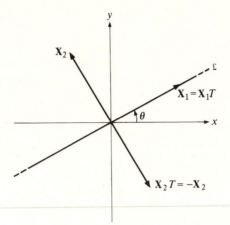

Fig. 35-1   Effect of the reflection in the line $\mathfrak{L}$ on a suitable basis.

To find the matrix P we need to know the coordinates of $\mathbf{X}_1$ and $\mathbf{X}_2$. The reader can easily verify that $\mathbf{X}_1 = (\cos\theta, \sin\theta)$ and $\mathbf{X}_2 = (-\sin\theta, \cos\theta)$, and therefore

$$P = \begin{bmatrix} \cos\theta & \sin\theta \\ -\sin\theta & \cos\theta \end{bmatrix}$$

Since P was formed from an orthonormal basis, its inverse is very easy to compute:

$$P = P^T = \begin{bmatrix} \cos\theta & -\sin\theta \\ \sin\theta & \cos\theta \end{bmatrix}$$

The usual matrix A for T is given by

$$A = P^{-1}BP$$
$$= P^T BP$$

$$= \begin{bmatrix} \cos^2\theta - \sin^2\theta & 2\sin\theta\cos\theta \\ 2\sin\theta\cos\theta & \sin^2\theta - \cos^2\theta \end{bmatrix}$$

$$= \begin{bmatrix} \cos 2\theta & \sin 2\theta \\ \sin 2\theta & -\cos 2\theta \end{bmatrix}$$

(The last equality follows from the two trigonometric identities $\sin(a+b) = \sin a \cos b + \cos a \sin b$ and $\cos(a+b) = \cos a \cos b - \sin a \sin b$.) Note that the matrix A allows us to compute $\mathbf{X}T$ for *any* reflection $T: \mathcal{R}^2 \to \mathcal{R}^2$, by using the appropriate value of θ.   ●

Example 35-2

Consider the problem of finding a simple matrix B for a rotation, in particular the $60°$ counterclockwise rotation about the axis in $\mathcal{R}^3$ $\{a(1, 2, 2) : a \in \mathcal{R}\}$. For this purpose, it is best that the new basis (a basis that the transformation has a simple effect on) be orthonormal. Clearly, one new basis vector should be on the axis of rotation, and the two others should lie on a plane perpendicular to this axis of rotation. The vector $(1, 2, 2)$ is on the axis. By Theorem 33-1, $(1, 2, 2)$ is perpendicular to the plane $x + 2y + 2z = 0$. We next find two basis vectors on that plane. Points on the plane are of the form $(-2a - 2b, a, b) = a(-2, 1, 0)$ $+ b(-2, 0, 1)$, and thus we have a basis $\mathbf{X}_1 = (-2, 1, 0)$, $\mathbf{X}_2 = (-2, 0, 1)$, $\mathbf{X}_3 = (1, 2, 2)$. This is almost what we want, except that it is not orthonormal. To make it orthonormal we first use the Gram-Schmidt process, obtaining an orthogonal basis $\mathbf{Y}_1 = (-2, 1, 0)$, $\mathbf{Y}_2 = (-\frac{2}{5}, -\frac{4}{5}, 1)$, $\mathbf{Y}_3 = (1, 2, 3)$. (It is no surprise that $\mathbf{Y}_3 = \mathbf{X}_3$ since $\mathbf{X}_3$ was perpendicular to $\mathbf{X}_1$ and $\mathbf{X}_2$ to begin with.) We then produce unit vectors by multiplying each $\mathbf{Y}_i$ by the reciprocal of its length. Our new orthonormal basis is thus $\mathbf{B}_1 = (1/\sqrt{5})(-2, 1, 0)$, $\mathbf{B}_2 = (5/3\sqrt{5})(-\frac{2}{5}, -\frac{4}{5}, 1)$, $\mathbf{B}_3 = (\frac{1}{3})(1, 2, 2)$. Figure 35-2 shows the orthonormal basis: $\mathbf{B}_1$ and $\mathbf{B}_2$ lie on the plane $x + 2y + 2z = 0$, and $\mathbf{B}_3$ is perpendicular to the plane.

Now that we have this carefully selected orthonormal basis, the effect

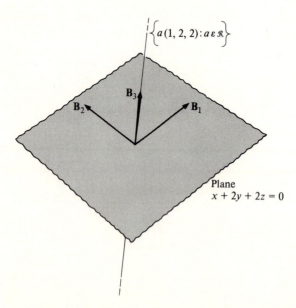

Fig. 35-2 An orthonormal basis for $\mathcal{R}^3$; $\mathbf{B}_1$ and $\mathbf{B}_2$ lie on the plane $x + 2y + 2z = 0$, and $\mathbf{B}_3$ is perpendicular to this plane.

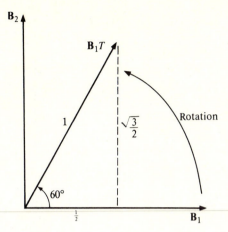

Fig. 35-3

of the rotation on the basis vectors is easy to determine. Of course, $\mathbf{B}_3$, which lies on the axis of rotation, is left unchanged. Thus

$$\mathbf{B}_3 T = \mathbf{B}_3$$

$\mathbf{B}_1$ and $\mathbf{B}_2$, which are perpendicular to each other, have unit length, and lie on the same plane, can be pictured much like the usual basis for $\mathcal{R}^2$. Instead of the x axis and the y axis we have a $\mathbf{B}_1$ axis and a $\mathbf{B}_2$ axis (Fig. 35-3). $\mathbf{B}_1 T$ is obtained by rotating $\mathbf{B}_1$ 60° in this plane. Thus the triangle in Fig. 35-3 is a 30°-60°-90° triangle, with sides $1/2$, $\sqrt{3}/2$, 1. It follows that

$$\mathbf{B}_1 T = \frac{1}{2}\mathbf{B}_1 + \frac{\sqrt{3}}{2}\mathbf{B}_2$$

A similar figure shows that

$$\mathbf{B}_2 T = \frac{-\sqrt{3}}{2}\mathbf{B}_1 + \frac{1}{2}\mathbf{B}_2$$

Hence the new matrix B relative to the basis $\mathbf{B}_1$, $\mathbf{B}_2$, $\mathbf{B}_3$ is

$$B = \begin{bmatrix} \dfrac{1}{2} & \dfrac{\sqrt{3}}{2} & 0 \\[2ex] -\dfrac{\sqrt{3}}{2} & \dfrac{1}{2} & 0 \\[2ex] 0 & 0 & 1 \end{bmatrix}$$

We can also find the usual matrix A for T. We know that $A = P^{-1}BP$, where

$$P = \begin{bmatrix} -\dfrac{2}{\sqrt{5}} & \dfrac{1}{\sqrt{5}} & 0 \\[2ex] -\dfrac{2}{3\sqrt{5}} & -\dfrac{4}{3\sqrt{5}} & \dfrac{5}{3\sqrt{5}} \\[2ex] \dfrac{1}{3} & \dfrac{2}{3} & \dfrac{2}{3} \end{bmatrix}$$

It would not be easy to compute P^{-1} directly, but, fortunately, we can use Corollary 35-2. The rows of P are formed from an orthonormal basis, hence $P^{-1} = P^T$, and P^{-1} is obtained at once. The computation of A is straightforward, if tedious, and we find that

$$A = \begin{bmatrix} \dfrac{5}{9} & \dfrac{3\sqrt{3}+1}{9} & \dfrac{-3\sqrt{3}+1}{9} \\[2ex] \dfrac{-3\sqrt{3}+1}{9} & \dfrac{13}{18} & \dfrac{3\sqrt{3}+4}{18} \\[2ex] \dfrac{3\sqrt{3}+1}{9} & \dfrac{-3\sqrt{3}+4}{18} & \dfrac{13}{18} \end{bmatrix}$$

●

In each of these two examples computations were simplified because we had matrices P whose inverses equaled their transposes. Matrices with this property arise often enough to warrant a special name.

Definition

A matrix M with the property that $M^{-1} = M^T$ is said to be **orthogonal.** Equivalently, M is orthogonal provided $MM^T = M^TM = I$.

In this terminology, Theorem 35-1 states that if the rows of a matrix form an orthonormal basis, the matrix is orthogonal. The converse is also true: if a matrix is orthogonal, then its rows form an orthonormal basis. The proof is left to the reader. Corollary 35-2 points out that if we switch to a new orthonormal basis, the simple matrix B satisfies $B = PAP^T$ since P is orthogonal.

It is not just in switching bases that orthogonal matrices arise—the matrix A or B itself may be orthogonal. (The reader may verify that the matrices A in both Examples 35-1 and 35-2 are orthogonal.) A transformation T that has an orthogonal matrix has various important properties, which we now explore.

Suppose that a linear transformation $T : \mathcal{R}^n \to \mathcal{R}^n$ has an orthogonal usual matrix A, where

$$
A = \begin{bmatrix}
\mathbf{X}_1 \\
\hdashline
\mathbf{X}_2 \\
\hdashline
\vdots \\
\hdashline
\mathbf{X}_n
\end{bmatrix}
$$

Then $(1, 0, \ldots, 0)T = \mathbf{X}_1$, $(0, 1, \ldots, 0)T = \mathbf{X}_2$, etc., since the rows of A give T's effect on the usual basis. Hence T takes the usual basis vectors (which are perpendicular and have length 1) to the basis vectors $\mathbf{X}_1, \mathbf{X}_2, \ldots, \mathbf{X}_n$ (which are also perpendicular and have length 1). This suggests that T does not alter the length of any vector and does not change the angle between any two vectors. This, and more, is the case, as we will now see.

Theorem 35-3

Let $T : \mathcal{R}^n \to \mathcal{R}^n$ be a linear transformation whose usual matrix A is orthogonal. Then

1   $|\mathbf{X}T| = |\mathbf{X}|$ for all $\mathbf{X} \in \mathcal{R}^n$. ($T$ leaves lengths unchanged.)

2   The angle between $\mathbf{X}$ and $\mathbf{Y}$ is equal to the angle between $\mathbf{X}T$ and $\mathbf{Y}T$ for all $\mathbf{X}, \mathbf{Y} \in \mathcal{R}^n$. ($T$ leaves angles unchanged.)

3   $\det A = \pm 1$. (Volumes or areas are left unchanged by T.)

Proof

(1)   Let $\mathbf{X} = (c_1, c_2, \ldots, c_n)$ be an arbitrary vector in $\mathcal{R}^n$. Denote the rows of A by $\mathbf{X}_1, \mathbf{X}_2, \ldots, \mathbf{X}_n$. The rows of A show what T does to the usual basis vectors, and thus $(1, 0, \ldots, 0)T = \mathbf{X}_1$, $(0, 1, \ldots, 0)T = \mathbf{X}_2$, etc. We therefore have

$$
\begin{aligned}
\mathbf{X}T &= (c_1, c_2, \ldots, c_n)T \\
&= c_1[(1, 0, \ldots, 0)T] + \cdots + c_n[(0, 0, \ldots, 1)T] \\
&= c_1\mathbf{X}_1 + c_2\mathbf{X}_2 + \cdots + c_n\mathbf{X}_n
\end{aligned}
$$

We now compare the length of $\mathbf{X}$ with the length of $\mathbf{X}T$. Clearly

$$
|\mathbf{X}| = \sqrt{c_1^2 + c_2^2 + \cdots + c_n^2}
$$

On the other hand,

$$
|\mathbf{X}T| = \sqrt{(c_1\mathbf{X}_1 + \cdots + c_n\mathbf{X}_n) \cdot (c_1\mathbf{X}_1 + \cdots + c_n\mathbf{X}_n)}
$$

When we carry out the dot product, each resulting term will be of the form $c_i c_j \mathbf{X}_i \cdot \mathbf{X}_j$, where $i = 1, 2, \ldots, n$ and $j = 1, 2, \ldots, n$. Since $\mathbf{X}_1, \mathbf{X}_2, \ldots, \mathbf{X}_n$ is orthonormal, $\mathbf{X}_i \cdot \mathbf{X}_j = 0$ except when $i = j$. Therefore,

$$
\begin{aligned}
|\mathbf{X}T| &= \sqrt{c_1^2(\mathbf{X}_1 \cdot \mathbf{X}_1) + c_2^2(\mathbf{X}_2 \cdot \mathbf{X}_2) + \cdots + c_n^2(\mathbf{X}_n \cdot \mathbf{X}_n)} \\
&= \sqrt{c_1^2(1) + c_2^2(1) + \cdots + c_n^2(1)} \\
&= \sqrt{c_1^2 + c_2^2 + \cdots + c_n^2}
\end{aligned}
$$

Hence $|\mathbf{X}| = |\mathbf{X}T|$.

(2) This is left as an exercise.

(3) We first note that $\det(A^T) = \det A$ for *any* $n \times n$ matrix A. This is evident in the 2×2 case since

$$
\det \begin{bmatrix} a & b \\ c & d \end{bmatrix} = ad - bc
$$

and

$$
\det \begin{bmatrix} a & c \\ b & d \end{bmatrix} = ad - cb = ad - bc
$$

For the 3×3 case, if we evaluate $\det A$ using the first row of A and if we evaluate $\det(A^T)$ using the first column of A^T, equal values will be obtained. The general $n \times n$ case is similar. For an orthogonal matrix we have $AA^T = I$, and therefore,

$$
\begin{aligned}
\det(AA^T) &= \det I \\
\det A \det(A^T) &= 1 \\
\det A \det A &= 1 \\
(\det A)^2 &= 1 \\
\det A &= \pm 1 \qquad \blacksquare
\end{aligned}
$$

In the vector spaces $\mathcal{R}^2$ and $\mathcal{R}^3$, the only ones that we can picture successfully, the properties listed in Theorem 35-3 suggest that T has the properties of a rotation or a reflection. The fact that the reflections and rotation of Examples 35-1 and 35-2 have orthogonal matrices supports this conjecture. In fact, it is true that in $\mathcal{R}^2$ and $\mathcal{R}^3$ any linear transformation with an orthogonal matrix is either a rotation, a reflection, or a product of these. The following theorem describes the 2×2 case.

Theorem 35-4   If the usual matrix for $T: \mathcal{R}^2 \to \mathcal{R}^2$ is orthogonal, then T is either a rotation or a reflection.

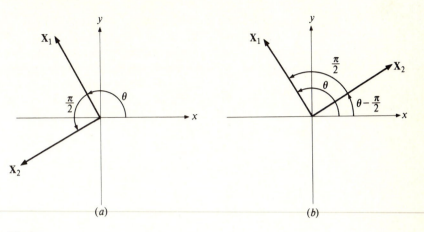

Fig. 35-4

Proof

Let A be the 2×2 usual matrix for T, and assume that A is orthogonal. We know that its rows form an orthonormal basis $\mathbf{X}_1, \mathbf{X}_2$ for $\mathcal{R}^2$. Suppose that $\mathbf{X}_1$ makes an angle of θ radians with the positive x axis. Then $\mathbf{X}_1 = (\cos \theta, \sin \theta)$. Since $\mathbf{X}_2$ is perpendicular to $\mathbf{X}_1$, $\mathbf{X}_2$ makes an angle of $\theta + \pi/2$ or $\theta - \pi/2$ radians with the positive x axis (Fig. 35-4). Each possibility is treated separately.

If $\mathbf{X}_2$ makes an angle of $\theta + \pi/2$ radians with the positive x axis (as in Fig. 35-4a), then $\mathbf{X}_2 = (\cos (\theta + \pi/2), \sin (\theta + \pi/2)) = (-\sin \theta, \cos \theta)$, using trigonometric identities. In this case

$$A = \begin{bmatrix} \cos \theta & \sin \theta \\ -\sin \theta & \cos \theta \end{bmatrix}$$

which, by Exercise 21-8, is the matrix for a rotation (counterclockwise, by θ radians).

If $\mathbf{X}_2$ makes an angle of $\theta - \pi/2$ radians with the positive x axis (Fig. 35-4b), then this vector $\mathbf{X}_2$ is the negative of the vector $\mathbf{X}_2$ in the last paragraph, and thus

$$A = \begin{bmatrix} \cos \theta & \sin \theta \\ \sin \theta & -\cos \theta \end{bmatrix}$$

which, by Example 35-1, is the matrix for a reflection (in the line through the origin making an angle of $\theta/2$ radians with the positive x axis).

Thus T is either a rotation or a reflection. ■

We have one more topic to consider concerning orthonormal bases.

In view of the way that orthonormal bases simplify certain computations, it would be especially desirable to be able, given a transformation with usual matrix A, to find an orthonormal basis of *eigenvectors*. In such a case we would have both the simplest possible matrix B (diagonal) and the comparatively simple relationship $B = PAP^T$. In order to obtain an orthonormal basis of eigenvectors, we need a special type of matrix A.

Definition

A matrix A is **symmetric** if $A = A^T$.

Thus A is symmetric when its rows are identical with its columns, taken in the same order. For example,

$$\begin{bmatrix} 1 & 3 & 6 \\ 3 & 2 & -1 \\ 6 & -1 & 4 \end{bmatrix}$$

is symmetric since row i and column i are identical, for $i = 1, 2, 3$. Equivalently, a matrix A is symmetric if its i, j entry is equal to its j, i entry for all i and j.

Symmetric matrices arise frequently in geometry, physics, and applied mathematics. As an illustration, see Example 25-2 where the i, j entry of the matrix is the number of edges joining vertex i to vertex j in a graph. Since every edge from vertex i to vertex j also goes from vertex j to vertex i, the i, j entry is equal to the j, i entry. In physical problems involving n objects, a matrix whose i, j entry is the magnitude of the force or displacement between objects i and j would be symmetric since the magnitude is the same between objects j and i as between objects i and j.

As the following theorem shows, symmetric matrices are similar to diagonal matrices. That is, if $T : \mathcal{R}^n \to \mathcal{R}^n$ has a symmetric matrix, then T has a diagonal matrix. Furthermore, the new basis of eigenvectors can be chosen so that they form an orthonormal basis.

Theorem 35-5

Let $T : \mathcal{R}^n \to \mathcal{R}^n$ be a linear transformation whose usual matrix A is symmetric. Then

1 T has a diagonal matrix $B = PAP^{-1}$, where the rows of P form a basis of eigenvectors for T.

2 The basis of eigenvectors in (1) can be chosen to be orthonormal; when this is done we have $P^{-1} = P^T$ and hence $B = PAP^T$.

Proof

We will sketch the proof for the 2×2 case, leaving some details as exercises. Let

$$A = \begin{bmatrix} r & s \\ s & t \end{bmatrix}$$

(the 1, 2 entry is equal to the 2, 1 entry since A is symmetric). The characteristic equation is then $\lambda^2 - (r + t)\lambda + (rt - s^2) = 0$, which has real roots λ_1 and λ_2 (proof in Exercise 35-5).

Let $\mathbf{X}_1$ be an eigenvector one unit long corresponding to λ_1. By the corollary to the Gram-Schmidt process, Corollary 34-3, we can find a vector $\mathbf{X}_2$ such that $\mathbf{X}_1$, $\mathbf{X}_2$ is an orthonormal basis for $\mathcal{R}^2$.

The new matrix B with respect to the new basis $\mathbf{X}_1$, $\mathbf{X}_2$ has first row $[\lambda_1 \quad 0]$ since $\mathbf{X}_1$ is an eigenvector. Say

$$B = \begin{bmatrix} \lambda_1 & 0 \\ p & q \end{bmatrix}$$

The matrix B must be symmetric. [To see this, first note that $B = PAP^T$, where P is the matrix with $\mathbf{X}_1$ and $\mathbf{X}_2$ as rows. Also, $B^T = (PAP^T)^T = PA^TP^T$, by Exercise 35-2. Since $A = A^T$, we have $B^T = PAP^T = B$. Thus B is symmetric.] Since B is symmetric, its 2, 1 entry is equal to its 1, 2 entry; that is, $p = 0$. Thus

$$B = \begin{bmatrix} \lambda_1 & 0 \\ 0 & q \end{bmatrix}$$

which is a diagonal matrix. (The entry q must equal λ_2 since according to the second row $\mathbf{X}_2 B = q\mathbf{X}_2$ and hence q is an eigenvalue.) ■

Exercises

35-1 For each of the following symmetric matrices A, find an orthogonal matrix P such that PAP^T is diagonal. (Hint: By Theorem 35-5 an orthonormal basis of eigenvectors exists. Find one.)

(a) $\begin{bmatrix} 2 & 1 \\ 1 & 2 \end{bmatrix}$ (b) $\begin{bmatrix} 5 & 2 \\ 2 & 2 \end{bmatrix}$

(c) $\begin{bmatrix} 0 & 1 & 1 \\ 1 & 0 & 0 \\ 1 & 0 & 0 \end{bmatrix}$ (d) $\begin{bmatrix} 1 & -1 & 0 \\ -1 & -1 & 1 \\ 0 & 1 & 1 \end{bmatrix}$

35-2 Suppose that A and B are arbitrary 2×2 matrices. Prove the following:

(a) $(A^T)^T = A$.
(b) $(AB)^T = B^T A^T$.

(c) $(A^{-1})^T = (A^T)^{-1}$ if A is invertible. [Hint: Show that $(A^{-1})^T$ acts like the inverse of A^T; that is, show that $A^T(A^{-1})^T = I$.]

(d) $(PAP^T)^T = PA^TP^T$ for any 2×2 matrix P. [Hint: Use (a) and (b).]

(Note: These four statements hold for $n \times n$ matrices also.)

35-3   Suppose that A and B are symmetric matrices. Which of the following are necessarily symmetric? Give reasons.

(a) $A + B$ 　　　　　　　　(b) A^2

(c) AB 　　　　　　　　　 (d) AA^T

(e) kA where k is a scalar

35-4   Repeat Exercise 35-3, replacing the word *symmetric* by the word *orthogonal*.

35-5   Suppose that $T : \mathcal{R}^2 \to \mathcal{R}^2$ has as its usual matrix the symmetric matrix

$$\begin{bmatrix} r & s \\ s & t \end{bmatrix}$$

Prove that the characteristic equation has real roots. (Hint: Use the quadratic formula.)

35-6   Let $\mathcal{S}$ be the set of all 2×2 symmetric matrices.

(a)   Prove that $\mathcal{S}$ is a subspace of $\mathcal{M}^{2 \times 2}$. (Hint: Exercise 35-5 shows what an arbitrary vector in $\mathcal{S}$ must look like.)

(b)   Find the dimension of $\mathcal{S}$.

35-7   Each of the following orthogonal matrices represents a linear transformation $T : \mathcal{R}^2 \to \mathcal{R}^2$, and thus T is either a rotation or a reflection. In each case where T is a reflection determine the line in which T reflects, and in each case where T is a rotation determine the angle of rotation. (Hint: Use Theorem 35-4.)

(a) $\begin{bmatrix} \dfrac{\sqrt{3}}{2} & \dfrac{1}{2} \\ -\dfrac{1}{2} & \dfrac{\sqrt{3}}{2} \end{bmatrix}$ (b) $\begin{bmatrix} \dfrac{\sqrt{3}}{2} & \dfrac{1}{2} \\ \dfrac{1}{2} & -\dfrac{\sqrt{3}}{2} \end{bmatrix}$

(c) $\begin{bmatrix} 0 & 1 \\ -1 & 0 \end{bmatrix}$ (d) $\begin{bmatrix} 0 & -1 \\ -1 & 0 \end{bmatrix}$

35-8(a) Verify that

$$\begin{bmatrix} \cos\theta & \sin\theta \\ -\sin\theta & \cos\theta \end{bmatrix} = \begin{bmatrix} \cos(-\theta) & \sin(-\theta) \\ \sin(-\theta) & -\cos(-\theta) \end{bmatrix}\begin{bmatrix} 1 & 0 \\ 0 & -1 \end{bmatrix}$$

(b) Determine whether the three matrices in (a) represent rotations or reflections.
(c) What does (a) say about the relationship between rotations and reflections in $\mathcal{R}^2$?

35-9(a) Let $X_1, X_2, \ldots, X_n$ be an orthonormal basis for $\mathcal{R}^n$. Let $X \in \mathcal{R}^n$. Prove that the coordinates of X relative to this basis are $X \cdot X_1, X \cdot X_2, \ldots, X \cdot X_n$. (Hint: We can write $X = a_1X_1 + a_2X_2 + \cdots + a_nX_n$. Then show that $a_1 = X \cdot X_1$, etc.)
(b) Find the coordinates of the vector $(6, 12, 3)$ relative to the basis $(\sqrt{2}/2, \sqrt{2}/2, 0), (-\sqrt{6}/6, \sqrt{6}/6, \sqrt{6}/3), (\sqrt{3}/3, -\sqrt{3}/3, \sqrt{3}/3)$ for $\mathcal{R}^3$. [Hint: The basis is orthonormal. Use the result of part (a).]

35-10 Let $T : \mathcal{R}^n \to \mathcal{R}^n$ be a linear transformation whose usual matrix A is orthogonal. Prove $XT \cdot YT = X \cdot Y$. [Hint: Let $X = (c_1, c_2, \ldots, c_n)$ and $Y = (d_1, d_2, \ldots, d_n)$. Use the ideas in the proof of part (1) of Theorem 35-3 to write XT and YT in terms of the orthonormal basis $X_1, X_2, \ldots, X_n$ that form the rows of A. Then form $XT \cdot YT$.]

35-11 Prove part (2) of Theorem 35-3. (Hint: Suppose that θ is the angle between X and Y. We know that $\cos\theta = X \cdot Y/|X||Y|$. Use Exercise 35-10 to show that this formula also gives the cosine of the angle between XT and YT.)

*35-12 Write a computer program that determines whether matrices are orthogonal.

36. INNER PRODUCT SPACES

In previous sections of this chapter we have seen how the dot product in $\mathcal{R}^n$ leads to the concepts of length and angle. Other vector spaces can also be provided with these concepts; what is needed in such cases is an *inner product*, a generalization of the dot product.

Recall that the dot product in $\mathcal{R}^n$ has the four properties of Theorem 32-1: $X \cdot Y = Y \cdot X$, $(X + Y) \cdot Z = X \cdot Z + Y \cdot Z$, $(rX) \cdot Y = r(X \cdot Y)$, and $X \cdot X > 0$ if and only if $X \neq 0$. There are other "products" in other vector spaces that satisfy these four properties; any such "product" is called an inner product.

Definition

Let $\mathcal{V}$ be a vector space. An **inner product** on $\mathcal{V}$ is a rule that assigns to each ordered pair of vectors $\mathbf{X}$, $\mathbf{Y}$ in $\mathcal{V}$ a real number $\mathbf{X} \cdot \mathbf{Y}$ satisfying the following four properties:

(a) $\mathbf{X} \cdot \mathbf{Y} = \mathbf{Y} \cdot \mathbf{X}$

(b) $(\mathbf{X} + \mathbf{Y}) \cdot \mathbf{Z} = \mathbf{X} \cdot \mathbf{Z} + \mathbf{Y} \cdot \mathbf{Z}$

(c) $r(\mathbf{X} \cdot \mathbf{Y}) = (r\mathbf{X}) \cdot \mathbf{Y}$

(d) $\mathbf{X} \cdot \mathbf{X} > 0$ if and only if $\mathbf{X} \neq \mathbf{0}$

When $\mathcal{V}$ has an inner product defined on it, $\mathcal{V}$ is called an **inner product space.**

Clearly, the dot product on $\mathcal{R}^n$ is an example of an inner product. Hence $\mathcal{R}^n$ is an inner product space. We now give other examples of inner product spaces.

Example 36-1

The vector space $\mathcal{C}[0, 1]$, consisting of all real-valued functions that are continuous on the interval $[0, 1]$, is an inner product space using the rule

$$f \cdot g = \int_0^1 f(x)g(x)\, dx$$

It is left to the reader to verify that this rule satisfies the four properties of an inner product. The proof amounts to a review of various properties of the integral. For example, to show property (c) of the definition we would have to verify that

$$r \int_0^1 f(x)g(x)\, dx = \int_0^1 (rf(x))g(x)\, dx$$

But the latter equation is well known from calculus: the integral of a constant times a function is equal to the constant times the integral of the function. ●

Example 36-2

The vector space $\mathcal{R}^2$ is an inner product space when we define

$$(x_1, y_1) \cdot (x_2, y_2) = 3x_1 x_2 + 2y_1 y_2$$

We will show that this rule satisfies property (b) of the definition:

REASONS

$$(\mathbf{X} + \mathbf{Y}) \cdot \mathbf{Z} \overset{?}{=} \mathbf{X} \cdot \mathbf{Z} + \mathbf{Y} \cdot \mathbf{Z}$$

Statement of property to be proved

$[(x_1, y_1) + (x_2, y_2)] \cdot (x_3, y_3)$	$(x_1, y_1) \cdot (x_3, y_3) + (x_2, y_2) \cdot (x_3, y_3)$	Rewriting, using appropriate notation for $\mathcal{R}^2$
$(x_1 + x_2, y_1 + y_2) \cdot (x_3, y_3)$		Definition of addition in $\mathcal{R}^2$
$3(x_1 + x_2)x_3 + 2(y_1 + y_2)y_3$	$(3x_1 x_3 + 2y_1 y_3) + (3x_2 x_3 + 2y_2 y_3)$	Using the rule for this inner product
$3x_1 x_3 + 3x_2 x_3 + 2y_1 y_3 + 2y_2 y_3$	$3x_1 x_3 + 3x_2 x_3 + 2y_1 y_3 + 2y_2 y_3$	Using various laws for real numbers

The proofs of the remaining three laws are left as exercises.

Note that this inner product for $\mathcal{R}^2$ is different from the dot product for $\mathcal{R}^2$. A vector space can have more than one inner product. ●

Example 36-3

Consider the vector space $\mathcal{P}^2$. Let $p(x) = a_2 x^2 + a_1 x + a_0$ and $q(x) = b_2 x^2 + b_1 x + b_0$ be arbitrary vectors in $\mathcal{P}^2$. Both the rule

$$p \cdot q = a_2 b_2 + a_1 b_1 + a_0 b_0$$

and the rule

$$p \cdot q = \int_0^1 p(x)q(x)\, dx$$

are inner products on $\mathcal{P}^2$. The proofs are left as exercises. ●

Since inner products have the same four properties that the dot product has, the concepts of length and angle that were defined using the dot product and the theorems concerning the dot product all carry over to *any* inner product space. This greatly extends the applicability of the key concepts of euclidean geometry.

Listed below are the key definitions and theorems, proved in previous sections for $\mathcal{R}^n$, rephrased in terms of inner product spaces. The rephrasing also serves as a review of the previous sections.

Definition

Let $\mathcal{V}$ be an inner product space and let **X** and **Y** be arbitrary vectors in $\mathcal{V}$.

(a)   The *length* $|\mathbf{X}|$ of **X** is defined by $|\mathbf{X}| = \sqrt{\mathbf{X} \cdot \mathbf{X}}$.

(b)   The *angle* between **X** and **Y** (neither of which is **0**) is the unique number θ between zero and π satisfying

$$\cos \theta = \frac{\mathbf{X} \cdot \mathbf{Y}}{|\mathbf{X}||\mathbf{Y}|}$$

(c)   **X** and **Y** are *perpendicular* if $\mathbf{X} \cdot \mathbf{Y} = 0$.

(d)   The *distance* between **X** and **Y** is $|\mathbf{X} - \mathbf{Y}|$.

(e)   The *projection* of **X** on **Y** (where $\mathbf{Y} \neq \mathbf{0}$) is

$$P_\mathbf{Y}(\mathbf{X}) = \left(\frac{\mathbf{X} \cdot \mathbf{Y}}{\mathbf{Y} \cdot \mathbf{Y}} \right) \mathbf{Y}$$

Example 36-4

Let $f(x) = x^2$ and $g(x) = x - \frac{3}{4}$ be vectors in the inner product space $\mathcal{C}[0, 1]$ of Example 36-1.

The length of f is

$$|f| = \left(\int_0^1 f(x)f(x)\, dx \right)^{1/2} = \left(\int_0^1 x^4\, dx \right)^{1/2} = \sqrt{\frac{1}{5}}$$

The length of g is

$$|g| = \left(\int_0^1 (x - \tfrac{3}{4})^2\, dx \right)^{1/2} = \sqrt{\frac{7}{48}}$$

(The reader should be warned that the term *length* used here does not mean arc length of the function. Here, *length* gives a measure of how far a function's graph deviates from that of the zero function. As usual, the length of a vector tells how far it is from the zero vector.)

The vectors $f(x)$ and $g(x)$ are perpendicular since

$$f \cdot g = \int_0^1 f(x)g(x)\, dx = \int_0^1 x^2(x - \tfrac{3}{4})\, dx = 0$$

In this case, perpendicularity can also be given a geometric interpretation: since the integral of the product of f and g equals zero, the area above the x axis bounded by the graph of $f(x)g(x)$ is the same as the area below the x axis bounded by the graph of $f(x)g(x)$.   ●

We now turn to the key theorems that hold in any inner product space.

Theorem 36-1

Let $\mathbf{X}$ and $\mathbf{Y}$ be nonzero vectors in an inner product space $\mathcal{V}$ and let $\mathbf{P}$ be the projection of $\mathbf{X}$ on $\mathbf{Y}$. Then the vector $\mathbf{X} - \mathbf{P}$ is perpendicular to $\mathbf{Y}$ and lies in the subspace of $\mathcal{V}$ spanned by $\mathbf{X}$ and $\mathbf{Y}$.

Theorem 36-2

The Gram-Schmidt process   Let $\mathbf{X}_1, \mathbf{X}_2, \ldots, \mathbf{X}_n$ be a basis for an n-dimensional inner product space $\mathcal{V}$. Then the n vectors $\mathbf{Y}_1, \mathbf{Y}_2, \ldots, \mathbf{Y}_n$ defined by

$$\mathbf{Y}_1 = \mathbf{X}_1$$
$$\mathbf{Y}_i = \mathbf{X}_i - P_{\mathbf{Y}_1}(\mathbf{X}_i) - P_{\mathbf{Y}_2}(\mathbf{X}_i) - \cdots - P_{\mathbf{Y}_{i-1}}(\mathbf{X}_i)$$

where $i = 2, 3, \ldots, n$, form an orthogonal basis for $\mathcal{V}$.

Corollary 36-3

Every n-dimensional inner product space has an *orthonormal* basis.

Theorem 36-4

Cauchy-Schwarz inequality   In any inner product space, $|\mathbf{X} \cdot \mathbf{Y}| \leq |\mathbf{X}||\mathbf{Y}|$ for all vectors $\mathbf{X}$ and $\mathbf{Y}$.

Theorem 36-5

Triangle inequality   If $\mathbf{X}$ and $\mathbf{Y}$ are vectors in an inner product space, then $|\mathbf{X} + \mathbf{Y}| \leq |\mathbf{X}| + |\mathbf{Y}|$.

The proofs of all five results are *identical* to the proofs given in the case of the dot product in $\mathcal{R}^n$.

We will now illustrate some of these theorems for the inner product space $\mathcal{P}^2$, with the inner product

$$p \cdot q = \int_0^1 p(x)q(x) \, dx$$

Example 36-5

We wish to find an orthonormal basis for $\mathscr{P}^2$. We begin with the basis $\mathbf{X}_1 = 1$, $\mathbf{X}_2 = x$, $\mathbf{X}_3 = x^2$. Using the Gram-Schmidt process to get an orthogonal basis, we have

$$\mathbf{Y}_1 = 1$$

$$\mathbf{Y}_2 = \mathbf{X}_2 - P_{\mathbf{Y}_1}(\mathbf{X}_2)$$

$$= x - \left(\frac{\mathbf{X}_2 \cdot \mathbf{Y}_1}{\mathbf{Y}_1 \cdot \mathbf{Y}_1}\right)\mathbf{Y}_1$$

$$= x - \left(\frac{\int_0^1 x \, dx}{\int_0^1 1 \, dx}\right)1$$

$$= x - \tfrac{1}{2}$$

$$\mathbf{Y}_3 = \mathbf{X}_3 - P_{\mathbf{Y}_1}(\mathbf{X}_3) - P_{\mathbf{Y}_2}(\mathbf{X}_3)$$

$$= x^2 - \left(\frac{\mathbf{X}_3 \cdot \mathbf{Y}_1}{\mathbf{Y}_1 \cdot \mathbf{Y}_1}\right)\mathbf{Y}_1 - \left(\frac{\mathbf{X}_3 \cdot \mathbf{Y}_2}{\mathbf{Y}_2 \cdot \mathbf{Y}_2}\right)\mathbf{Y}_2$$

$$= x^2 - \left(\frac{\int_0^1 x^2 \, dx}{\int_0^1 1 \, dx}\right)1 - \left(\frac{\int_0^1 x^2(x - \tfrac{1}{2}) \, dx}{\int_0^1 (x - \tfrac{1}{2})^2 \, dx}\right)\left(x - \frac{1}{2}\right)$$

$$= x^2 - \tfrac{1}{3} - 1(x - \tfrac{1}{2})$$

$$= x^2 - x + \tfrac{1}{6}$$

Thus the vectors 1, $x - \tfrac{1}{2}$, $x^2 - x + \tfrac{1}{6}$ form an orthogonal basis.

To obtain an ortho*normal* basis we divide each vector by its length. Since $|\mathbf{Y}_1| = 1$, $|\mathbf{Y}_2| = \sqrt{\tfrac{1}{12}}$, $|\mathbf{Y}_3| = \sqrt{\tfrac{1}{180}}$, we have the orthonormal basis 1, $\sqrt{12}(x - \tfrac{1}{2})$, $\sqrt{180}(x^2 - x + \tfrac{1}{6})$. ●

Example 36-6

If $p(x)$ and $q(x)$ are vectors in $\mathscr{P}^2$, then by the Cauchy-Schwarz inequality we have

$$|p(x) \cdot q(x)| \leq |p(x)|\,|q(x)|$$

Thus

$$\left|\int_0^1 p(x)q(x) \, dx\right| \leq \left(\int_0^1 p(x)^2 \, dx\right)^{1/2}\left(\int_0^1 q(x)^2 \, dx\right)^{1/2}$$

Squaring both sides, we obtain the inequality

$$\left(\int_0^1 p(x)q(x)\, dx\right)^2 \le \left(\int_0^1 p(x)^2\, dx\right)\left(\int_0^1 q(x)^2\, dx\right)$$

The reader should note how easily this inequality was obtained using the inner product. No other derivation of this inequality could be so simple. ●

Exercises

36-1   Prove that the following is an inner product for $\mathcal{R}^2$:

$$(x_1,\, y_1)\cdot(x_2,\, y_2) = x_1 x_2 + 2y_1 y_2$$

36-2   Complete the verification that the rule of Example 36-2 is an inner product.

36-3   Which of the following are inner products for $\mathcal{R}^3$?

(a) $(x_1,\, y_1,\, z_1)\cdot(x_2,\, y_2,\, z_2) = x_1 x_2 + 3y_1 y_2 + 2z_1 z_2$
(b) $(x_1,\, y_1,\, z_1)\cdot(x_2,\, y_2,\, z_2) = x_1 x_2 y_1 y_2 z_1 z_2$
(c) $(x_1,\, y_1,\, z_1)\cdot(x_2,\, y_2,\, z_2) = x_1 x_2 - y_1 y_2 + z_1 z_2$
(d) $(x_1,\, y_1,\, z_1)\cdot(x_2,\, y_2,\, z_2) = x_1 y_2 + y_1 z_2 + z_1 x_2$

36-4   For vectors in $\mathcal{M}^{2\times 2}$, define $A\cdot B = \det(AB)$. Is this an inner product for $\mathcal{M}^{2\times 2}$? Explain.

36-5(a)   Show that the rule defined in Example 36-1 satisfies properties (a) and (b) for an inner product.
(b)   Show that the rule defined in Example 36-1 satisfies property (d) for an inner product. {Hint: To say that $f = 0$ means that $f(x) = 0$ for all numbers x in $[0, 1]$. Argue in terms of area, since integrals have that interpretation.}

36-6   Suppose that an inner product rule for $\mathcal{P}^2$ is given by

$$p\cdot q = \int_{-1}^1 p(x)q(x)\, dx$$

(a)   Beginning with the basis $\frac{1}{2}$, $4x$, x^2, use the Gram-Schmidt process to find an orthogonal basis.
(b)   Change the basis found in part (a) to an orthonormal basis.

36-7   Suppose that f is a function continuous on $[0, 1]$. Prove that

$$\left(\int_0^1 f(x)\, dx \right)^2 \le \int_0^1 (f(x))^2\, dx$$

(Hint: Use the Cauchy-Schwarz inequality on the functions f and 1.)

37.   APPLICATIONS

Example 37-1

Analysis of Conic Sections

The conic sections—circles, ellipses, parabolas, hyperbolas—are graphs of equations such as $x^2 + y^2/4 = 1$, $x^2 - y^2 = 1$, etc. More generally, they are graphs of equations of the form $ax^2 + bxy + cy^2 + dx + ey = f$. Such equations lend themselves to the use of matrices, and the simplification of their matrices aids in analyzing their graphs.

Rather than deal with such graphs with full generality, we will restrict ourselves to the problem of analyzing graphs of equations of the form $ax^2 + bxy + cy^2 = 1$. We can express this equation in matrix notation as

$$[xy] \begin{bmatrix} a & \dfrac{b}{2} \\ \dfrac{b}{2} & c \end{bmatrix} \begin{bmatrix} x \\ y \end{bmatrix} = 1$$

(The reader can check this by carrying out the matrix multiplication on the left side.) The matrix

$$A = \begin{bmatrix} a & \dfrac{b}{2} \\ \dfrac{b}{2} & c \end{bmatrix}$$

is symmetric, and by Theorem 35-5 is therefore similar to a diagonal matrix

$$B = \begin{bmatrix} \lambda_1 & 0 \\ 0 & \lambda_2 \end{bmatrix}$$

By that same theorem we may choose the new basis to be orthonormal, in which case $A = P^T B P$ since $P^{-1} = P^T$ for new orthonormal bases. Thus we can transform

$$[xy]A\begin{bmatrix} x \\ y \end{bmatrix} = 1 \tag{4}$$

to

$$[xy]P^T B P \begin{bmatrix} x \\ y \end{bmatrix} = 1$$

By Theorem 27-1, the product $[xy]P^T$ is the new coordinate matrix $[(x, y)]_{\text{new}}$ and $P\begin{bmatrix} x \\ y \end{bmatrix}$ is $[(x, y)]^T_{\text{new}}$.† Thus we have transformed (4) to

$$[(x, y)]_{\text{new}} B \, [(x, y)]^T_{\text{new}} = 1$$

To simplify notation, let $[X \quad Y]$ be the new coordinate matrix; we therefore have

$$[X \quad Y]B\begin{bmatrix} X \\ Y \end{bmatrix} = 1$$

or

$$\lambda_1 X^2 + \lambda_2 Y^2 = 1$$

This last equation is easy to analyze since it is in a standard form:

If $\lambda_1 = \lambda_2 > 0$, it is the equation of a circle.
If $\lambda_1 \neq \lambda_2$ and both are positive, it is the equation of an ellipse.
If one λ is positive and the other is negative, it is the equation of a hyperbola.

(Other possibilities for the λ's result in degenerate conics.)

We now use this technique on the specific example $4x^2 + 2\sqrt{3}\, xy + 2y^2 = 1$. In matrix notation this is

$$[xy]\begin{bmatrix} 4 & \sqrt{3} \\ \sqrt{3} & 2 \end{bmatrix}\begin{bmatrix} x \\ y \end{bmatrix} = 1$$

---

† $[(x, y)]^T_{\text{new}} = ([x \quad y]P^T)^T = P^{TT}[x \quad y]^T = P\begin{bmatrix} x \\ y \end{bmatrix}$

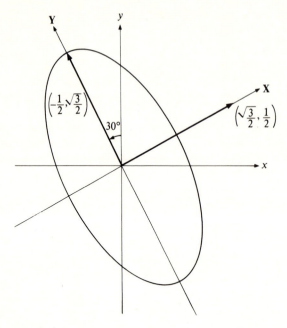

Fig. 37-1

Here

$$A = \begin{bmatrix} 4 & \sqrt{3} \\ \sqrt{3} & 2 \end{bmatrix}$$

The characteristic equation $\det (A - \lambda I) = \lambda^2 - 6\lambda + 5 = 0$ has the roots $\lambda_1 = 5$ and $\lambda_2 = 1$. The new orthonormal basis of eigenvectors can be chosen to be $(\sqrt{\tfrac{3}{2}}, \tfrac{1}{2})$, $(-\tfrac{1}{2}, \sqrt{\tfrac{3}{2}})$. Thus the transformed equation is

$$[X \quad Y] \begin{bmatrix} 5 & 0 \\ 0 & 1 \end{bmatrix} \begin{bmatrix} X \\ Y \end{bmatrix} = 1$$

or $5X^2 + Y^2 = 1$, where $[X \quad Y]$ is the coordinate matrix relative to the new basis above. Since $5X^2 + Y^2 = X^2/(1/\sqrt{5})^2 + Y^2/1^2$, we recognize the transformed equation as that of an ellipse centered at the origin, with minor axis along the X axis, and with endpoints that are $1/\sqrt{5}$ units and 1 unit from the origin, as in Fig. 37-1. The ellipse is in standard position relative to the *new* basis vectors and their corresponding coordinate axes.

Thus we have identified the graph as that of an ellipse, tilted 30° counterclockwise from the vertical. ●

Example 37-2

International Standard Book Numbers

In 1969 a method was established for assigning an identification number to every book published. A unique ten-digit number, called the International Standard Book Number (ISBN), is assigned to each book. This number consists of four parts separated by hyphens—for example, 0-865-40615-4—and appears on the back of the title page or at the bottom of the back cover. The first part, 0, indicates the country of publication;† the second part, 865, denotes the publisher; the third part, 40615, is the number assigned to the particular book; and the last part, 4, is called the *check digit*. The check digit is used to detect errors that might be made in copying the ISBN. We will see how the check digit is determined.

Suppose that *a-bcd-efghi* is the ISBN except for the check digit. Rewrite these nine digits as a vector in $\mathcal{R}^9$:

$$\mathbf{A} = (a, b, c, d, e, f, g, h, i)$$

and let

$$\mathbf{B} = (1, 2, 3, 4, 5, 6, 7, 8, 9)$$

Form the dot product

$$\mathbf{B} \cdot \mathbf{A} = 1a + 2b + 3c + 4d + 5e + 6f + 7g + 8h + 9i = z$$

and then divide z by 11 to obtain a remainder r. This remainder r is the check digit. Thus the check digit is an integer from zero to ten, with the roman numeral X used in place of ten.

For example, to determine the check digit for the number 0-865-40615 we form the dot product

$$(1, 2, 3, 4, 5, 6, 7, 8, 9) \cdot (0, 8, 6, 5, 4, 0, 6, 1, 5) = 169$$

and divide by 11, yielding 4 as remainder. Hence 4 is the check digit, and the ISBN is 0-865-40615-4.

The chief use of the check digit is to detect errors in copying an ISBN. Many errors in copying the ISBN can be uncovered by again forming the dot product $\mathbf{B} \cdot \mathbf{A}$, dividing by 11, and checking to see that the remainder

---

† A first digit of 0, which indicates publication in the United States, is often omitted. If this digit is omitted it must be supplied before carrying out the method of this example.

is the same as the check digit. For example, the number 0-301-21401-6 must have been incorrectly copied since

$$(1, 2, 3, 4, 5, 6, 7, 8, 9) \cdot (0, 3, 0, 1, 2, 1, 4, 0, 1) = 63$$

and when 63 is divided by 11 we have a remainder of 8, not the desired remainder 6.

The majority of errors made in copying ISBN's involve either miscopying one digit or else in transposing two digits (for example, writing 27 rather than 72). The check-digit method guarantees that both types of errors will be detected.   ●

Example 37-3

Expected Value

When two dice are rolled, the sum of the spots showing will be a number between 2 and 12. The following table gives the probabilities of the possible outcomes. To see how the probabilities are arrived at, note that there are 36 ways in which the two dice can land: 1-1, 1-2, ..., 1-6, 2-1, ..., 2-6, ..., 6-6. All 36 possibilities are assumed equally likely (that is, the probability of each is assumed to be $\frac{1}{36}$). It follows that, for example, the probability of rolling a 4 is $\frac{3}{36}$ because a 4 can be obtained by rolling 1-3, 2-2, or 3-1.

Outcome	2	3	4	5	6	7	8	9	10	11	12
Probability	$\frac{1}{36}$	$\frac{2}{36}$	$\frac{3}{36}$	$\frac{4}{36}$	$\frac{5}{36}$	$\frac{6}{36}$	$\frac{5}{36}$	$\frac{4}{36}$	$\frac{3}{36}$	$\frac{2}{36}$	$\frac{1}{36}$

If the dice are rolled over and over, what number can we expect to have as the average of all the numbers we roll? To answer this question we note that the outcome 2 can be expected to occur once in 36 rolls, the outcome 3 twice in 36 rolls (since it could occur as either 1-2 or 2-1), the outcome 4 three times in 36 rolls, etc. Thus to find the average of the outcomes we would find the average of one 2, two 3s, three 4s, ..., two 11s, and one 12. Doing this, we obtain

$$\frac{2 + 3 + 3 + 4 + 4 + 4 + \cdots + 11 + 11 + 12}{36} = 7 \qquad (5)$$

This number is called the *expected value* (or mean, or weighted average).

We can simplify the above calculation by introducing vector notation and the dot product. We form the *outcome vector* $\mathbf{A} = (2, 3, 4, ..., 12)$ and the *probability vector* $\mathbf{B} = (\frac{1}{36}, \frac{2}{36}, \frac{3}{36}, \ldots, \frac{1}{36})$, using the possible outcomes as the entries of $\mathbf{A}$ and their corresponding probabilities as entries of $\mathbf{B}$. Forming the dot product, we have

$$\mathbf{A} \cdot \mathbf{B} = 2(\tfrac{1}{36}) + 3(\tfrac{2}{36}) + 4(\tfrac{3}{36}) + \cdots + 12(\tfrac{1}{36}) = 7$$

which is equivalent to (5) above.

In fact, the dot product is the basis for the definition of expected value in probability theory. In general, if $\mathbf{X} = (x_1, x_2, \ldots, x_n)$ is the outcome vector for an experiment and if $\mathbf{P} = (p_1, p_2, \ldots, p_n)$ is the probability vector (that is, if the entries of $\mathbf{P}$ are the probabilities of the corresponding possible outcomes), then $\mathbf{X} \cdot \mathbf{P}$ is the expected value of the outcome.

The expected value is the keystone of any "successful" gambling game, for example, roulette. A roulette wheel has 38 slots numbered 1, 2, ..., 36, 0, 00. A player can bet on any of the 38 numbers. If the ball, when spun in the roulette wheel, stops on that number, the player wins 35 times his bet;† otherwise, the player loses his bet.

For each dollar a player bets, how much can he be expected to win? In this case the outcome vector $\mathbf{X} = (35, -1)$. The first entry shows the amount he wins if the ball lands on his number, and the second entry shows the amount he "wins" if the ball lands on any other number (a "win" of -1 is actually a *loss* of 1). The probability vector $\mathbf{P} = (\frac{1}{38}, \frac{37}{38})$ lists the probabilities (a player will win $\frac{1}{38}$ of the time and will lose the remaining $\frac{37}{38}$ of the time). Thus the expected value is

$$\mathbf{X} \cdot \mathbf{P} = (35, -1) \cdot (\tfrac{1}{38}, \tfrac{37}{38})$$

$$= -\tfrac{2}{38}$$

$$\approx -\$0.05$$

Therefore, after a long sequence of $1 bets, a player can expect his losses to average about 5 cents per bet. This 5 cents is the "edge" for the house.

●

Exercises

37-1 In a certain game a player tosses two coins. If they both show heads he wins $5, if they both show tails he wins $3, and if they show one head and one tail he loses $4. Verify that the expected value is zero. (Note: A game whose expected value is zero is called a *fair* game since in the long run the player will neither win nor lose.)

---

† That is, he gets back his bet plus 35 times the amount bet.

37-2 A box has 100 balls in it, 80 labeled zero, 15 labeled 50 cents, 4 labeled $1, and 1 labeled $10. A player pays 25 cents, blindly draws one ball, and is paid the amount shown on that ball. How much can he be expected to gain (or lose) each time he plays? (Remember that, even though he may win, it has cost him 25 cents to play.)

37-3 Determine which of these numbers could be correct ISBN's:

(a) 0-23-515920-1   (b) 1-357-88174-6
(c) 0-07-492231-X (d) 0-25-616200-7

37-4 Verify the accuracy of the ISBN for this book.

37-5 Various types of roulette bets can be made. The type of bet described in Example 37-3 is called a *straight* bet. Each of the following parts (a) to (d) describes another type of roulette bet. Find the expected value for a $1 bet.

(a) A *high* bet: you win $1 if any of the numbers 19 to 36 shows.
(b) A *street* bet: you bet on three numbers of the form 1, 2, 3, or 4, 5, 6, etc., and win $11 if any one of the three shows.
(c) An *even* bet: you win $1 if any of the numbers 2, 4, 6, ..., 36 shows.
(d) A *dozen* bet: you bet on 1 to 12, or 13 to 24, or 25 to 36, and win $2 if any number in that dozen shows.
(e) Which is the best bet to make: straight, high, street, even, or dozen?

37-6 Analyze the graph of $2x^2 - 6xy + 2y^2 = 1$. Include a sketch showing the new X and Y axes, properly positioned (you must find the new basis), and the graph of the transformed equation in X and Y.

37-7 Answer Exercise 37-6 for the graph whose equation is $2\sqrt{2}\,xy - y^2 = 1$.

37-8 Suppose that *a-bcd-efghi-j* is a correct ISBN. Suppose that two unequal adjacent digits, say *f* and *g*, are accidentally transposed. Show that the check digit for this incorrect number will not be *j*. (Thus the check-digit method detects a transposition error.)

Review of Chapter 6

1 Let $X = (x_1, x_2, \ldots, x_n)$ and $Y = (y_1, y_2, \ldots, y_n)$ be vectors in $\mathcal{R}^n$. The dot product $X \cdot Y =$ _____. The dot product satisfies four properties:

(a) $\mathbf{X} \cdot \mathbf{Y} = $ _____

(b) $(\mathbf{X} + \mathbf{Y}) \cdot \mathbf{Z} = $ _____

(c) $r(\mathbf{X} \cdot \mathbf{Y}) = $ _____

(d) $\mathbf{X} \cdot \mathbf{X} > 0$ if and only if _____.

2   The dot product is the key to defining length, distance, and angle in any $\mathcal{R}^n$. If $\mathbf{X} = (x_1, x_2, \ldots, x_n)$ is a vector in $\mathcal{R}^n$, the length of $\mathbf{X}$ is defined by $|\mathbf{X}| = $ _____. The distance between vectors $\mathbf{X}$ and $\mathbf{Y}$ in $\mathcal{R}^n$ is defined to be _____. If $\mathbf{X}$ and $\mathbf{Y}$ are nonzero vectors in $\mathcal{R}^n$, the angle between $\mathbf{X}$ and $\mathbf{Y}$ is the unique number θ between zero and π, such that $\cos \theta = $ _____. In particular, $\mathbf{X}$ and $\mathbf{Y}$ are perpendicular (or orthogonal) if $\theta = \pi/2$, that is, if $\mathbf{X} \cdot \mathbf{Y} = $ ____.

3   If $\mathbf{X}$ is a nonzero vector in $\mathcal{R}^n$, the projection of $\mathbf{X}$ on $\mathbf{Y}$ is the vector $P_\mathbf{Y}(\mathbf{X}) = $ _____. The projection of one vector on another is the starting point for proving the following results:

(a)   The Cauchy-Schwarz inequality: $|\mathbf{X} \cdot \mathbf{Y}| \le $ _____.
(b)   The Gram-Schmidt process for constructing an orthonormal basis: we begin with any basis $\mathbf{X}_1, \mathbf{X}_2, \ldots, \mathbf{X}_n$ for $\mathcal{R}^n$ and define $\mathbf{Y}_1 = \mathbf{X}_1$, $\mathbf{Y}_2 = $ _____, $\mathbf{Y}_3 = $ _____, etc. The vectors $\mathbf{Y}_1, \mathbf{Y}_2, \ldots, \mathbf{Y}_n$ form an orthogonal basis. Furthermore, the vectors $(1/|\mathbf{Y}_1|)\mathbf{Y}_1$, $(1/|\mathbf{Y}_2|)\mathbf{Y}_2, \ldots, (1/|\mathbf{Y}_n|)\mathbf{Y}_n$ form an orthonormal basis, that is, a basis of orthogonal unit vectors.

(c)   The triangle inequality: _____ $\le$ _____.

4   It is often easiest to work with an orthonormal basis $\mathbf{X}_1, \mathbf{X}_2, \ldots, \mathbf{X}_n$ for $\mathcal{R}^n$. For such a basis, $\mathbf{X}_i \cdot \mathbf{X}_i = $ ____ and $\mathbf{X}_i \cdot \mathbf{X}_j = $ ____ if $i \ne j$. If we use these basis vectors to form the rows of a matrix P, then $P^{-1} = $ ____.
The coordinates $c_1, c_2, \ldots, c_n$ of a vector $\mathbf{X}$ with respect to the orthonormal basis $\mathbf{X}_1, \mathbf{X}_2, \ldots, \mathbf{X}_n$ are easy to compute: $c_1 = $ _____, $c_2 = $ _____, $\ldots, c_n = $ ____. (Refer to Exercise 35-9.)

5   A matrix with the property that $AA^T = I$ is said to be orthogonal. If $T: \mathcal{R}^n \to \mathcal{R}^n$ has an orthogonal usual matrix A, then:

(a) T leaves lengths unchanged; that is, $|\mathbf{X}T| = $ ____.
(b) T leaves angles unchanged; that is, the angle between $\mathbf{X}$ and $\mathbf{Y}$ equals _____.
(c) $\det A = $ ____.

6 A symmetric matrix A is a matrix such that $A = $ _____. If $T : \mathcal{R}^n \to \mathcal{R}^n$ has a symmetric usual matrix A, then there is an orthonormal basis of _____ for T, and thus T has a diagonal matrix B. If P is the matrix whose rows are the new orthonormal basis, we have $B = $ ____.

7 If $\mathcal{V}$ is a vector space, an inner product on $\mathcal{V}$ is any rule assigning a real number to each ordered pair of vectors that satisfies the following four properties:

(a) _____

(b) _____

(c) _____

(d) _____

The definitions and theorems stated in terms of the dot product for $\mathcal{R}^n$ can be restated to apply to any inner product space.

Review Exercises

1 Let $\mathbf{X}_1 = (2, 0, -1)$, $\mathbf{X}_2 = (3, 1, 6)$, $\mathbf{X}_3 = (3, \sqrt{10}, 1)$.

(a) Find the length of each vector.
(b) Find the distance between each pair of vectors.
(c) Find the angle between $\mathbf{X}_1$ and $\mathbf{X}_2$; find the angle between $\mathbf{X}_1$ and $\mathbf{X}_3$.

2 Let $\mathbf{X} = (2, 5, 1, 0)$, $\mathbf{Y} = (-3, 1, 1, -1)$, $\mathbf{Z} = (1, -1, 1, -2)$.

(a) Find the projection of $\mathbf{X}$ on $\mathbf{Y}$ and on $\mathbf{Z}$.
(b) Find a vector perpendicular to $\mathbf{X}$.
(c) Find a vector perpendicular to both $\mathbf{Y}$ and $\mathbf{Z}$.

3 Find an orthonormal basis for $\mathcal{R}^2$ where one of the vectors is on the line $y = -2x$.

4 Use the Gram-Schmidt process on $\mathbf{X}_1 = (1, 3, 0)$, $\mathbf{X}_2 = (1, 0, 4)$, $\mathbf{X}_3 = (2, 1, -1)$ to find an orthogonal basis for $\mathcal{R}^3$.

5 Suppose that $T : \mathcal{R}^3 \to \mathcal{R}^3$ is the perpendicular projection onto the plane $x - 2y = 0$.

(a) Find an orthonormal basis for $\mathcal{R}^3$ where two of the three vectors are on the plane $x - 2y = 0$.
(b) Find the matrix for T relative to this basis.
(c) Find the usual matrix for T.

6 Suppose that $S : \mathcal{R}^3 \to \mathcal{R}^3$ is the perpendicular projection onto the plane $x - 2y - 2z = 0$. Find $(1, 0, 3)S$.

7 Suppose that $U : \mathcal{R}^2 \to \mathcal{R}^2$ is the reflection in the line $y = -3x$. Find $(1, 1)U$.

8 Suppose that $T : \mathcal{R}^3 \to \mathcal{R}^3$ is the reflection in the plane $x - 2y = 0$ (the plane of Exercise 5). Find the usual matrix for T.

9 Find the inverse of the matrix

$$
A = \begin{bmatrix} \dfrac{1}{\sqrt{6}} & \dfrac{2}{\sqrt{6}} & \dfrac{1}{\sqrt{6}} \\[2mm] -\dfrac{2}{\sqrt{5}} & \dfrac{1}{\sqrt{5}} & 0 \\[2mm] -\dfrac{1}{\sqrt{30}} & -\dfrac{2}{\sqrt{30}} & \dfrac{5}{\sqrt{30}} \end{bmatrix}
$$

(Hint: First verify that the rows of the matrix form an orthonormal basis for $\mathcal{R}^3$.)

10 Show directly that if T has the matrix A of Exercise 9, then $|(1, 2, 0)T| = |(1, 2, 0)|$. (Compute both sides separately and compare.)

11 Suppose that $S : \mathcal{R}^3 \to \mathcal{R}^3$ is the 60° clockwise rotation about the axis $\{a(1, 2, 2) : a \in \mathcal{R}\}$. Find $(1, 0, 0)S$. (Hint: Find the relationship between this linear transformation and the linear transformation of Example 35-2.)

12 Let

$$
A = \begin{bmatrix} 1 & 0 & 0 \\ 0 & 3 & 2 \\ 0 & 2 & 0 \end{bmatrix}
$$

Find a diagonal matrix B similar to A, and also find a new orthonormal basis of eigenvectors.

ANSWERS TO SELECTED EXERCISES

CHAPTER 1

Section 1

1-1 (3, 7)
1-3 (2, 1/2)
1-5 (4, 1, 2)
1-7 $(1/2, 0, -1, 3)$
1-9 (7, 2)
1-11 26 and 14

Section 2

2-1
$$\begin{bmatrix} 1 & 0 & 0 & 0 \\ 0 & 1 & 0 & 0 \\ 0 & 0 & 1 & 0 \\ 0 & 0 & 0 & 1 \end{bmatrix}$$

2-2
$$\begin{bmatrix} 1 & 5 & 0 & 0 \\ 0 & 0 & 1 & 0 \\ 0 & 0 & 0 & 0 \\ 0 & 0 & 0 & 0 \end{bmatrix}$$

2-3
$$\begin{bmatrix} 1 & 0 & 0 \\ 0 & 1 & 0 \\ 0 & 0 & 1 \\ 0 & 0 & 0 \end{bmatrix}$$

2-6 $(32, -6, -6)$
2-9 $(0, 0, 0)$
2-11 $(2, 0, 6, 1/2)$

Section 3

3-1 $\{a(3, -2, 1): a \in \mathcal{R}\}$
3-3 $\{(2, 1, 4, 0) + a(-9, 1, -3, 1): a \in \mathcal{R}\}$
3-5 $\{(3, 0, -1) + a(-4, 1, 0): a \in \mathcal{R}\}$
3-7 $\{(7, 5, -9, 0, 0) + a(1, -5, -2, 1, 0) + b(-4, -3, -3, 0, 1): a, b \in \mathcal{R}\}$
3-9 $\{(0, 1, 0, -2, 0, 0) + a(1, 0, 0, 0, 0, 0) + b(0, -2, 1, 0, 0, 0)$
$+ c(0, -2, 0, -3, 0, 1): a, b, c \in \mathcal{R}\}$
3-11 $\{a(-3/2, -1/2, 1): a \in \mathcal{R}\}$
3-13 No solution
3-15 $\{a(1, 1, 0) + b(-2, 0, 1): a, b \in \mathcal{R}\}$
3-17 $(-2a + 3b, a - b)$

Section 4

4-1 (a) A line in $\mathcal{R}^3$ passing through the point $(0, 1, 2)$, parallel to the line passing through $(0, 0, 0)$ and $(-2, -2, 0)$
(b) A plane in $\mathcal{R}^3$ passing through $(1, 3, 1)$, parallel to the plane through $(0, 0, 0)$, $(1, 0, 0)$, $(0, 2, -4)$
4-2 (a) $\{a(-4, -3, 1): a \in \mathcal{R}\}$, a line in $\mathcal{R}^3$ passing through $(0, 0, 0)$ and $(-4, -3, 1)$
(c) $\{(3, 0) + a(5, 1): a \in \mathcal{R}\}$, a line in $\mathcal{R}^2$ passing through $(3, 0)$, parallel to the line through $(0, 0)$ and $(5, 1)$
(e) $\{(5, 3, 0) + a(-2, 3, 1): a \in \mathcal{R}\}$, a line in $\mathcal{R}^3$ passing through $(5, 3, 0)$, parallel to the line through $(0, 0, 0)$ and $(-2, 3, 1)$
4-3 $\{(0, 11/4, 1/4) + a(1, 0, 0): a \in \mathcal{R}\}$, a line in $\mathcal{R}^3$ passing through $(0, 11/4, 1/4)$, parallel to the line through $(0, 0, 0)$ and $(1, 0, 0)$
4-5 The single point $(5/3, 2/3, 2/3)$

Section 5

5-1 $A = \begin{bmatrix} 3 & 4 & 5 & 6 & \vdots & 7 \\ 0 & 5 & 8 & 9 & \vdots & 2 \end{bmatrix}$, infinitely many solutions

$C = \begin{bmatrix} 86.27 & 6.381 & 0 \\ 0 & 19.8 & .37 \end{bmatrix}$, infinitely many solutions

$E = \begin{bmatrix} 2 & 5 & -4 & 6 & -3 & 2 \\ 0 & 0 & 2 & 5 & -4 & -2 \\ 0 & 0 & 0 & 4 & 4 & 4 \\ 0 & 0 & 0 & 0 & 0 & 0 \end{bmatrix}$, infinitely many solutions

$$G = \begin{bmatrix} 46 & 5.3 & 1.1 & \vdots & 8 \\ 0 & 0 & \underline{0.2} & \vdots & 0 \\ 0 & 0 & 3 & \vdots & \underline{5} \end{bmatrix}, \text{ no solution}$$

5-2 (a) -8
5-3 (a) No   (b) Yes   (c) Yes
5-4 (a) Yes   (b) No   (c) Yes

Section 6

6-1 75% must be economy car owners and 25% must be full-size car owners
6-2 (a) Twice as many suburban dwellers as urban dwellers should be chosen
(b) Ten years ago: 8900 urban dwellers, 800 suburban dwellers; ten years from now: 6010 urban dwellers, 3690 suburban dwellers
6-3 $3Cl_2 + 6KOH = 5KCl + KClO_3 + 3H_2O$
6-4 $4C_3H_5(NO_3)_3 = 10H_2O + 12CO + 6N_2 + 7O_2$
6-5 (a) Total output of A $= 8\frac{1}{2}$, total output of B $= 14$
(b) Total output of A $= 8$, total output of B $= 12$

Review

1 (a) $\{(-1, -5, 0) + a(-3, 1, 1) : a \in \mathcal{R}\}$
(b) $\{a(-3, 1, 1) : a \in \mathcal{R}\}$
(c) $\{(-10/21, -13/21, 25/21)\}$
(d) $\{a(3/2, 1, 0, 0) + b(1/2, 0, -1, 1) : a, b \in \mathcal{R}\}$
(e) $\{a(-2, 1, 0, 0, 0) + b(-1, 0, 1, 0, 0) + c(-3, 0, 0, -2, 1) : a, b, c \in \mathcal{R}\}$
(f) No solution
(g) $\{(0, 0, 0)\}$
2 (a) $a = 0$ (b) $a = 15/2$
3 (b) $p(x) = 2x^2 + 5x - 3$
4 A $= 3$, B $= -3$, C $= -1$; $3 \log (x - 1) - \frac{3}{2} \log (x^2 + 1) - \arctan x + $ constant
5 $\{(5/4, 1/2, 0) + a(1/4, 3/2, 1) : a \in \mathcal{R}\}$, a line in $\mathcal{R}^3$ passing through the point $(5/4, 1/2, 0)$, parallel to the line through $(0, 0, 0)$ and $(1/4, 3/2, 1)$
7 4 cars of type I, 6 cars of type II, 2 cars of type III
8 The system of equations (where $N =$ number of nickels, $D =$ number of dimes, $Q =$ number of quarters) is

$$N + \quad D + \quad Q = 17$$
$$D - \quad 2Q = 0$$
$$5N + 10D + 25Q = 185$$

and this system has no solutions where N, D, and Q are positive integers
9 $(2, 3), (2, -3), (-2, 3), (-2, -3)$
10 (i) no, no, yes
(ii) yes, no, yes
(iii) no, yes, yes
(iv) yes, yes, yes

CHAPTER 2

Section 7

7-1 Arbitrary vectors: (a, b, c), (x, y, z), (x_1, x_2, x_3), etc.
Zero vector: $(0, 0, 0)$
Inverses: $(-a, -b, -c)$, $(-x, -y, -z)$, $(-x_1, -x_2, -x_3)$, etc.

7-3 Arbitrary vectors: $\begin{bmatrix} a & b \\ c & d \end{bmatrix}$, $\begin{bmatrix} w & x \\ y & z \end{bmatrix}$, $\begin{bmatrix} a_{11} & a_{12} \\ a_{21} & a_{22} \end{bmatrix}$, etc.

Zero vector: $\begin{bmatrix} 0 & 0 \\ 0 & 0 \end{bmatrix}$

Inverses: $\begin{bmatrix} -a & -b \\ -c & -d \end{bmatrix}$, $\begin{bmatrix} -w & -x \\ -y & -z \end{bmatrix}$, $\begin{bmatrix} -a_{11} & -a_{12} \\ -a_{21} & -a_{22} \end{bmatrix}$, etc.

7-6 Arbitrary vectors: $(x, 0, 0)$, $(a, 0, 0)$, $(b, 0, 0)$, etc.
Zero vector: $(0, 0, 0)$
Inverses: $(-x, 0, 0)$, $(-a, 0, 0)$, $(-b, 0, 0)$, etc.

7-9 Arbitrary vectors: $(-3a, a, a)$, $(-3z, z, z)$, $(3v, -v, -v)$, etc.
Zero vector: $(0, 0, 0)$
Inverses: $(3a, -a, -a)$, $(3z, -z, -z)$, $(-3v, v, v)$, etc.

Section 9

9-1 Arbitrary vectors: $ax^2 + bx + c$, $a_2 x^2 + a_1 x + a_0$, $b_2 x^2 + b_1 x + b_0$, etc.
Zero vector: the zero polynomial, that is, $0x^2 + 0x + 0$
Inverses: $-ax^2 - bx - c$, $-a_2 x^2 - a_1 x - a_0$, $-b_2 x^2 - b_1 x - b_0$, etc.

9-3 Arbitrary vectors: $(-a, -2a, a)$, $(-b, -2b, b)$, $(a, 2a, -a)$, etc.
Zero vector: $(0, 0, 0)$
Inverses: $(a, 2a, -a)$, $(b, 2b, -b)$, $(-a, -2a, a)$, etc.

9-5 Arbitrary vectors: f, g, h, etc.
Zero vector: the function f whose rule is $f(x) = 0$ for all real numbers x
Inverses: $-f$, $-g$, $-h$ (where $-f$ is the function whose rule is given by $(-f)(x) = -(f(x))$, and similarly for $-g$ and $-h$)

9-7 Arbitrary vectors: $(2a - b, a, b)$, $(2c - d, c, d)$, $(2r - s, r, s)$, etc.
Zero vector: $(0, 0, 0)$
Inverses: $(-2a + b, -a, -b)$, $(-2c + d, -c, -d)$, $(-2r + s, -r, -s)$, etc.

Section 10

10-1	Subspace	**10-3**	Not a subspace
10-5	Subspace	**10-7**	Subspace
10-9	Subspace	**10-11**	Not a subspace

10-13   Subspace   10-15   Subspace
10-17   Not a subspace
10-21   (b) Plane   (c) Line

Section 11

11-3   Ground speed is 120 mi/h
11-4   If **F** is the vector whose entries measure temperatures in degrees Fahrenheit, then $\mathbf{F} = \frac{9}{5}\mathbf{H} + 32$

Review

2   (a)   Subspace
    (b)   Not a subspace
    (c)   Not a subspace
10   (7, 1, 3) is not in the given subspace
11   (b)   No
12   (a)   Not a subspace
    (b)   Subspace
    (c)   Subspace
    (d)   Subspace
    (e)   Not a subspace

CHAPTER 3

Section 12

12-6   (1, 3, 0), (0, 2, 1)
12-7   (−1, 1)
12-11   Yes
12-13   No
12-15   (a) Plane   (b) Line (c) Line   (d) $\mathcal{R}^3$

Section 13

13-1   Independent   13-3   Dependent
13-5   Independent   13-7   Independent
13-11   (a)   Either (2, 1) or (10, 5) can be discarded
       (c)   (1, 2, 13) and (10, 4, 2) can be discarded

Section 14

14-3 (a) Basis   (c) Not a basis
14-9 2

Section 15

15-1 (a) $\begin{bmatrix} 4 & 0 & 23 & 78 \\ 0 & 1 & 2 & 10 \\ 0 & 0 & 0 & 7 \end{bmatrix}$; rank = 3; set of solutions has the form $a(\quad)$

(c) $[3 \quad 3 \quad -6/7 \quad -5/2 \mathrel{\vdots} 3]$; rank = 1; set of solutions has the form $(\quad) + a(\quad) + b(\quad) + c(\quad)$

(e) $\begin{bmatrix} \underline{4} & 89 & 65 \\ 0 & \underline{11.3} & 501 \\ 0 & 0 & \underline{7.1} \end{bmatrix}$; rank = 3; unique solution (0, 0, 0)

15-3 (a) Two-dimensional; $(-3, 1, 0)$, $(8, 0, 1)$
(c) Two-dimensional; $(1, -8, 1, 0)$, $(1, -9, 0, 1)$

Section 16

16-1 (a) $3a - 2b, b - a$
(c) 0, 0
(e) $0, -4$
16-2 (a) $3a/4 + c/4, b/2, a/4 - c/4$
(c) $-1, 1, 1$
(e) $64, -63/2, 66$
16-3 (a) 1, 3, 0
(c) 1, 1, 0
16-4 (b) 2, 0, 1, 0
16-5 (a) $a, -8a + 4b - c, 6a - 3b + c$
(c) $2, -19, 14$
(e) $2000, -19000, 14000$
16-7 $0, 3, 0, -2$

Section 17

17-1 (a) $[1, \infty)$
(d) The line $y = 2x$
17-2 (a) $F(0) = F(1) = 0$
(c) $f(0) = f(2\pi) = 1$
(e) $H(1, 0, 0) = H(2, 0, 0) = (0, 0)$
17-4 (a) Any number less than -3
(c) Any vector in $\mathcal{R}^2$ whose second entry is negative

Section 19

19-3 (b) Two dimensional

Review

1 (a) Not a basis   (c) Not a basis
   (b) Not a basis   (d) Not a basis
2 (a) Not a basis   (c) Basis
   (b) Not a basis   (d) Basis
3 (a) Not a basis   (c) Basis
   (b) Basis   (d) Not a basis
4 (a) Basis   (c) Basis
   (b) Not a basis   (d) Not a basis
5 (a) No; 2   (b) Yes; 4   (c) Yes; 3
6 (a) 3   (b) 0   (c) 0   (d) 3   (e) 3   (f) 2
   (g) 3
9 (a) $(1, 1, 1), (1, 0, 0)$
   (b) $(-4, 1, 0), (1, 0, 1)$
   (c) $(9/5, 7/5, 1)$
   (d) $(-1, 0, 1)$
   (e) $(1, 1, 0), (-4, 0, 1)$
   (f) $(\frac{1}{3}, 1)$
   (g) $(9/5, 7/5, 1)$
10 (a) $-1, 2, 0, 4$
    (b) $0, 2, 4, -1$
    (c) $-1, 3, 17/3, -4/3$
11 $1/2, -1/2$

CHAPTER 4

Section 20

20-3 $(x, y)T = (x, 0)$ **20-5** $(x, y)S = (y, -x)$
20-7 $(x, y)T = (x - 2y, x + y)$
20-9 (a) $\{(0, 0, 0)\}$ (e) $\langle(2, 2, 0)\rangle$
20-10 (a) $\langle(0, -2, 0)\rangle$ (c) $\mathcal{R}^2$
20-13 (a) $\{(a, 0) : a \in \mathcal{R}\}$ (c) $\{(0, 0, 0)\}$

Section 21

21-1 $\begin{bmatrix} 0 & 0 & 1 \\ 0 & 2 & 0 \\ 0 & 0 & 3 \end{bmatrix}$ **21-3** $\begin{bmatrix} 1 & 0 & 0 & -1 \\ 1 & 0 & 3 & 1 \end{bmatrix}$

21-5 $\begin{bmatrix} -1 & 0 \\ 0 & 1 \end{bmatrix}$ **21-7** $\begin{bmatrix} \sqrt{3}/2 & 1/2 \\ -1/2 & \sqrt{3}/2 \end{bmatrix}$ **21-9** $\begin{bmatrix} 1/2 & \sqrt{3}/2 & 0 \\ -\sqrt{3}/2 & 1/2 & 0 \\ 0 & 0 & 1 \end{bmatrix}$

21-11 $\begin{bmatrix} 0 & 1 \\ 1 & 0 \end{bmatrix}$, $(1, -2)U = (-2, 1)$, $(x, y)U = (y, x)$

21-13 $(\sqrt{3} - 2, 1 + 2\sqrt{3})$

21-15 T is the projection of $\mathcal{R}^3$ onto the xz plane

Section 22

22-1 (a) $AB = \begin{bmatrix} 4 & 2 & -4 \\ 12 & 7 & -3 \end{bmatrix}$, $BD = I$, $B^2 = \begin{bmatrix} 4 & 2 & -1 \\ -5 & -3 & 2 \\ 0 & -1 & 1 \end{bmatrix}$

    (b) BA, C^2, IA, and CI do not exist

22-3 (a) $(x, y, z)ST = (6x + y, 2x)$

22-5 (a) $A = \begin{bmatrix} 0 & 0 \\ 0 & 1 \end{bmatrix}$, B $\begin{bmatrix} 0 & 1 \\ -1 & 0 \end{bmatrix}$

    (b) $AB = \begin{bmatrix} 0 & 0 \\ -1 & 0 \end{bmatrix}$, $(x, y)ST = (-y, 0)$

    (c) $(x, y)TS = (0, x)$

22-7 (a) $A = \begin{bmatrix} 1 & 1 \\ 0 & 1 \end{bmatrix}$, $B = \begin{bmatrix} 0 & 0 \\ 0 & 2 \end{bmatrix}$

    (b) $C = \begin{bmatrix} 0 & 2 \\ 0 & 2 \end{bmatrix}$

22-9 (a) $\begin{bmatrix} 1/\sqrt{2} & 1/\sqrt{2} & 0 \\ -1/\sqrt{2} & 1/\sqrt{2} & 0 \\ 0 & 0 & 0 \end{bmatrix}$;

    $(x, y, z)ST = (x/\sqrt{2} - y/\sqrt{2}, x/\sqrt{2} + y/\sqrt{2}, 0)$

    (b) Same as (a)

22-11 ST is a 90° clockwise rotation

    TS is a 90° counterclockwise rotation

Section 23

23-1 (a) $\begin{bmatrix} 1 & -2/5 \\ 0 & 1/5 \end{bmatrix}$ (c) Not invertible

    (d) $\begin{bmatrix} 1 & -1 & 2 & -3 \\ -1 & 2 & -3 & 9/2 \\ 2 & -4 & 8 & -23/2 \\ -1 & 2 & -4 & 6 \end{bmatrix}$

23-6 (a) $(1, -1/2, -1)$

    (c) $(1/8, 131/48, 5/24)$

23-9 T^{-1} does not exist
23-11 (a) T has no inverse
    (b) T has no inverse
    (c) T^{-1} exists
23-15 For $(x, y, z)T = (x, 0, z)$, $\mathcal{K} = \{(0, a, 0) : a \in \mathcal{R}\}$
    For $(x, y, z)T = (x - y, y + z, x + z)$, $\mathcal{K} = \{(-a, -a, a) : a \in \mathcal{R}\}$

Section 24

24-1 (a)   140   (c) -1 (e)   0
24-3 (b)   det $A = -6$, det $B = -36$
24-8 0
24-11 (a) -1

Section 25

25-1 (a)   51   (b)   819
25-2 (a) $\begin{bmatrix} 0 & 1 & 0 & 0 & 0 \\ 0 & 0 & 1 & 1 & 0 \\ 0 & 0 & 0 & 0 & 1 \\ 1 & 1 & 0 & 0 & 1 \\ 0 & 1 & 1 & 0 & 0 \end{bmatrix}$ (b)   3   (c)   0   (d)   1

25-3 Adding the entries in the rows of $A + A^2$, player 1 has a score of 7, while players 3 and 4 each have scores of 5
25-4 (a)   4; 1 and 6 tied; 2 and 5 tied; 3
    (b)   6; 4; 1; 2; 3 and 5 tied
25-5 (a)   2 and 6 tied; 4; 1, 3, and 5 tied
    (b)   2; 1; 6; 3 and 4 tied; 5

Review

1 a, d, e, f, g are linear transformations
2   (a) $\begin{bmatrix} 1 & 1 \\ -1 & 2 \end{bmatrix}$ (d) $\begin{bmatrix} 0 & 1 \\ 0 & 1 \\ 0 & 1 \end{bmatrix}$ (e) $\begin{bmatrix} 0 & 0 & 1 \\ 1 & 0 & 0 \end{bmatrix}$

    (f) $\begin{bmatrix} 3 & 0 \\ 0 & -1 \end{bmatrix}$ (g) $\begin{bmatrix} 3 & 1 & 2 \\ -5 & -1 & -4 \\ 0 & 0 & 0 \end{bmatrix}$

3   (a) a and f are invertible
    (b) $a: (x, y)T^{-1} = (2x/3 + y/3, -x/3 + y/3)$
        $f: (x, y)T^{-1} = (x/3, -y)$

4 $\begin{bmatrix} 1/2 & -1/2 \\ -1/2 & 1/2 \end{bmatrix}$

5 (a) $(1, 2, -3)T = (1, -1/\sqrt{2}, -5/\sqrt{2})$
$(1, 2, -3)T^{-1} = (1, 5/\sqrt{2}, -1/\sqrt{2})$

(b) $(1, 2, -3)T = \left(1 - \dfrac{3\sqrt{3}}{2}, 1, -\sqrt{3} - \dfrac{3}{2}\right)$

$(1, 2, -3)T^{-1} = \left(2, \dfrac{1 + 3\sqrt{3}}{2}, \dfrac{\sqrt{3} - 3}{2}\right)$

(c) $(1, 2, -3)T = (-6, -4, 14)$
$(1, 2, -3)T^{-1} = (1/4, -1/2, 5/8)$

6 (a) $(x, y, z)ST = (x/2 + y/2, x/2 + y/2, z)$

(b) $(x, y, z)ST = (9y + 5z, x - 9y + 3z, 3y - 9z)$

7 (a)
$$\begin{bmatrix} 34/77 & -37/77 & 12/11 & 1/7 \\ -24/77 & 93/154 & -15/22 & -1/14 \\ -17/77 & 37/154 & -1/22 & -1/14 \\ 4/77 & 23/154 & -3/22 & -1/14 \end{bmatrix}$$
(b) Not invertible

(c)
$$\begin{bmatrix} -6/41 & -4/41 & 24/41 \\ 48/41 & 32/41 & -28/41 \\ 2/41 & 15/41 & -8/41 \end{bmatrix}$$
(d) Not invertible

8 (a)
$$B = \begin{bmatrix} 1 & 1 & 1 \\ 1 & 1 & 1 \\ -1 & -1 & -1 \end{bmatrix}$$
(b) No

9
$$\begin{bmatrix} 1/2 & -1 & 1 \\ 1 & -1/6 & -1/6 \end{bmatrix}$$

12 $\det A = -4$, $\det B = 0$, $\det AB = 0$
13 0
14 b, d, and e are always true
15 $-7/10$
16 The line $\langle (1, 1, 2) \rangle$
17 $45°$ counterclockwise rotation about the z axis (as viewed from the positive side of the z axis)

CHAPTER 5

Section 26

26-1 (a)
$$\begin{bmatrix} 1 & 0 \\ 0 & -1 \end{bmatrix}$$
(b) $(1, 5), (-5, 1)$

26-3 (a) If X_1 and X_2 lie on the plane and X_3 is perpendicular to the plane, then $B = \begin{bmatrix} 1 & 0 & 0 \\ 0 & 1 & 0 \\ 0 & 0 & 0 \end{bmatrix}$

(b) $(3, 1, 0), (-2, 0, 1), (1, -3, 2)$

26-5 If X_1 lies on the line and X_2 and X_3 are perpendicular to the line (and $X_2 S = X_3$), then $B = \begin{bmatrix} 1 & 0 & 0 \\ 0 & 0 & 1 \\ 0 & -1 & 0 \end{bmatrix}$

26-7 Perpendicular projection onto the line

26-9 (a) $\begin{bmatrix} 2 & 0 & 0 \\ 0 & 2 & 7 \\ 0 & 0 & -5 \end{bmatrix}$ (b) $\begin{bmatrix} 2 & 0 & 0 \\ 0 & 2 & 0 \\ 0 & 0 & -5 \end{bmatrix}$

Section 27

27-1 (a) $[2/11 \;\; -7/11]$ (b) $[-1 \;\; 2/5]$

27-3 Coordinates of $(2, 3, 0, 4)$: $2, 3/2, -1, 1$
Coordinates of $(1, 1, 1, -1)$: $1, 1/2, 1/4, -1$
Coordinates of (a, b, c, d): $a, b/2, a/8 - b/4 + c/4 - d/8, d/2 - a/2$

Section 28

28-1

(a) $A = P^{-1}BP = \begin{bmatrix} 6/29 & -12/29 & -3/29 \\ 25/29 & 8/29 & 2/29 \\ 8/29 & 13/29 & -4/29 \end{bmatrix}$

$\times \begin{bmatrix} 1 & 0 & 0 \\ 0 & 1 & 0 \\ 0 & 0 & -1 \end{bmatrix} \begin{bmatrix} 2/3 & 1 & 0 \\ -4/3 & 0 & 1 \\ -3 & 2 & -4 \end{bmatrix}$

$= \begin{bmatrix} 11/29 & 12/29 & -24/29 \\ 12/29 & 21/29 & 16/29 \\ -24/29 & 16/29 & -3/29 \end{bmatrix}$

(b) $(35/29, 54/29, 8/29)$

28-3 (a)

$A = P^{-1}BP = \begin{bmatrix} 1/2 & 0 & 1/2 \\ 1/2 & 0 & -1/2 \\ 0 & 1 & 0 \end{bmatrix} \begin{bmatrix} 1 & 0 & 0 \\ 0 & 1 & 0 \\ 0 & 0 & 0 \end{bmatrix} \begin{bmatrix} 1 & 1 & 0 \\ 0 & 0 & 1 \\ 1 & -1 & 0 \end{bmatrix} =$

$\begin{bmatrix} 1/2 & 1/2 & 0 \\ 1/2 & 1/2 & 0 \\ 0 & 0 & 1 \end{bmatrix}$

(b) $(3/2, 3/2, 1)$

28-5 The usual matrix is $\begin{bmatrix} -3/5 & -4/5 \\ -4/5 & 3/5 \end{bmatrix}$, hence $(1, 5)T = (-23/5, 11/5)$

28-7 (a)
$$A^6 = P^{-1}B^6P = \begin{bmatrix} 2/3 & -5/3 & 0 \\ -1/3 & 4/3 & 0 \\ 1 & 6 & -1 \end{bmatrix} \begin{bmatrix} 64 & 0 & 0 \\ 0 & 1 & 0 \\ 0 & 0 & 0 \end{bmatrix} \begin{bmatrix} 4 & 5 & 0 \\ 1 & 2 & 0 \\ 10 & 17 & -1 \end{bmatrix} =$$

$$\begin{bmatrix} 169 & 210 & 0 \\ -84 & -104 & 0 \\ 262 & 332 & 0 \end{bmatrix}$$

(b) det $A = 0$
(c) A is not invertible

Section 29

29-1 (a) 2   (b) -5 (c) $0, -5$
29-3 $(1, 0)$ and $(0, 1)$ are eigenvectors, with corresponding eigenvalues -1 and 1
29-6 1

Section 30

30-1 (a) $(5 - \lambda)(2 - \lambda) = 0$
    (b) 5, 2
    (c) $(3, 4), (0, 1)$

    (d) $\begin{bmatrix} 5 & 0 \\ 0 & 2 \end{bmatrix}$ relative to the basis $(3, 4), (0, 1)$

30-3 (a) $(1 - \lambda)^2 = 0$ (b) 1   (c) $(0, 1)$
    (d) T has no diagonal matrix
30-5 (a) $(1 - \lambda)(2 - \lambda)(4 - \lambda) = 0$
    (b) 1, 2, 4
    (c) $(-3, 8, 1), (0, 1, 0), (0, 1, 2)$
    (d) $\begin{bmatrix} 1 & 0 & 0 \\ 0 & 2 & 0 \\ 0 & 0 & 4 \end{bmatrix}$ relative to the basis in (c)
30-7 (a) $(-\lambda)^3 + \lambda = 0$
    (b) $0, 1, -1$
    (c) $(0, 1, 0), (1, 0, 1), (-1, 0, 1)$
    (d) $\begin{bmatrix} 0 & 0 & 0 \\ 0 & 1 & 0 \\ 0 & 0 & -1 \end{bmatrix}$ relative to the basis in (c)

30-9 (a) $-\lambda^3 + 3\lambda^2 - 4 = 0$
    (b) $-1, 2$
    (c) $(1, -1, 0), (2, 1, 0)$
    (d) No diagonal matrix

30-11 (a) $-\lambda^3 + 5\lambda^2 - 7\lambda + 3 = 0$
    (b) $1, 3$
    (c) $(1, 1, 0), (-1, 0, 1); (0, 1, 3)$
    (d) $\begin{bmatrix} 1 & 0 & 0 \\ 0 & 1 & 0 \\ 0 & 0 & 3 \end{bmatrix}$ relative to the basis in (c)

30-13 The eigenvalues are a, d, f

30-16 $(0, 0, 0, 1, -1), (0, 0, 0, 0, 1)$

$$B = \begin{bmatrix} 2 & 1 & 0 & \vdots & 0 & 0 \\ 0 & 2 & 1 & \vdots & 0 & 0 \\ 0 & 0 & 2 & \vdots & 0 & 0 \\ \hline 0 & 0 & 0 & \vdots & 3 & 1 \\ 0 & 0 & 0 & \vdots & 0 & 3 \end{bmatrix}$$

Section 31

31-1 (a) $x_0 = 1, x_1 = 1, x_n = x_{n-1} + 2x_{n-2}$ if $n \geq 2$
    (b) $x_n = 2^{n+1}/3 + (-1)^n/3$

31-2 (a) $\begin{bmatrix} 0 & 2/3 & 0 \\ 0 & 0 & 1/2 \\ 3 & 0 & 0 \end{bmatrix}$ (b) $(3, 2, 1)$

31-3 (a) $X_0 = (.5, .5)$

$$Q = \begin{bmatrix} .8 & .2 \\ .1 & .9 \end{bmatrix}$$

    (b) $7/10, 1$, with corresponding eigenvectors $(-1, 1), (1, 2)$

    (c) $Q^n = \begin{bmatrix} \dfrac{2}{3}\left(\dfrac{7}{10}\right)^n + \dfrac{1}{3} & -\dfrac{2}{3}\left(\dfrac{7}{10}\right)^n + \dfrac{2}{3} \\ -\dfrac{1}{3}\left(\dfrac{7}{10}\right)^n + \dfrac{1}{3} & \dfrac{1}{3}\left(\dfrac{7}{10}\right)^n + \dfrac{2}{3} \end{bmatrix}$

    (d) $X_n = X_0 Q^n = \left(\dfrac{1}{6}\left(\dfrac{7}{10}\right)^n + \dfrac{1}{3}, -\dfrac{1}{6}\left(\dfrac{7}{10}\right)^n + \dfrac{2}{3}\right)$

    (e) X_n approaches $(1/3, 2/3)$

31-4 5/9 should be placed in the feeding compartment and 4/9 in the exercise compartment

31-5 $f(x) = ae^{5x} + be^x, g(x) = 3ae^{5x} - be^x$

31-6 $f(x) = ae^x + 3be^{2x}, g(x) = 3be^{2x}, h(x) = ae^x + 2be^{2x} + ce^{-x}$

Review

1 (a) $\begin{bmatrix} 1 & 0 \\ 0 & -1 \end{bmatrix}$ (b) (1, 1), (1, 3)

2 No diagonal matrix
3 No diagonal matrix
4 (a) $\begin{bmatrix} -1 & 0 & 0 \\ 0 & 4 & 0 \\ 0 & 0 & 4 \end{bmatrix}$ (b) (1, −2, 0), (2, 1, 0), (0, 0, 1)

5 No diagonal matrix
6 (a) $\begin{bmatrix} 1 & 0 & 0 \\ 0 & 2 & 0 \\ 0 & 0 & -2 \end{bmatrix}$ (b) (0, 1, 0), (2, 4, 1), (−6, 4, 3)

7 (a) $\begin{bmatrix} 1 & 0 & 0 \\ 0 & 1 & 0 \\ 0 & 0 & -2 \end{bmatrix}$ (b) (−1, 0, 1), (1, 1, 0), (1, 0, 1)

8 (a) $\begin{bmatrix} 1 & 0 \\ 0 & -1 \end{bmatrix}$ relative to the basis (1, 3), (3, −1)

   (b) $\begin{bmatrix} -4/5 & 3/5 \\ 3/5 & 4/5 \end{bmatrix}$ (c) (−2, −1)

9 (a) $\begin{bmatrix} 1 & 0 \\ 0 & 0 \end{bmatrix}$ relative to the basis (1, −4), (4, 1)

   (b) $\begin{bmatrix} 1/17 & -4/17 \\ -4/17 & 16/17 \end{bmatrix}$ (c) (1/17, −4/17)

10 (a) $\begin{bmatrix} 1 & 0 & 0 \\ 0 & 1 & 0 \\ 0 & 0 & -1 \end{bmatrix}$ relative to the basis (5, 1, 0), (−1, 0, 1), (1, −5, 1)

   (b) $\begin{bmatrix} 25/27 & 10/27 & -2/27 \\ 10/27 & -23/27 & 10/27 \\ -2/27 & 10/27 & 25/27 \end{bmatrix}$ (c) (10/27, −23/27, 10/27)

11 (a) $\begin{bmatrix} 1 & 0 & 0 \\ 0 & 1 & 0 \\ 0 & 0 & 0 \end{bmatrix}$ relative to the basis (1, 0, 1), (0, 1, 0), (1, 0, −1)

   (b) $\begin{bmatrix} 1/2 & 0 & 1/2 \\ 0 & 1 & 0 \\ 1/2 & 0 & 1/2 \end{bmatrix}$ (c) (3, −2, 3)

12 (a) $\begin{bmatrix} 1 & 0 & 0 & 0 & 0 \\ 0 & 2 & 0 & 0 & 0 \\ 0 & 0 & -2 & 0 & 0 \\ 0 & 0 & 0 & 3 & 0 \\ 0 & 0 & 0 & 0 & -3 \end{bmatrix}$

(b) $\begin{bmatrix} 1 & 0 & 0 & 0 & 0 \\ 0 & 1 & 0 & 0 & 0 \\ 0 & 0 & 1 & 0 & 0 \\ 0 & 0 & 0 & 2 & 0 \\ 0 & 0 & 0 & 0 & -2 \end{bmatrix}$ (c) $\begin{bmatrix} 2 & 0 & 0 & 0 & 0 \\ 0 & 2 & 0 & 0 & 0 \\ 0 & 0 & 3 & 0 & 0 \\ 0 & 0 & 0 & 3 & 0 \\ 0 & 0 & 0 & 0 & 3 \end{bmatrix}$ (d) No

13 (a) $\begin{bmatrix} 2 & 0 & 0 \\ 0 & 2 & 0 \\ 0 & 0 & 3 \end{bmatrix}$ (b) $\begin{bmatrix} 1 & 1 & 0 \\ 0 & 1 & 1 \\ 1 & 2 & 4 \end{bmatrix}$

(c) $\begin{bmatrix} 7/3 & 2/3 & 4/3 \\ -1/3 & 4/3 & -4/3 \\ 1/3 & 2/3 & 10/3 \end{bmatrix}$

15 (a) 0 and 3; corresponding eigenvectors $(0, 1, 0)$ and $(1/3, 7/9, 1)$

(b) No diagonal matrix exists

16 (a) 0 and 6; eigenvectors $(-2, 1, 0)$ and $(-3, 0, 1)$ correspond to 0, $(1, 1, 1)$
corresponds to 6

(b) $\begin{bmatrix} 0 & 0 & 0 \\ 0 & 0 & 0 \\ 0 & 0 & 6 \end{bmatrix}$

17 (a) $\begin{bmatrix} -127/2 & 0 & -129/2 \\ 129/2 & 1 & 129/2 \\ -129/2 & 0 & -127/2 \end{bmatrix}$ (b) $\begin{bmatrix} 1/4 & 0 & -3/4 \\ 3/4 & 1 & 3/4 \\ -3/4 & 0 & 1/4 \end{bmatrix}$

18 $\begin{bmatrix} 1 & 0 \\ 0 & -1 \end{bmatrix}$ relative to the basis $(1, 1), (-1, 1)$

19 $\begin{bmatrix} 1 & 0 \\ 0 & -1 \end{bmatrix}$ relative to the basis $(1 + \sqrt{2}, 1), (1 - \sqrt{2}, 1)$

CHAPTER 6

Section 32

32-1 (a) $\sqrt{2}$ (c) 1   (e) 1   (g) $\sqrt{13}$
32-2 (a) $\sqrt{2}$ (c) $\sqrt{62}$ (e) 3
32-4 (a) $(3/5, 4/5)$ (c) $(1/\sqrt{30}, 2/\sqrt{30}, 5/\sqrt{30})$
(e) $(1/2, 1/2, 1/2, 1/2)$

Section 33

33-1 (a) $\pi/2$ (c) $\pi/3$ (e) $\pi/4$
33-3 (a) $(3, 0, -1)$ for example
(c) $(-b, a)$, or any non-zero multiple of $(-b, a)$
33-5 (a) $(2/\sqrt{21}, -1/\sqrt{21}, 4/\sqrt{21})$
(b) $(3/\sqrt{22}, 2/\sqrt{22}, 3/\sqrt{22})$
33-7 $(-1/2, 5/6, 1)$

Section 34

34-1 (a) $P_Y(X) = (24/29, 56/29)$
$P_X(Y) = (24/5, 8/5)$

(c) $P_Y(X) = (-2, 0, -4, 2)$
$P_X(Y) = (0, -2/5, 2, -4/5)$

34-2 (a) $(9/17, 36/17)$ (c) $(1, 2, 3)$

34-3 (a) $(1/\sqrt{17}, 4/\sqrt{17})$ (c) $(1/\sqrt{14}, 2/\sqrt{14}, 3/\sqrt{14})$

34-5 (a) Using the Gram-Schmidt process on $(5, 1, 0)$, $(2, 0, 1)$ we have the orthogonal basis $(5, 1, 0)$, $(1/13, -5/13, 1)$. Hence an orthonormal basis is $(5/\sqrt{26}, 1/\sqrt{26}, 0)$, $(1/\sqrt{195}, -5/\sqrt{195}, 13/\sqrt{195})$

(b) $(5/\sqrt{26}, 1/\sqrt{26}, 0)$, $(1/\sqrt{195}, -5/\sqrt{195}, 13/\sqrt{195})$, $(1/\sqrt{30}, -5/\sqrt{30}, -2/\sqrt{30})$

34-7 Using the Gram-Schmidt process on $(2, 0, 1)$, $(0, 1, 0)$, $(0, 0, 1)$, we have the orthogonal basis $(2, 0, 1)$, $(0, 1, 0)$, $(-2/5, 0, 4/5)$. Hence an orthonormal basis is $(2/\sqrt{5}, 0, 1/\sqrt{5})$, $(0, 1, 0)$, $(-1/\sqrt{5}, 0, 2/\sqrt{5})$

34-9 $(1/\sqrt{2}, 1/\sqrt{2}, 0)$, $(1/\sqrt{11}, -1/\sqrt{11}, 3/\sqrt{11})$, $(-3/\sqrt{22}, 3/\sqrt{22}, 2/\sqrt{22})$

Section 35

35-1 (a) $\begin{bmatrix} 1/\sqrt{2} & 1/\sqrt{2} \\ 1/\sqrt{2} & -1/\sqrt{2} \end{bmatrix}$ (c) $\begin{bmatrix} 0 & 1/\sqrt{2} & -1/\sqrt{2} \\ 1/\sqrt{2} & 1/2 & 1/2 \\ 1/\sqrt{2} & -1/2 & -1/2 \end{bmatrix}$

35-7 (a) Counterclockwise rotation by $\pi/6$ radians

(c) Counterclockwise rotation by $\pi/2$ radians

35-9 (b) $9\sqrt{2}, 2\sqrt{6}, -\sqrt{3}$

Section 36

36-3 (a) Inner product

(c) Not an inner product

36-4 No

36-6 (a) $1/2, 4x, x^2 - 1/3$

(b) $\sqrt{2}/2, \sqrt{3}x/\sqrt{2}, \dfrac{3\sqrt{5}}{2\sqrt{2}}\left(x^2 - \dfrac{1}{3}\right)$

Section 37

37-2 He can expect to lose $3\frac{1}{2}$¢ each time he plays

37-3 Only b is correct

37-5 All have the same expected value, approximately $-\$.05$

37-6 Hyperbola

37-7 Hyperbola

Chapter 6

Review

1 (a) $|\mathbf{X}_1| = \sqrt{5}, \ |\mathbf{X}_2| = \sqrt{46}, \ |\mathbf{X}_3| = 2\sqrt{5}$

   (b) $|\mathbf{X}_1 - \mathbf{X}_2| = \sqrt{51}, \ |\mathbf{X}_1 - \mathbf{X}_3| = \sqrt{15}, \ |\mathbf{X}_2 - \mathbf{X}_3| = \sqrt{36 - 2\sqrt{10}}$

   (c) $\pi/2, \ \pi/3$

2 (a) $(0, 0, 0, 0), \ (-2/7, 2/7, -2/7, 4/7)$

   (b) $(0, 0, 0, 1)$ for example

   (c) $(1, 2, 1, 0)$ for example

3 $(1/\sqrt{5}, -2/\sqrt{5}), \ (2/\sqrt{5}, 1/\sqrt{5})$

4 $(1, 3, 0), \ (9/10, -3/10, 4), \ (276/169, -92/169, -69/169)$

5 (a) $(2/\sqrt{5}, 1/\sqrt{5}, 0), \ (0, 0, 1), \ (1/\sqrt{5}, -2/\sqrt{5}, 0)$

   (b) $\begin{bmatrix} 1 & 0 & 0 \\ 0 & 1 & 0 \\ 0 & 0 & 0 \end{bmatrix}$

   (c) $\begin{bmatrix} 4/5 & 2/5 & 0 \\ 2/5 & 1/5 & 0 \\ 0 & 0 & 1 \end{bmatrix}$

6 $(14/9, -10/9, 17/9)$

7 $(-7/5, 1/5)$

8 $\begin{bmatrix} 3/5 & 4/5 & 0 \\ 4/5 & -3/5 & 0 \\ 0 & 0 & 1 \end{bmatrix}$

9 $A^{-1} = A^T$

11 $\left(5/9, \ \dfrac{-3\sqrt{3} + 1}{9}, \ \dfrac{3\sqrt{3} + 1}{9}\right)$

12 $\begin{bmatrix} 1 & 0 & 0 \\ 0 & -1 & 0 \\ 0 & 0 & 4 \end{bmatrix}$

relative to the orthonormal basis $(1, 0, 0), \ (0, 1/\sqrt{5}, -2/\sqrt{5}), \ (0, 2/\sqrt{5}, 1/\sqrt{5})$

INDEX

INDEX